MW01031445

SIMPLIFIED DESIGN OF WOOD STRUCTURES

SIMPLIFIED DESIGN OF WOOD STRUCTURES

Sixth Edition

JAMES AMBROSE
University of Southern California

and

PATRICK TRIPENY
University of Utah

John Wiley & Sons, Inc.

This book is printed on acid-free paper. ♾

Published by John Wiley & Sons, Inc., Hoboken, New Jersey
Published simultaneously in Canada

Library of Congress Cataloging-in-Publication Data:

Ambrose, James E.
Simplified design of wood structures / James Ambrose and Patrick Tripeny.—6th ed.
p. cm.
Includes bibliographical references and index.
ISBN 978-0-470-18784-5 (cloth)
1. Building, Wooden. 2. Structural design. I. Tripeny, Patrick. II. Title.
TH1101.P37 2009
721′.0448—dc22

2008018990

Printed in the United States of America

10 9 8 7 6 5 4 3 2 1

CONTENTS

Preface to the Sixth Edition xi

Preface to the First Edition xv

Introduction 1

1 Structural Uses of Wood 12

1.1 Sources of Wood / 13
1.2 Tree Growth / 13
1.3 Density of Wood / 14
1.4 Defects in Lumber / 15
1.5 Seasoning of Wood / 16
1.6 Nominal and Dressed Sizes / 17
1.7 Use Classification of Structural Lumber / 17
1.8 Grading of Structural Lumber / 18
1.9 Fabricated Wood Products / 18

2 Design Issues and Methods **20**

2.1 Design Goals / 20
2.2 Methods of Investigation and Design / 23
2.3 Choice of Design Method / 25

3 Structural Investigation **27**

3.1 General Concerns / 27
3.2 Forces and Loads / 28
3.3 Direct Stress / 30
3.4 Kinds of Stress / 31
3.5 Deformation / 32
3.6 Elastic Response and Limit / 32
3.7 Inelastic Behavior and Ultimate Strength / 33
3.8 Modulus of Elasticity / 33
3.9 Permissible Values for Design / 35

4 Design Data and Criteria **36**

4.1 General Concerns / 36
4.2 Reference Design Values for Allowable Stress Design (ASD) / 37
4.3 Adjustment of Design Values / 41
4.4 Modification for Loading with Relation to Grain Direction / 45
4.5 Design Controls for LRFD / 47

5 Beam Functions **50**

5.1 General Considerations / 50
5.2 Moments / 52
5.3 Beam Loads and Reaction Forces / 57
5.4 Beam Shear / 60
5.5 Bending Moment / 65
5.6 Tabulated Values for Beam Behavior / 78
5.7 Multiple-Span Beams / 82

6 Behavior of Beams **86**

6.1 Shear in Beams / 86
6.2 Bending in Beams / 93
6.3 Deflection / 97

6.4 Bearing / 101
6.5 Buckling of Beams / 103
6.6 Unsymmetrical Bending / 105
6.7 Behavior Considerations for LRFD / 109

7 Design of Beams **117**

7.1 Design Procedure / 117
7.2 Beam Design Examples / 118
7.3 Joists and Rafters / 121
7.4 Alternative Spanning Elements / 126

8 Wood Decks **129**

8.1 Board Decks / 129
8.2 Wood Fiber Decks / 132
8.3 Plywood Decks / 132
8.4 Spanning Capability of Decks / 133

9 Wood Columns **135**

9.1 Slenderness Ratio for Columns / 135
9.2 Compression Capacity of Simple Solid Columns / 136
9.3 Column Load Capacity, LRFD / 145
9.4 Round Columns / 147
9.5 Stud Wall Construction / 148
9.6 Spaced Columns / 150
9.7 Built-Up Columns / 154
9.8 Columns with Bending / 155

10 Connections for Wood Structures **166**

10.1 Bolted Joints / 166
10.2 Nailed Joints / 176
10.3 Screws / 179
10.4 Mechanically Driven Fasteners / 181
10.5 Shear Developers / 181
10.6 Split-Ring Connectors / 182
10.7 Formed Steel Framing Elements / 189
10.8 Concrete and Masonry Anchors / 192
10.9 Plywood Gussets / 192
10.10 Investigation of Connections, LRFD / 194

11 Trusses **196**

11.1 General Considerations / 196
11.2 Types of Trusses / 199
11.3 Bracing for Trusses / 199
11.4 Loads on Trusses / 201
11.5 Investigation for Internal Forces in Planar Trusses / 203
11.6 Design Forces for Truss Members / 221
11.7 Combined Actions in Truss Members / 221
11.8 Truss Members and Joints / 222
11.9 Timber Trusses / 223
11.10 Manufactured Trusses / 224

12 Miscellaneous Wood Products and Elements **226**

12.1 Engineered Wood Products / 226
12.2 Glued Laminated Structural Members / 227
12.3 Structural Composite Lumber / 229
12.4 Wood Structural Panels / 231
12.5 Plywood / 232
12.6 Prefabricated Wood I-Joists / 237
12.7 Built-Up Panel and Lumber Beams / 238
12.8 Flitched Beams / 239
12.9 Pole Structures / 244

13 Wood Structures for Lateral Bracing **246**

13.1 Application of Wind and Earthquake Forces / 247
13.2 Horizontal Diaphragms / 250
13.3 Vertical Diaphragms (Shear Walls) / 262
13.4 Investigation and Design of Wood-Framed Shear Walls / 270
13.5 Trussed Bracing for Wood Frames / 277
13.6 Special Lateral Bracing / 284

14 General Considerations for Building Structures **288**

14.1 Choice of Building Construction / 288
14.2 Structural Design Standards / 289
14.3 Loads for Structural Design / 289
14.4 Dead Loads / 290
14.5 Building Code Requirements for Structures / 292
14.6 Live Loads / 294

14.7 Lateral Loads (Wind and Earthquake) / 297
14.8 Load Combinations and Factors / 301
14.9 Determination of Design Loads / 302
14.10 Structural Planning / 302
14.11 Building Systems Integration / 303
14.12 Economics / 303

15 Building Design Examples **306**

15.1 Building One: Single-Story Light Wood Frame / 307
15.2 Building Two: Multistory Light Wood Frame / 327
15.3 Building Three: Masonry and Timber Structure / 334
15.4 Building Four: Steel and Wood Structure / 351

Appendix A: Properties of Sections **354**

Appendix B: Study Aids **369**

Appendix C: Answers to Problems **381**

Glossary **387**

References **390**

Index **393**

PREFACE TO THE SIXTH EDITION

Publication of this book presents an opportunity for yet another generation of students of building design to access the subject of wood structures. The particular focus of this work is a concentration on widely-used, simple, and ordinary methods of construction. In addition, the effort has been made to keep mathematical work at a low level, in order to emphasize the accessibility of the work by untrained persons.

The basic purpose of this "simplified" work is well expressed in the preface to the first edition by the originator of the simplified series of books, the late Professor Harry Parker of the University of Pennsylvania. Excerpts from Professor Parker's preface to the first edition follow this preface. To the extent possible, we have adhered to the spirit expressed by Professor Parker.

Of course, structural engineering is no longer really simple. Utilization of all of the available resources for investigation and design of structures is a complex and exhaustive task. Professional engineers must climb this notable learning curve in pursuit of credibility as professionals. We do not mean to belittle the work of serious engineers by our simplified approach. Still, practical means for construction and the use of very ordinary structural products provides the resource for the vast amount of building construction—even in the computer age. It is possible, therefore, to present real structural solutions for ordinary structural tasks with a minimum of complexity. And it is also possible to present the design process for these solutions in relatively elementary form.

Readers of this book should obtain a useful overview of the field of wood structures and of the means for their design. Those wishing to pursue the study to more advanced levels can find many publications and educational opportunities for their study. For many, seeking mostly only a general view of the topic and an understanding of basic design processes, this book may suffice well.

This topic is supported by an amazing archive of reference materials, including publications and computer programs. We have used a few essential sources for this work, a primary one being the *National Design Specification for Wood Construction* (Ref. 1), generally referred to as the NDS, published by the American Forest and Paper Association (AFPA). To the extent possible, the work presented here conforms to the specifications of the latest edition of that work. We are grateful to the AFPA for permission to use some materials from the NDS for this work. We have also used materials from several other sources, whose permissions are acknowledged in the book.

Preparation of instructional material needs to include the testing of the materials in classroom situations. The authors of this book have had extensive opportunities to utilize classroom experience for development of the materials in this book. We are considerably in debt to the many students who have sat in our classes, and from whom we have undoubtedly learned more than they have from our efforts. We are also indebted to the schools that have provided our teaching opportunities, especially the University of Utah and the University of Southern California. We are very grateful for the support and encouragement provided by these schools to classroom teachers.

We are very appreciative for the support of our publisher, John Wiley and Sons. This publisher has a long history of maintaining a strong catalog of publications in the fields of architecture and construction and we thank them for that continuing effort. We are especially grateful to our publisher, Amanda Miller, our editor, Paul Drougas and his assistant Sadie Abuhoff, and our production liaison, Diana Cisek. This book would indeed not have been produced without their significant contributions and support.

Finally, we need to express the gratitude we have to our families. Writing work, especially when added to an already full-time occupation, is very time consuming. We thank our spouses and children for their patience, endurance, support, and encouragement in permitting us to achieve this work.

James Ambrose

Patrick Tripeny

PREFACE TO THE FIRST EDITION

(The following is an excerpt from Professor Parker's preface to the first edition.)

This volume is the fifth of a series of elementary books relating to the design of structural members in buildings. The author has endeavored to explain as simply as possible the methods commonly used in determining proper timber sizes. This book deals primarily with wood members that support loads in buildings.

The material is arranged not only for classroom use but also for the many young architects and builders who desire a guide for home study. With this in mind, a major portion of the book is devoted to the solution of practical problems, which are followed by problems to be solved by the student. In addition to explanations of basic principles of mechanics involved in the design of members, numerous safe load tables have been included. These tables will enable one to select promptly members of proper size to use for given conditions.

Stress tables, properties of sections, and tabulations of technical information pertinent to timber construction are included, thus making reference books unnecessary.

It is assumed that those who use this book have had no previous training. As in the previous books of the series, the use of advanced mathematics has been avoided, and a knowledge of high school arithmetic and algebra is all that is necessary to understand the mathematics involved.

Harry Parker
High Hollow, Southampton, Bucks County, Pennsylvania
1948

INTRODUCTION

Design for the use of wood structures in buildings involves a broad range of considerations. These include basic properties of the materials, forms of common industrialized products, and common usages for typical building construction. However, the topic also embraces concerns for building design in general, regulatory codes, and commonly used methods of professional designers. This chapter briefly considers some of these very basic issues.

USE OF WOOD FOR BUILDING STRUCTURES

This book deals in general with the common uses of wood for the structures of ordinary buildings. If fully considered, this involves a considerable range of usage, as wood is used in one form or another in many structures of small to moderate size. It is the most widely used structural material in the United States. The topic of wood structures as developed here is limited essentially to situations in which wood is used as the major material for the primary structure. Exceptions to this are the use of foundations and vertical bearing structures of concrete or masonry and the use of composite spanning elements of steel and wood. Erection of frames of wood requires the use of nails, screws, bolts, anchorage devices, and various metal connectors—all ordinarily made of steel.

METHODS OF INVESTIGATION AND DESIGN

For the determination of design load conditions and the establishment of structural resistance of wood structures, there are currently two different methods in use. These are:

1. *Load and Resistance Factor Design (LRFD).* This method is based on use of ultimate strength resistance of structural members and determination of failure load levels. Design loads are determined by multiplying the true (service) loads by adjustment factors, usually resulting in an increase in the load. Member resistances are determined and then multiplied by adjustment factors, usually resulting in a reduction of the member resistance for design. The resulting two-shot adjustment is manipulated to achieve a desired level of safety.
2. *Allowable Stress Design (ASD).* Also known as Working Stress Design or simply Stress Design. This method uses the true (service) loads and sets limits on stresses in members under these load conditions. This is the classic method developed and used for wood structures in the 19th century and most of the 20th century.

The LRFD method is now used by many structural engineering firms and it is the principal focus of the current design aids provided by the industry. Structural design software is based on this method, and some building codes either require its exclusive use or grant an exception to it in some cases. Factored loads are based on statistical likelihood, and member resistances are based on evaluations backed by laboratory tests to failure. It is unquestionably the more accurate method of design.

The old ASD method is still used by many engineers and by architects who do little structural design work. It is also mostly used for teaching structures to architecture students and in general for teaching where short time is available for the subject (only part of a whole semester course). Longstanding and easy-to-use design aids support the design work providing shortcuts that are easily learned and used. It is not, however, as closely related to the research and development of design produced over the past 50 years and is sure to steadily fade from use.

While choice of methods used by practicing engineers and architects can only be modified over time, the LRFD method will certainly eventually prevail. Anyone expecting to participate in design of structures in the future should learn the LRFD method. It is not that hard to learn, although it involves a few more steps in the design process. Analyses for resistance for some modes of failure are quite complex, but then they are also complex with the stress method. For most ordinary cases, design is simple, and direct and design aids can be used to shortcut the process.

Of course, teaching old dogs new tricks is certain to be met with resistance, especially with persons long experienced with the stress method. On the other hand, teaching students or other inexperienced persons is not really more difficult, as they have no point of comparison and no attachment to other methods. It can easily be accomplished in the same amount of classroom time.

We have chosen to demonstrate the use of both ASD and LRFD for wood design in this book since it is important that both methods be understood at the present time. The usual presentation method consists of first explaining the ASD method and then explaining what is different about using LRFD for the same task. In design work there is only a small fraction of the effort involved with computation of structural behaviors. It is an essential part of design work, but many other issues and circumstances must also be dealt with. Planning of structural systems, development of construction details, and integration of the building structure with the rest of the building construction and services involve much greater time and attention.

Inspection of this book will show that considerable space is devoted to issues other than computation of structural behaviors. Furthermore, when actual design work is performed for building cases, use is made wherever possible of shortcuts and available design aids. This is how professional designers work, not by spending the majority of their time in cranking out data from complex equations. It helps in the development of understanding to know about the complex equations, but the shortcuts are much more practical for design work.

The procedures of the LRFD method are described more fully in Chapter 2, and its application is illustrated for various situations of structural investigation and design throughout the book.

REFERENCE SOURCES FOR DESIGN

Information about wood structures is forthcoming from a number of sources. These range from relatively unbiased textbooks and research reports to clearly biased promotional materials from the producers and suppliers of construction materials, equipment, and services. Bias is not to be considered as evil, it is merely to be acknowledged in evaluating the potential for completeness and neutrality in the materials presented. One cannot really expect people in the business of selling wood and fabricated wood products to provide unbiased information about the use of wood, especially information about any of its drawbacks or its true competitiveness with other materials.

The published information about wood and all of the issues relating to it would fill a large building. Some is essential and general in nature; most is highly detailed and narrowly directed. The material presented in this book is general in nature, specific to the interests of building designers, and represents a small, distilled essence of a number of general publications. Principal references used for this book are listed in the References section following the text.

There are several industry-wide organizations in the United States that have a relation to wood structures. While these organizations are largely industry-supported, they do represent major sources of design codes and standards, as well as information about wood products. Materials from several of these sources are presented in this book, with the publications from which they are taken noted.

Widely used sources for information for design of wood structures are the publications of the American Forest and Paper Association (AFPA). This organization deals with both sawn-wood products and engineered wood products. For design of wood structures in general, the *National Design Specification for Wood Construction* (Ref. 1), commonly referred to as the NDS, is published by the AFPA and is used as a reference by most building codes.

Although the AFPA is the principal industry-sponsored organization in the area of wood structures, many other industry and professional organizations also provide materials for designers. Some of these are the following:

The American Society for Testing and Materials (ASTM). This organization provides widely used standards for all sorts of materials, including many structural products. Just about every structural product is produced with some ASTM specification.

International Code Council (ICC). This is the publisher of the number one model building code, *International Building Code* (Ref. 4), commonly referred to as the IBC.

American Society of Civil Engineers (ASCE). This organization includes a division which is a major affiliation of structural engineers, and is the sponsor of many publications on structural design. Their publication, *Minimum Design Loads for Buildings and Other Structures* (Ref. 3.), commonly referred to as ASCE 2006, is generally used by most building codes.

APA—The Engineered Wood Association. This organization publishes standards for classification and grading of structural plywood products as well as other engineered wood products. It also publishes many design guides for structural applications of plywood and other products.

American Institute of Timber Construction (AITC). This organization deals primarily with timber structures (timber meaning wood pieces with thickness greater than 4 in.) consisting of both sawn-wood and glued-laminated wood elements. Their publication, *Timber Construction Manual* (Ref. 2), is a major reference source for design work in these topic areas.

There is in fact an industry or trade organization for just about every type of product used for construction. Any of them may be the source of useful information, and the ones referred to here are just a sampling of the mountain of available information.

Anyone intending to pursue the study of this subject beyond the scope of the treatment in this book should obtain access to a basic text, such as those used for courses in civil engineering schools.

UNITS OF MEASUREMENT

Previous editions of this book have used U.S. units (feet, inches, pounds, etc.) for the basic presentation. In this edition the basic work is developed with U.S. units with equivalent metric unit values in brackets [thus]. While the building industry in the United States is still in process of changing to metric units, our decision for the presentation here is a pragmatic one. Most of the references used for this book are still developed primarily in U.S. units.

Table 1 lists the standard units of measurement in the U.S. system with the abbreviations used in this work and a description of common

TABLE 1 Units of Measurement: U.S. System

Name of Unit	Abbreviation	Use in Building Design
Length		
Foot	ft	Large dimensions, building plans, beam spans
Inch	in.	Small dimensions, size of member cross sections
Area		
Square feet	ft^2	Large areas
Square inches	$in.^2$	Small areas, properties of cross sections
Volume		
Cubic yards	yd^3	Large volumes, of soil or concrete (commonly called simply "yards")
Cubic feet	ft^3	Quantities of materials
Cubic inches	$in.^3$	Small volumes
Force, Mass		
Pound	lb	Specific weight, force, load
Kip	kip, k	1000 pounds
Ton	ton	2000 pounds
Pounds per foot	lb/ft, plf	Linear load (as on a beam)
Kips per foot	kips/ft, klf	Linear load (as on a beam)
Pounds per square foot	lb/ft^2, psf	Distributed load on a surface, pressure
Kips per square foot	k/ft^2, ksf	Distributed load on a surface, pressure
Pounds per cubic foot	lb/ft^3	Relative density, unit weight
Moment		
Foot-pounds	ft-lb	Rotational or bending moment
Inch-pounds	in.-lb	Rotational or bending moment
Kip-feet	kip-ft	Rotational or bending moment
Kip-inches	kip-in.	Rotational or bending moment
Stress		
Pounds per square foot	lb/ft^2, psf	Soil pressure
Pounds per square inch	$lb/in.^2$, psi	Stresses in structures
Kips per square foot	$kips/ft^2$, ksf	Soil pressure
Kips per square inch	$kips/in.^2$, ksi	Stresses in structures
Temperature		
Degree Fahrenheit	°F	Temperature

TABLE 2 Units of Measurement: SI System

Name of Unit	Abbreviation	Use in Building Design
Length		
Meter	m	Large dimensions, building plans, beam spans
Millimeter	mm	Small dimensions, size of member cross sections
Area		
Square meters	m^2	Large areas
Square millimeters	mm^2	Small areas, properties of member cross sections
Volume		
Cubic meters	m^3	Large volumes
Cubic millimeters	mm^3	Small volumes
Mass		
Kilogram	kg	Mass of material (equivalent to weight in U.S. units)
Kilograms per cubic meter	kg/m^3	Density (unit weight)
Force, Load		
Newton	N	Force or load on structure
Kilonewton	kN	1000 Newtons
Stress		
Pascal	Pa	Stress or pressure (1 pascal = 1 N/m^2)
Kilopascal	kPa	1000 pascals
Megapascal	MPa	1,000,000 pascals
Gigapascal	GPa	1,000,000,000 pascals
Temperature		
Degree Celsius	°C	Temperature

usage in structural design work. In similar form, Table 2 gives the corresponding units in the metric system. Conversion factors to be used for shifting from one unit system to the other are given in Table 3. Direct use of the conversion factors will produce what is called a *hard conversion* of a reasonably precise form.

In the work in this book many of the unit conversions presented are *soft conversions*, meaning ones in which the converted value is rounded off to produce an approximate equivalent value of some slightly more

TABLE 3 Factors for Conversion of Units

To Convert from U.S. Units to Metric Units, Multiply by:	U.S. Unit	SI Unit	To Convert from SI Units to U.S. Units, Multiply by:
25.4	in.	mm	0.03937
0.3048	ft	m	3.281
645.2	in.^{2}	mm^2	1.550×10^{-3}
16.39×10^3	in.^{3}	mm^3	61.02×10^{-6}
416.2×10^3	in.^{4}	mm^4	2.403×10^{-6}
0.09290	ft^2	m^2	10.76
0.02832	ft^3	m^3	35.31
0.4536	lb (mass)	kg	2.205
4.448	lb (force)	N	0.2248
4.448	kip (force)	kN	0.2248
1.356	ft-lb (moment)	N-m	0.7376
1.356	kip-ft (moment)	kN-m	0.7376
16.0185	lb/ft^3 (density)	kg/m^3	0.06243
14.59	lb/ft (load)	N/m	0.06853
14.59	kip/ft (load)	kN/m	0.06853
6.895	psi (stress)	kPa	0.1450
6.895	ksi (stress)	MPa	0.1450
0.04788	psf (load or pressure)	kPa	20.93
47.88	ksf (load or pressure)	kPa	0.02093
$0.566 \times (°F - 32)$	°F	°C	$(1.8 \times °C) + 32$

relevant numerical significance to the unit system. Thus a wood 2 × 4 (actually 1.5 × 3.5 inches in the U.S. system) is precisely 38.1 mm × 88.9 mm in the metric system. However, the metric equivalent ″2 × 4″ is more likely to be made 40 × 90 mm: close enough for most purposes in construction work.

ACCURACY OF COMPUTATIONS

Structures for buildings are seldom produced with a high degree of dimensional precision. Exact dimensions of some parts of the construction—such as window frames and elevator rails—must be reasonably precise; however, the basic structural framework is ordinarily achieved with only a very limited dimensional precision. Add this to considerations for the lack of precision in predicting loads for any structure, and the significance of highly precise structural computations becomes moot. This is not to be used for an argument to justify sloppy mathematical work, sloppy construction, or use of vague theories of investigation of behaviors. Nevertheless, it makes a case for not being highly concerned with any numbers beyond about the second digit (103 or 104; who cares?).

While most professional design work these days is likely to be done with computer support, most of the work illustrated here is quite simple and was actually performed with a hand calculator (the 8-digit, scientific type is adequate). Rounding off of even these primitive computations is sometimes done with no apologies.

SYMBOLS

The following shorthand symbols are frequently used:

Symbol	Reading
$>$	is greater than
$<$	is less than
$\geq$	equal to or greater than
$\leq$	equal to or less than
$6'$	6 feet
$6''$	6 inches
Σ	the sum of
ΔL	change in L

NOTATION

Notation used in this book complies generally with that used in the wood industry and the latest editions of standard specifications. The following list includes the notation used in this book and is compiled and adapted from more extensive lists in the references:

A = area, general
A_g = gross area of a section, defined by the outer dimensions
A_n = net area (gross area less area removed by holes or notches)
C = compressive force
C_D = load duration factor
C_F = size factor for sawn lumber
C_M = wet (moisture) service factor
C_P = column stability factor
C_T = buckling stiffness factor for dimension lumber
C_r = repetitive member factor for dimension lumber
D = diameter

E = reference modulus of elasticity
E' = adjusted modulus of elasticity
E_{min} = modulus of elasticity for stability investigation
F_b = reference bending design value
F'_b = adjusted bending design value
F_c = reference compressive design value parallel to the grain, due to axial load only
F'_c = adjusted compressive design value parallel to the grain, due to axial load only
F_{cE} = design value for critical buckling in compression members
$F_{c\perp}$ = reference compression design value perpendicular to the grain
$F'_{c\perp}$ = adjusted compression design value perpendicular to the grain
F_v = reference shear design value parallel to the grain
F'_v = adjusted shear design value parallel to the grain
G = specific gravity
I = moment of inertia, or importance factor (wind and earthquakes)
L = length (usually of a span), or unbraced height of a column
M = bending moment
M_r = reference design moment
M'_r = adjusted design moment
P = concentrated load or axial compression load
Q = statical moment of an area about the neutral axis
R = radius of curvature
S = section modulus
T = temperature in degrees Fahrenheit, or tension force
V = shear force, or vertical component of a force
V_r = reference design shear
V'_r = adjusted design shear
W = total gravity load, or weight, or dead load of an object, or total wind load force, or total of a uniformly distributed load or pressure due to gravity
b = width or breadth of bending member
c = in bending: distance from extreme fiber stress to the neutral axis
d = overall beam depth, or pennyweight of nail
e = eccentricity of a non-axial load, from the point of application of the load to the centroid of the section
f_b = actual computed bending stress
f_c = actual computed compressive stress due to axial load

f'_c = specified compressive strength of concrete
f'_m = specified compressive strength of masonry
f_p = actual computed bearing stress
f_t = actual computed stress in tension parallel to the grain
f_v = actual computed shear stress
r = radius of gyration
s = spacing of objects, center to center
t = thickness, general
w = unit of a distributed load on a beam (lb/ft, etc.)
Δ = deflection, usually maximum vertical deflection of a beam (also used to indicate "change of" in mathematical expression)
Σ = sum of
λ = time effect factor (LRFD), or adjustment factor for building height (wind)
ϕ = resistance factor (LRFD)

1

STRUCTURAL USES OF WOOD

Wood from trees has a long history of usage for structural purposes, most notably in regions where large stands of trees exist. At the time of the early colonization of the United States, vast areas of this country were covered with forests. It was, indeed, a major problem for early settlers of the eastern, southeastern, and midwestern areas. Travel was difficult because of the dense growth, and up to the middle of the nineteenth century, was mostly accomplished by using the many navigable rivers. As in many other countries today, land for cultivation of crops or grazing of animals had to be claimed by burning off or otherwise destroying forest lands.

While much of that early dense forest was lost—most notably vast stands of hardwood trees—a considerable amount of timber was used for construction. Thus a heritage of wood construction was developed and an extensive industry was established. This industry extends to today, with wood remaining as a major source for building construction uses.

We no longer build extensively with construction that directly utilizes the source. Log cabins, roughly hacked boards, and pole construction with peeled logs do not account for the majority of buildings. Today, wood as a building material is treated as an industrialized product, receiving considerable processing on the way to the construction site.

Still, a major use—and one treated extensively in this book—is that of the lightly processed pieces of wood that are cut directly from the logs, smoothed up a bit, and used as quickly as possible in their solid-sawn form. This product is what we generally refer to as *lumber*, and the lumberyard is still a major business in almost every large community in the United States. Wood is indeed the all-American building material.

This chapter deals with some of the basic issues concerning the use of wood, with concentration on the direct usage for structural lumber.

1.1 SOURCES OF WOOD

The particular type of tree from which wood comes is called the *species*. Although there are thousand of species of trees, most structural wood comes from a few dozen species that are selected for commercial processing.

The two groups of trees used for building purposes are the *softwoods* and *hardwoods*. Most softwood trees like pine and spruce are coniferous, or cone bearing, whereas hardwood trees have broad leaves exemplified by oak and maple. Softwoods are mostly indeed softer than hardwoods, although there are other properties that define the types.

The two species of trees used most extensively for structural wood in the United States are Douglas Fir and Southern Pine. However, several other species are also used, depending partly on regional availability. Although the terms *timber* and *lumber* are often used interchangeably, current industry usage tends to reserve timber for structural wood members of large cross-sectional area.

1.2 TREE GROWTH

The trees used for lumber in the United States are exogenous; that is, they increase in size by growth of new wood on the outer surface under the bark. The cross section of a tree trunk reveals the layers of new wood that are formed annually. These layers, called *annual rings*, are typically composed of alternating light and dark material. In most areas, the lighter, more porous layers are grown in the warmer months of the year (spring and summer in the northern hemisphere), and the denser, darker layers are grown in the colder months.

The number of layers of annual rings at the base of the tree trunk indicates the age of a tree. To build up a trunk large enough to saw structural lumber from requires several years of growth, the number of years depending on the climate and the type of tree. In a real sense, however, no matter how many years it takes, wood is a renewable source of building materials.

The youngest band of annual rings at the outer edge of the tree is called the *sapwood*. This is usually lighter in color than wood at the center of the log, which is called the *heartwood*. For specific purposes either the sapwood or the heartwood may be the more desirable material. However, how an individual piece of lumber is cut from the log with respect to orientation to the general pattern of the annual rings is often of greater concern.

The structure of wood is composed primarily of long and slender cells called *fibers*. These cells have a hollow, tubular form with an orientation of their lengths in the longitudinal direction of the log (up the tree for transporting of water and nutrients during growth). This gives cut pieces of wood a character described as its *grain*, with the grain being directed along the length of cut pieces of lumber. This in turn provides a reference for viewing various structural actions as relating to the grain; that is, as being *parallel to the grain, perpendicular to the grain*, or at some *angle to the grain*.

The fibrous, tubular cells of the wood are composed primarily of *cellulose* and the material that binds the cells is called *lignin*. These two materials are the main chemical components of wood.

1.3 DENSITY OF WOOD

The difference in the arrangement and size of the cell cavities and the thickness of the cell walls determine the specific gravity, or relative *density*, of various species of wood. The strength of wood is closely related to its density. The term *close grained* refers to wood with narrow, closely spaced annual rings. Certain woods, such as Douglas Fir and Southern Yellow Pine, show a distinct contrast between the springwood and summerwood, and the proportion of summerwood affords a visual basis for approximating strength and density. The solid material in wood is about 1.53 times the weight of water, but the wood cells contain open spaces in varying degrees. These spaces are typically filled partly with air and partly with water. The weight of wood varies with regard to the amount of open cell space and the amount of trapped water. For purposes of computation in this book, the average weight of structural softwood is taken as 35 lb per cu ft (pcf).

1.4 DEFECTS IN LUMBER

Any irregularity in wood that affects its strength or durability is called a *defect*. Because of the natural characteristics of the material, several common defects are inherent in wood. The most common are described here.

A *knot* is a portion of a branch or limb that has been surrounded by subsequent growth of the tree. There are several types of knots, and the the strength of a structural member is affected by the size and location of those it may contain. Grading rules for structural lumber are specific concerning the number, size, and position of knots, and their presence is considered when establishing design values for structural response.

A *shake* is a separation along the grain, principally between annual rings. The cross section of a shake is shown in Figure 1.1*a*. Shakes reduce the resistance to shear, and consequently members subjected to bending are directly affected by their presence. The strength of members in longitudinal compression (directed parallel to the wood grain), such as columns and posts, is not greatly affected by shakes.

A *check* is a separation along the grain, the greater part of which occurs across the annual rings (Figure 1.1*b*). Checks generally develop during the process of seasoning (drying out from the green condition). Like shakes, checks also reduce the resistance to shear.

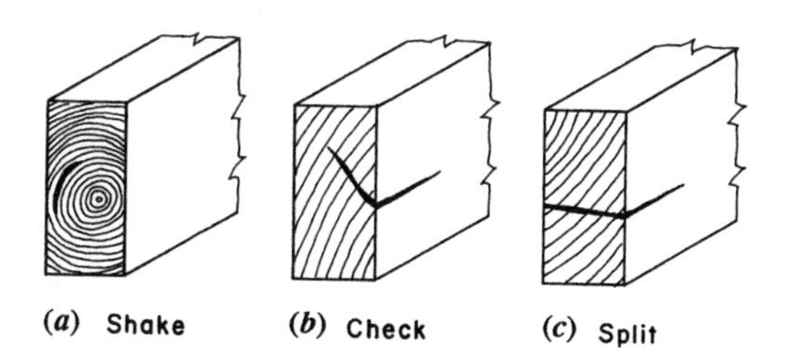

Figure 1.1 Common defects in structural lumber.

A *split* is shown in Figure 1.1*c*. It is a lengthwise separation of the wood that extends through the piece from one surface to another. Splits obviously have a major effect in reducing shear resistance.

A *pitch pocket* is an opening parallel to the annual rings that contains pitch, either solid or liquid.

Logs are typically tapered in form, and when a long piece of lumber is sawn from a relatively short tree trunk, or from a log that is not held straight during sawing, a condition may occur that is described as having a piece of lumber with a *slope of grain*. This has some direct effects on certain structural uses, and is one of many properties of an individual piece of lumber that is noted when the piece is evaluated for structural applications.

A major concern for wood for construction is the general problem of *decay* of the wood. This is actually a natural process for the organic (once living) material, and preserving the wood is literally a nature-defying effort. Some decay occurs within the tree even during its growth period and pockets of decay are another form of defect of the sawn lumber pieces. Old decay may be arrested or suspended by treatment of the wood, or simply be eliminated by cutting out the decayed portions. Often of greater concern is continuing or new decay, which is a major problem for the general development of the construction. Numerous treatments are possible, including the impregnation of the wood mass with chemicals to arrest future decay. Wood generally untreated and exposed to the weather is especially vulnerable in this regard.

1.5 SEASONING OF WOOD

All wood contains moisture, and the serviceability of wood for construction is generally improved by reduction of the amount of moisture below the content in the *freshly-cut* pieces, referred to as *green* wood. The process of removing moisture from green wood is known as *seasoning*; it is accomplished by exposing the wood to relatively drier air for an extended period of time or by heating it in kilns to drive out the moisture. Whether *air dried* or *kiln dried*, seasoned wood is generally stiffer, stronger, and less subject to shape changes.

Drying out of the wood results in shrinkage of the cellular structure of the material. This occurs differently in the three primary directions: along the grain, parallel to the annual rings, and perpendicular to the annual rings. This is where the orientation of the grain, as well as its relative uniformity and absence of large defects, becomes quite important. Shape and dimensional changes of some degree are to be expected, affecting a property described as the *dimensional stability* of the wood. It is quite important that as much of this change as is possible should occur before installation of the wood in the construction assemblage.

The *moisture content* of wood is the ratio of the weight of contained water in a piece of wood to the weight of an oven-dried (zero moist) sample, expressed as a percentage. Specific limits are set for this value for structural applications.

1.6 NOMINAL AND DRESSED SIZES

A piece of structural lumber is designated by its nominal cross-sectional dimensions. As an example, a designation of 6 by 12 (written 6 × 12) indicates a member with a width of 6 in. and a depth of 12 in. However these are actually *nominal dimensions* which roughly describe the shape as sawn from the log. When the member faces are smoothed by planing and/or sanding, the member is said to be *dressed* or *surfaced*. The 6 × 12 when dressed will have actual dimensions of 5.5 × 11.5 in. Standard lumber sizes produced by the lumber industry are listed in Table A.3, which yields both the nominal and actual dimensions of the pieces.

1.7 USE CLASSIFICATION OF STRUCTURAL LUMBER

Because the effects of natural defects on the strength of lumber vary with the type of loading, and specific types of defects relate to structural usage of members, structural lumber is classified according to both *size* and *use*. All dimensions used in classification are nominal dimensions and not actual dimensions. The four principal classifications are as follows:

Dimension Lumber. Rectangular cross sections with nominal dimensions that range from 2 in. to 4 in. in thickness and 2 in. or more in width. A further distinction is made between *light framing* 2 in. to 4 in. wide, and *joists* and *planks* 5 in. and wider.

Beams and Stringers. Rectangular cross sections 5 in. or more thick and with a width at least 2 in. greater than the thickness, graded for strength in bending when loaded on the narrow face.

Posts and Timbers. Square or nearly square cross sections with nominal dimensions 5 in. by 5 in. and larger, and with width not more than

2 in. greater than thickness, are graded primarily for use as posts or columns or other uses where bending strength is not a major concern.

Decking. This consists of lumber 2 in. to 4 in. thick and 4 in. or more wide. Jointing of members is provided by tongue and groove edges or edges splined (slotted) for interlocking on the narrow face. Decking is graded for use with the wide face placed flatwise in contact with supporting members.

1.8 GRADING OF STRUCTURAL LUMBER

Grading is necessary to establish the quality of lumber. Structural grades are identified in relation to strength properties and use classifications. Structural grading may be done by mechanical testing of wood samples but is most often accomplished by visual inspection. Once graded, a piece of structural lumber is stamped with the grade and the grading authority or standard used. Use of specific grades is established by the structural designer for each type of structural member, and is ordinarily designated for use on the structural construction documents. The topic of grading is discussed in Chapter 4 as it relates to the process of structural design work.

1.9 FABRICATED WOOD PRODUCTS

Wood is indeed used in solid sawn form for many structures. However, many products—structural and other—are fabricated from wood in some reconstituted form. Reduced to fiber form, wood is used widely—a major use being for paper. Newspapers, bathroom tissue, paper bags, cardboard boxes, and the pages of this book are produced primarily from wood fiber. This constitutes *the* major use of wood by volume.

With wood fiber production in place for many years, a natural outgrowth has been the production of structural products. Wood fiber panels have been produced for many years and are now used extensively for furniture and in many situations in building construction. With the wood reduced to a fine fiber form, strength is limited, so uses are mostly in construction with low structural tasks.

A more recent product is OSB (for oriented strand board), which is produced with wood in small thin pieces laid on top of each other. The grain direction of individual pieces is random with respect to other pieces, which gives the end product a two-way grain effect, similar to that produced with plywood, where alternating plies have their grain directions perpendicular to adjacent plies. This product now sees major use for wall paneling and roof decks, due to its considerable strength.

Another recent product is that produced with shredded strands of wood adhered with the grain direction the same. This produces something very similar to solid-sawn wood, and is used as a replacement for smaller sizes of lumber.

Another fabricated product in wide use is that produced with wood pieces adhered in laminations. This includes thin plywood panels and large forms for girders and columns. This process can also be used to produce curved forms. Laminated products are discussed in Chapter 12.

Fabricated products also include many assembled components, such as trusses and boxed elements. These may use various combinations of solid sawn wood, fiber products, and possibly elements of metal. Some of these products are discussed in Chapter 12. A product of particular note is the truss produced with wood chords and steel web members, as discussed in Section 11.8.

One reason for the expanding use of fiber, particle, and strand products is that the wood source for these products can be from slow-growth trees and parts of trees not usable for sawn lumber. With the steady loss of old-growth forests, this is a major consideration for the fuller use of forest resources.

2

DESIGN ISSUES AND METHODS

The primary aim of this book is to deal with the design of structures for buildings that use wood products. It is also intended to be used as an introductory study reference for the general topic of wood structures. For the latter purpose, it is necessary to consider many phenomena and issues in isolation in order to concentrate learning activities. Design, however, is typically a widebased, comprehensive activity. The design process in professional work often works backwards in comparison to typical programmed learning processes, proceeding ordinarily from a first consideration of the whole building, and then of its component parts. This chapter presents some general concerns for design work with specific application to the situation encountered in designing building structures with wood products.

2.1 DESIGN GOALS

Design of any building structure usually has some simple goals. Their ranked order of importance may vary in different situations, but ordinary goals are the following:

To provide a structure with adequate safety for anticipated loading conditions. Safety here referring to life safety for the users of buildings.

To shape and arrange the structure so as to accommodate the other elements of the building construction with greatest ease and least interference.

To provide a completed structure at the lowest cost, in terms of options with equivalent performance capabilities.

To satisfy the current standards of design practice as promulgated by industry and professional organizations and as regulated by enforceable building codes.

To facilitate the achievement of the various requisite aims of the architectural planning of the building.

Accomplishment of these goals involves considerable professional judgment and often some subjective evaluation. However, safety and cost (dollars spent) can be reasonably accurately measured. Safety is understandably a high priority of the community, while cost is usually high on the list for the building owners and investors. Without prejudice as to ranked order, we will consider these two concerns in some detail.

Cost

Dollar cost for building construction is a design concern for all buildings, large and small, utilitarian and luxurious. Almost universally there is a desire for the lowest cost possible for a solution with a defined set of requirements. Final cost is hard to accurately predict in early design stages, but experience and an awareness of general cost factors are essential for the designer. General concerns for cost of construction are discussed in Section 14.11.

Safety

Life safety is a major concern in building structures. Two primary considerations are for resistance to fires and for a low likelihood of collapse under anticipated loads. Both considerations strongly affect choices facing structural designers. Major elements of fire resistance are:

Combustibility of the Structure. If structural materials can burn, they will contribute fuel to the fire as well as hasten the collapse of the structure.

Loss of Strength at High Temperatures. This sets up a race against time, from the moment of inception of the fire to the failure of the structure, with a long interval representing a better chance for occupants to escape the burning building.

Containment of the Fire. Fires usually start at a single location, and prevention of their spread is highly desirable. Walls, floors, and roofs should have construction that resists burn-through by a fire.

Major portions of building code regulations have to do with aspects of fire safety. Materials, systems, and details of construction are rated for fire safety on the basis of experience and tests. These regulations constitute constraints on building design.

Building fire safety is much more than a structural design concern. Clear exit paths for occupants, proper exits, detection and alarm systems, firefighting devices (sprinklers, standpipes, hose cabinets, etc.), and lack of toxic or highly combustible materials are also important. All of these factors can contribute to the race with time. The nature of this race and basic means for controlling it are symbolized in Figure 2.1.

The building structure must, of course, also sustain loads. Safety in this case is represented by the structure's extra margin of capacity beyond that strictly required for the specific magnitude of the loads. This margin may be visualized in the form of a *safety factor*, SF, defined

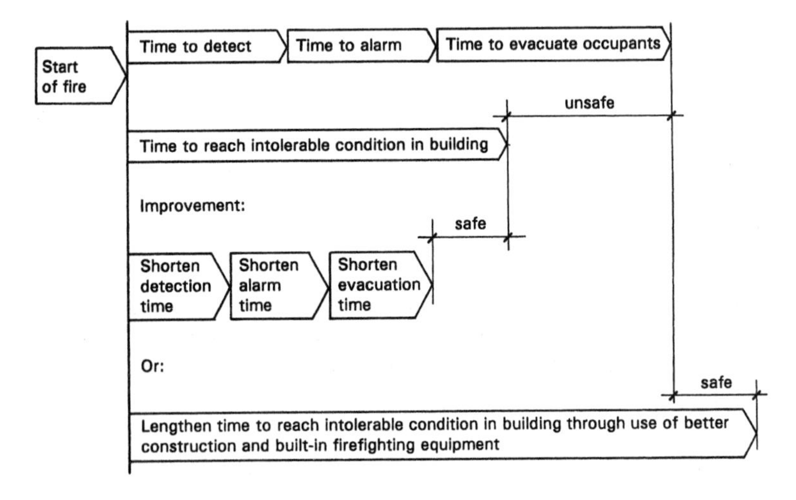

Figure 2.1 Concept of fire safety.

as follows:

$$SF = \frac{\text{Actual capacity of the structure}}{\text{Required capacity for determined loads}}$$

The user of a structure may find comfort in a safety factor as high as 10, but the cost or simply the gross size of such a structure may be unreasonable. For many years, building structures have been designed for an average safety factor of 2. There is no particular reason for this other than experience. Regulation of structural safety is now accomplished using statistical and empirical evaluations, done in various ways during the processes of investigation and design work.

2.2 METHODS OF INVESTIGATION AND DESIGN

Traditional structural design was developed primarily with a method now referred to as stress design. This method utilizes basic relationships derived from classic theories of elastic behavior of materials, and the adequacy or safety of designs is measured by comparison with two primary limits: an acceptable level for maximum stress and a tolerable limit for the extent of deformation (deflection, stretch, etc.). These limits are calculated as they occur in response to the service loads; that is, the loads caused by the normal usage conditions visualized for the structure. This method is also called the working stress method, the stress limits are called allowable working stresses, and the tolerable movements are called allowable deflection, allowable elongation, etc. This design method is now designated as *allowable stress design*, abbreviated ASD.

In order to convincingly establish both stress and strain limits, it was necessary to perform tests on actual structures. This was done extensively, in both the field (on real structures) and in testing laboratories (on specimen prototypes or models). When nature provides its own tests in the form of structural failures, forensic studies are typically made extensively by various people—for research or establishment of liability.

Testing has helped to prove, or disprove, the design theories and to provide data for the shaping of the design process into an intelligent operation. The limits for stress levels and for magnitudes of deformation—essential to the ASD method—have been established in this manner. Thus, although a difference is clearly seen between the stress and strength methods, they are actually both based on evaluations of the total capacity of structures tested to their failure limits. The difference is not minor, but it is really mostly one of procedure.

Allowable Stress Design (ASD)

The ASD method generally consists of the following:

1. The service (working) load conditions are visualized and quantified as intelligently as possible. Adjustments may be made here by the determination of various statistically likely load combinations (dead load plus live load plus wind load, etc.), by consideration of load duration, and so on.
2. Stress, stability, and deformation limits are set by standards for the various responses of the structure to the loads: in tension, bending, shear, buckling, deflection, and so on.
3. The structure is then evaluated (investigated) for its adequacy or is proposed (designed) for an adequate response.

An advantage obtained in working with the ASD method is that the real usage condition (or at least an intelligent guess about it) is kept continuously in mind. The principal disadvantage comes from its detached nature regarding real failure conditions, since most structures develop much different forms of stress and strain as they approach their failure limits.

Strength Design (LRFD)

In essence, the stress method consists of designing a structure to work at some established appropriate percentage of its total capacity. The strength method, now called the LRFD method (for *load and resistance factor design*), consists of designing a structure to fail, but at a load condition well beyond what it should have to experience in use. A major reason for favoring of strength methods is that the failure of a structure is relatively easily demonstrated by physical testing. What is truly appropriate as a working condition, however, is pretty much a theoretical speculation. In any event, the strength method is now largely preferred in professional design work. It was first mostly developed for design of reinforced concrete structures, but is now generally taking over all areas of structural design work.

Nevertheless, it is considered necessary to study the classic theories of elastic behavior as a basis for visualization of the general ways that structures work. Ultimate responses are usually some form of variant from the classic responses (because of inelastic materials, secondary effects, multi-mode responses, etc.). In other words, the usual study procedure is to first consider a classic, elastic response, and then to observe (or speculate about) what happens as failure limits are approached.

For the strength method, the process is as follows:

1. The service (working, usage) loads are quantified as in step (1) of the ASD method, and then are multiplied by an adjustment factor (essentially a safety factor) to produce the *factored load.* See discussion in Section 14.8.
2. The form of response of the structure is visualized and its ultimate (maximum, failure) resistance is quantified in appropriate terms (resistance to compression, to buckling, to bending, etc.). This quantified resistance is also subject to an adjustment factor called the *resistance factor.* See discussion in Section 4.5.
3. The useable resistance of the structure is then compared to the ultimate resistance required (an investigation procedure), or a structure with an appropriate resistance is proposed (a design procedure).

When the design process using the strength method employs both load and resistance factors, it is now called load and resistance factor design (abbreviated LRFD). The procedures used in this method are presented in the example problems throughout this book.

2.3 CHOICE OF DESIGN METHOD

Application of design procedures in the working stress method tend to be simpler and more direct-appearing than in the strength methods. For example, the design of a beam may amount to simply inverting a few basic stress or strain equations to derive some required properties (section modulus for bending, area for shear, moment of inertia for deflection, etc.). Strength method applications tend to be more obscure, mostly because the mathematical formulations and data are usually more complex and less direct-appearing.

Extensive experience with either method will eventually produce some degree of intuitive judgment, allowing designers to make quick approximations even before the derivation of any specific requirements. Thus an experienced designer can look merely at the basic form, dimensions, and general usage of a structure and quickly determine an approximate solution—probably quite close in most regards to what a highly exact investigation will produce. Having designed many similar structures before is the essential basis for this quick solution. This can be useful when some error in computation causes the exact design process to produce a weird answer.

Extensive use of strength methods has required careful visualizations and analyses of modes of failures. After many such studies, one develops some ability to quickly ascertain the single, major response characteristic that is the primary design determinant for a particular structure. Concentration on those selected, critical design factors helps to quickly establish a reasonable design, which can then be tested by many basically routine investigative procedures for other responses.

Use of multiple load combinations, multi-mode structural failures, multiple stress and strain analyses, and generally complex investigative or design procedures is much assisted by computers in most professional design work. A treacherous condition, however, is to be a slave to the computer, accepting its answers with no ability to judge their true appropriateness or correctness. Grinding it out by hand—at least a few times—helps one to appreciate the process.

3

STRUCTURAL INVESTIGATION

A critical part of all structural design work is the necessity to understand and evaluate the physical performance of the structure in the work of resisting the loads it must carry. Some amount of mathematical work must usually be done to support this investigation. The work in this chapter presents some of the fundamentals of applied mechanics that are applied to this investigative work. In the end, it is the design work that must be accomplished, but the structural behaviors described here must be addressed in that work.

3.1 GENERAL CONCERNS

As a prelude to discussing the strength and behavior of structural wood under load, it is necessary to establish a clear understanding of the concept of *unit stress*. Throughout this book there will be many technical terms which may or may not be familiar, depending on the extent of the reader's experience in structural work. For those who have had some preparation in structural mechanics (statics and strength of materials), the material in this chapter will serve as a review. In any event, it is important that these terms and the concepts to which they apply be understood.

3.2 FORCES AND LOADS

A *force* is defined in mechanics as that which tends to change the state of rest or motion of a body. It may be considered as pushing or pulling a body at a definite point and in a definite direction. Such a force tends to give motion to a body at rest, but this tendency may be neutralized by the action of other forces. In building construction the primary concern is with forces that exist in a state of *equilibrium*; that is, with bodies that remain at rest. Motion produced by the applied forces thus consists of deformations of the bodies acted upon.

The units of force are pounds, tons, newtons, and so on, and in engineering practice the term *kip* is used, designating 1000 pounds or literally a *kilopound.* The weight of a body is a vertical, downward force due to gravity.

A *load* is a force generated by an external effect on a structure, such as those due to gravity or wind. As applied to a structure, most loads are *concentrated* at some point or *distributed* in a line or generally on a surface. If a distributed load is the same magnitude at all points, it is said to be *uniformly distributed.* A floor joist that supports a deck is an example of a member supporting a uniformly distributed load that is applied in a linear form and its magnitude is expressed as lb/ft, kip/ft, and so on. The supported deck usually is loaded with a distributed load on its surface and its magnitude is expressed as lb/sq ft (also psf), kip/sq ft (also ksf), and so on.

Concentrated loads mostly consist of supported loads; that is, of load from one member transferred to some supporting member. A girder in a framing system that supports a series of spaced beams is said to be loaded with concentrated loads.

The term *dead load* is applied to the weight of the materials of construction that are supported by a structural member, including the weight of the member itself. For design purposes, the dead load is considered to be permanent for the life of the structure. The term *live load,* on the other hand, is used to designate loads that are not permanent, such as the weight of occupants, furniture, or snow, or the pressure of a gust of wind. In practice the term live load is used exclusively

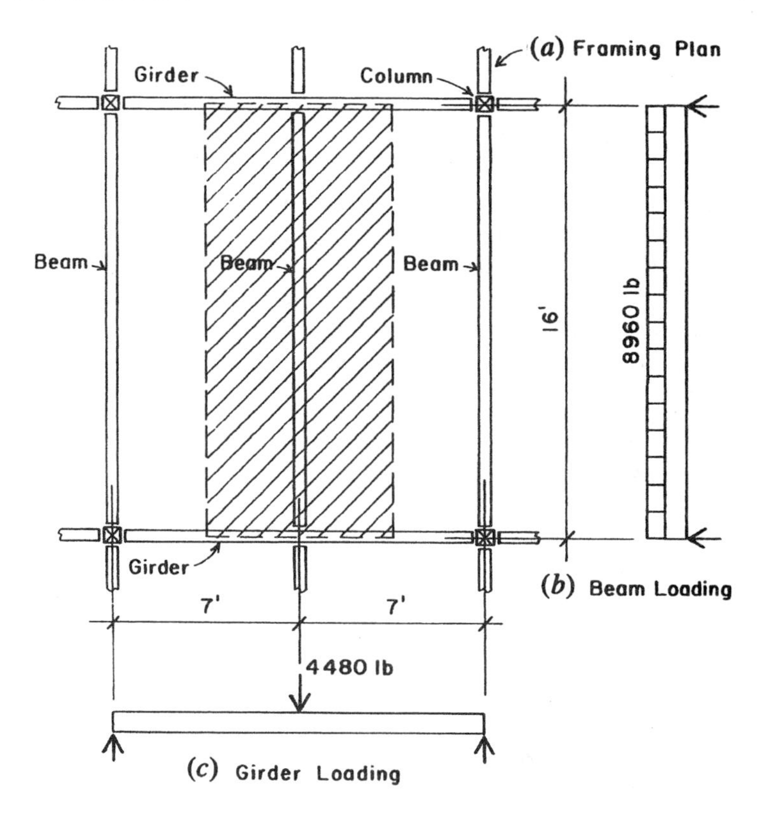

Figure 3.1 Determination of loading of members in a framed system.

to describe non-permanent gravity loads. Other loads are specifically named—wind, earthquake (seismic), fluid pressure, and so on.

Figure 3.1*a* represents a plan of the floor framing of one plan unit (sometimes described as a *bay*). The system shown consists of a deck that is supported by a spaced series of beams. A major plan arrangement consists of the supporting columns which are spaced at 14 ft in one direction and 16 ft in the other direction. The columns thus support every other beam, while the remaining beams are supported by girders between the columns. Assuming a total distributed load, dead plus live, of 80 lb/sq ft (or psf), the area of the deck supported by a single beam is $7 \times 16 = 112\,\text{ft}^2$ (shown as the hatched area in Figure 3.1*a*), and the total load on the beam is $112 \times 80 = 8960\,\text{lb}$. Figure 3.1*b* is the conventional diagram that represents the action of load on the beam.

In this text the designation of the unit of distributed load is w (in lb/ft), and the total distributed load is designated as W (in lb). In this example $W = 8960$ lb and $w = 8960/16 = 560$ lb/ft. For design computations, the load on the beam would be increased by the dead load of its own weight.

The loading from one supported beam on the girder is shown in Figure 3.1*c*. Of course, the girder as shown in the plan actually supports two beams so the total load is twice this value. Furthermore, the girder also carries its own weight as a distributed force along its length.

The loads on the columns in Figure 3.1 are equal to the end support loads of two girders plus the end support loads of two beams; again, plus the weight of the framing members.

3.3 DIRECT STRESS

The term *stress* is used to describe an internal resistance to some external loading on a structure. When the external effect consists of a force that acts along the axis of a linear member (such as a column), the effect is described as an *axial force* or an *axial load*. In Figure 3.2*b* the wood post B supports an axial load equal to the force P. The load exerts a compressive force on the post, tending to shorten it, and this tendency is resisted by the compressive stress developed in the post. The stress produced in this case is called *direct stress*.

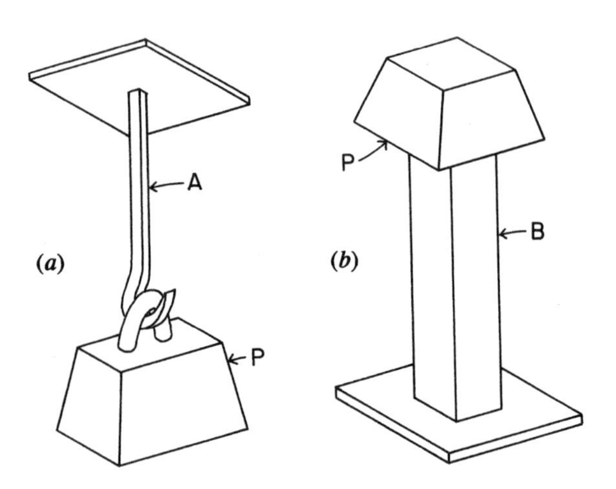

Figure 3.2 Development of direct stress.

A characteristic of direct stress is that the internal resistance may be considered to be evenly distributed over the cross-sectional area of the post. Thus, if P in Figure 3.2*b* is 6400 lb, and B is a nominal 6×6 (actually 5.5×5.5 with an area of $30.25\,\text{in.}^2$), each square inch of the post is stressed to $6400/30.25 = 212$ pounds per square inch (psi). This stress per unit area is called the *unit stress*, to distinguish it from the internal force of 6400 lb. By calling the force P, the area of the cross section A, and the unit stress f, this fundamental relationship between load and stress may be stated as:

$$f = \frac{P}{A} \quad \text{or} \quad P = fA \quad \text{or} \quad A = \frac{P}{f}$$

When using this equation, remember the two assumptions on which it is based: the loading is axial and the stress is distributed evenly over the cross section. All three forms shown are used for different situations. The first yields the stress value for a defined load and area; the second is used to find the useable load for a defined area at a given limiting stress; the third is a design formula used to find the cross-sectional area required for a defined load and a given limiting stress.

In Figure 3.2*b* the post is subjected to a compression force and the resulting stress on the post cross section is one of compression. In Figure 3.2*a* the load on the hanger produces tension stress in the cross section of the hanger shaft. Tension stress and compression stress are opposite in sense, but they are similar in being the result of direct force. The same formulas are used to express the relation between force and stress.

3.4 KINDS OF STRESS

The three kinds of stress are *compression*, *tension*, and *shear*. Compression and tension (Figure 3.2) are opposite in sign; that is, they are the reverse of each other and act at right angles to a section. The deformation they produce is one of shortening or lengthening of the stressed member.

Shear is a different form of stress that acts in the plane of a cross section and causes a deformation which is diagonal in form. The form of shear stress may be visualized as comparable to friction, although it occurs inside the material, not on a surface.

Figure 3.3*a* represents a beam with a uniformly distributed load. There is a tendency for the beam to drop down at the faces of the

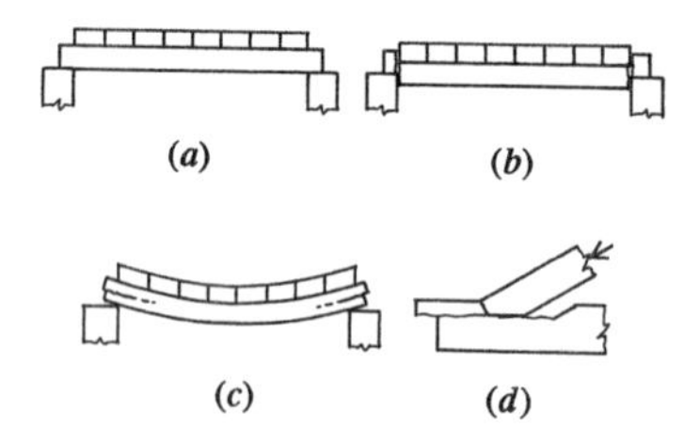

Figure 3.3 Development of shear stress.

supports, which is produced by what is called *vertical shear force*, as indicated in Figure 3.3*b*. Figure 3.3*c* shows an exaggerated bending action of the beam and a failure consisting of upper and lower portions sliding horizontally with respect to each other at the beam's ends; this action is called *horizontal shear failure*. Horizontal shear is critical for solid wood members, as it produces a splitting of the wood along its grain direction, which is one of the weakest stress resistances of wood. Another form of shear failure is shown in Figure 3.3*d*, which indicates a splitting of the wood.

Bending and shear in beams are discussed in Chapter 6. These actions produce the three basic kinds of stress—compression, tension, and shear—but the stresses are not uniformly distributed on the beam cross sections.

3.5 DEFORMATION

Whenever a body is subjected to force, and the force action produces stresses, the body is deformed by the stresses. In a soft sponge or a rubber band these deformations are readily seen. Although they are not often observable, the same deformations occur in building structures. Incremental deformations due to stress accumulate in the stressed body, producing shortening, lengthening, or—in the case of beams—a change of curvature described as deflection. The deformation of a unit of material is called *strain* and it is the accumulation of strain that produces deformations. Some of the relationships between stress and strain are described in the following discussions.

3.6 ELASTIC RESPONSE AND LIMIT

Elasticity is the property of a material that enables it to return to its original size and shape when a stress-producing force action is removed.

Although materials may continue to offer resistance beyond the elastic limit, a condition called *permanent set* occurs, meaning that some of the load-produced deformation will be permanent. The elastic limit is of great concern in steel, but not so significant in wood. Within the elastic range of stress, defined relations between stress and strain may be observed. A perfectly elastic material is one in which stress and strain maintain a constant proportionality throughout a particular range of stress magnitude. In the allowable stress design (ASD) method, this proportionality of stress to strain is usually assumed through the range of stress up to the limiting (allowable) stress.

3.7 INELASTIC BEHAVIOR AND ULTIMATE STRENGTH

At the upper limits of stress capability of most materials the simple proportionality of stress and strain no longer occurs. This behavior is described as an *inelastic* condition. While not so significant for the ASD method, inelastic conditions are the general case in the strength design method (LRFD). It is important to recognize both elastic and inelastic responses in defining the allowable load (at service levels) and the ultimate strength of structural members. Use of these conditions for various force actions is treated in Chapter 6, where both the ASD and LRFD methods are explained.

3.8 MODULUS OF ELASTICITY

Stress and strain are related not only in the basic forms they take, but in their actual magnitudes. Figure 3.4 shows the relation between stress and strain magnitudes for direct stress for a number of different materials. The form of such a graph illustrates various aspects of the nature of structural behavior of the materials.

Curves 1 and 2 represent materials with a constant proportionality of the stress and strain magnitudes throughout the range of their structural actions. Such materials are sometimes described as being perfectly elastic. For these materials, a quantified relationship between stress and strain can be simply defined in terms of the slope or angle of the straight line graph. This relationship is usually expressed as the tangent of the angle of the graph and is called the *modulus of elasticity* of the material. The higher the value of this modulus—that is, the steeper the slope of its graph—the stiffer the material. Thus, the material represented by curve 1 in Figure 3.4 is stiffer (more resistant to deformation) than that represented by curve 2.

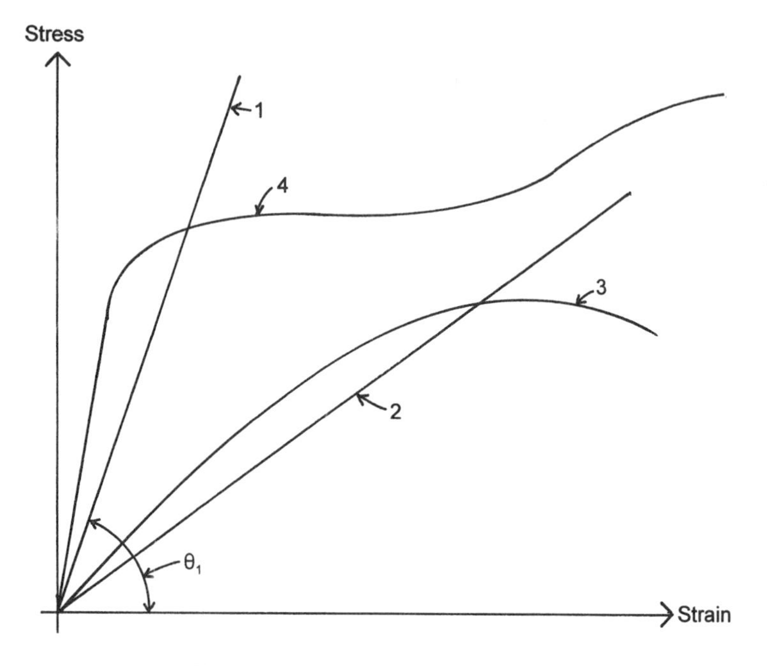

Figure 3.4 Relation of stress and strain.

For direct stress of tension or compression, the strain is measured as a linear shape change, and the modulus is called the direct stress modulus of elasticity. For shear stress, strain is measured as an angular change, and the resulting modulus is called the shear modulus of elasticity.

Some materials, such as glass and very high-strength steel, have a constant modulus of elasticity for just about the full range of stress up to failure of the material. This is the behavior exhibited by the curves labeled 1 and 2 in Figure 3.4. Other materials, such as wood, concrete, and plastic, have a curved form for the stress-strain graph (curve 3 in Figure 3.4). The curved graph indicates that the relationship between stress and strain varies throughout the range of response. However, if the value of stress permitted for response to service loads is a small fraction of the material's actual ultimate strength, and thus only the lower portion of the graph is used, it is reasonable to assume a constant value for the modulus of elasticity, which is typically done for determination of deformations at service load levels.

Stress is defined with a unit of force per unit area (such as lb/in.^{2} or psi). Unit strain, on the other hand, is expressed as a percentage or a constant of proportionality. Thus the modulus of elasticity, designated E, is defined as

$$E = \frac{\text{stress}}{\text{strain}}$$

And its dimensional unit is the same as the stress (psi, etc.), since the strain is a dimensionless number. For structural steel E = 29,000,000 psi [200 GPa], and for wood, depending on its species and grade, it varies from something less than 1,000,000 psi to about 1,900,000 psi [713 GPa]. This indicates that—as everyone knows—steel is considerably stiffer than wood.

As described earlier, the value for E is assumed to remain constant through the range of acceptable stress for the ASD method. For consideration of failure load levels in the LRFD method, however, an acknowledgment must be made of the nonlinear character of the stress/strain relationship. Since buckling of slender members is essentially a failure condition, a different value is used for E, defined as E_{min}. Use of this modified value is demonstrated in the investigation of columns in Chapter 9.

3.9 PERMISSIBLE VALUES FOR DESIGN

For structural design an important piece of data that must be established is the maximum values to be used for the structural limits of the material being used. For wood structures these are defined as *reference design values*. Although utilized somewhat differently, these values are used for both the ASD and LRFD methods. The values defined consist of the modulus of elasticity and the various kinds of stress. The source for these values and their determination for specific design conditions are described in Chapter 4.

4

DESIGN DATA AND CRITERIA

A major consideration for design of any wood structure is the determination of design values for stresses and the modulus of elasticity to be used for any given piece of wood. This chapter treats the issues involved in establishing these values, with special concern for sawn lumber.

4.1 GENERAL CONCERNS

There are many factors to be considered in determining the unit stresses to be used for design of wood structures. Extensive testing has produced values known as *reference design values*. To obtain values for design work, the base values are modified for considerations such as loss of strength from defects, the size and position of knots, the size of standard dimension members, the degree of density of the wood, and the condition of seasoning or specific value of moisture content of the wood at the time of use. For specific uses of the structure, modifications may be made for considerations such as load duration, direction of stress with respect to the wood grain, and the type of structural element. Separate groups of design values are established for decks, closely-spaced rafters and joists, large-dimension beams, and posts (columns).

4.2 REFERENCE DESIGN VALUES FOR ALLOWABLE STRESS DESIGN (ASD)

Because the relative effects of natural defects vary with the size of sawn pieces and the usage application, structural lumber is classified with respect to its size and use. Incorporating these and other considerations, the four major classifications are (note that *nominal dimensions*, as explained in the Appendix, are used here):

1. *Dimension Lumber.* Sections with thickness of 2 to 4 in. and width of 2 in. or more (includes most studs, rafters, joists, and planks).
2. *Beams and Stringers.* Rectangular sections 5 in. or more in thickness with width 2 in. or more greater than thickness, graded for strength in bending when loaded on the narrow face.
3. *Posts and Timbers.* Square or nearly square sections, 5 × 5 or larger, with width not more than 2 in. greater than thickness, graded primarily for use as compression elements where bending strength is not a major concern.
4. *Decking.* Lumber from 2 to 4 in. thick, tongued and grooved or splined on the narrow face, graded for flat application (mostly as plank deck).

Design data and standards for structural lumber are provided in the publications of the wood industry, such as the *NDS* (Ref. 1).

A broad grouping of trees identifies them as *softwoods* or *hardwoods*. Softwoods such as pine, cypress, and redwood, mostly come from trees that are coniferous or cone-bearing, whereas hardwoods come mostly from trees that have broad leaves, as exemplified by oaks and maples. Two species of trees used extensively for structural lumber in the United States are Douglas fir and Southern pine, both of which are classified among softwoods.

Dimensions

As discussed in Chapter 1, structural lumber is described in terms of a nominal size which is slightly larger than the true dimensions of pieces. However, properties for structural computations, as given in Appendix

Table A.3, are based on the true dimensions which are also listed in the table. For sake of brevity, we have omitted metric units from the text and the tabular data.

Reference Design Value Tables

Table 4.1 gives reference design values to be used for ordinary allowable stress design. It is adapted from the *NDS* (Ref. 1) and gives data for one popular wood species: Douglas fir-larch. To obtain values from the table the following is determined:

1. *Species.* The *NDS* publication lists values for several different species, only one of which is included in Table 4.1.
2. *Moisture Condition at Time of Use.* The moisture condition corresponding to the table values is given with the species designation in the table. Adjustments for other conditions are described in the table footnotes or in various specifications in the *NDS.*
3. *Grade.* This is indicated in the first column of the table and is based on visual grading standards.
4. *Size and Use.* The second column of the table identifies size ranges or usages of the lumber.
5. *Structural Function.* Individual columns in the table yield values for various stress conditions. The last two columns yield the material modulus of elasticity.

In the reference document there are extensive footnotes for this table. Data from Table 4.1 is used in various example computations in this book, and some issues treated in the document footnotes are explained. In many situations there are modifications (or adjustments, as they are called in the *NDS*) to the design values, as will be explained later. Referred to in a footnote to Table 4.1, Table 4.2 yields adjustment factors for dimension lumber and decking based on the dimensions of the piece.

Bearing Stress

There are various situations in which a wood member may develop a contact bearing stress: essentially a surface compression stress. Some examples are the following:

At the base of a wood column supported in direct bearing. This is a case of bearing stress that is in a direction parallel to the grain.

TABLE 4.1 Reference Design Values for Visually-Graded Lumber of Douglas Fir-Larch[a] (Values in psi)

Species and Commercial Grade	Member Size and Use Classification	Bending, F_b	Tension Parallel to Grain, F_t	Shear Parallel to Grain, F_v	Compression Perpendicular to Grain, $F_{c\perp}$	Compression Parallel to Grain, F_c	Modulus of Elasticity	
							E	E_{min}
Dimension Lumber 2 to 4 in. thick								
Select structural	2 in. & wider	1500	1000	180	625	1700	1,900,000	690,000
No. 1 & better		1200	800	180	625	1550	1,800,000	660,000
No. 1		1000	675	180	625	1500	1,700,000	620,000
No. 2		900	575	180	625	1350	1,600,000	580,000
No. 3		525	325	180	625	775	1,400,000	510,000
Stud		700	450	180	625	850	1,400,000	510,000
Timbers								
Dense select structural	Beams & Stringers	1900	1100	170	730	1300	1,700,000	620,000
Select structural		1600	950	170	625	1100	1,600,000	580,000
Dense No. 1		1550	775	170	730	1100	1,700,000	620,000
No. 1		1350	675	170	625	925	1,600,000	580,000
No. 2		875	425	170	625	600	1,300,000	470,000
Dense select structural	Posts & Timbers	1750	1150	170	730	1350	1,700,000	620,000
Select structural		1500	1000	170	625	1150	1,600,000	580,000
Dense No. 1		1400	950	170	730	1200	1.700,000	620,000
No. 1		1200	825	170	625	1000	1,600,000	580,000
No. 2		750	475	170	625	700	1,300,000	470,000
Decking								
Select Dex	2–4 in. thick, 6–8 in. wide	1750	—	—	625	—	1,800,000	660,000
Commercial Dex		1450	—	—	625	—	1,700,000	620,000

[a] Values listed are for reference in determining design use values subject to various modifications. See Table 4.2 for size adjustment factors for dimension lumber (2–4 in. thick). For design purposes, values are subject to various modifications.

Source: Data adapted from *National Design Specification® for Wood Construction, 2005 edition* (Ref. 1), with permission of the publishers, American Forest & Paper Association. The tables in the reference document have data for many other species and also have extensive footnotes.

TABLE 4.2 Size Adjustment Factors (C_F) for Dimension Lumber and Decking

Grades	Width (depth)	Thickness (breadth), F_b		F_t	F_c
		2-in. & 3-in.	4-in.		
Select structural, No. 1 and better, No. 1, No. 2, No. 3	2, 3 and 4-in.	1.5	1.5	1.5	1.15
	5-in.	1.4	1.4	1.4	1.1
	6-in.	1.3	1.3	1.3	1.1
	8-in.	1.2	1.3	1.2	1.05
	10-in.	1.1	1.2	1.1	1.0
	12-in.	1.0	1.1	1.0	1.0
	14-in. and wider	0.9	1.0	0.9	0.9
Stud	2, 3, and 4-in.	1.1	1.1	1.1	1.05
	5 and 6-in.	1.0	1.0	1.0	1.0
	8-in. and wider	Use No. 3 grade tabulated design values and size factors			
Decking		2-in.	3-in.		
		1.10	1.04		

Source: Data adapted from *National Design Specification® for Wood Construction, 2005 edition* (Ref. 1), with permission of the publishers, American Forest & Paper Association.

At the end of a beam that is supported by bearing on a support. This is a case of bearing stress that is perpendicular to the grain.

Within a bolted connection at the contact surface between the bolt and the wood at the edge of the bolt hole.

In a timber truss where a compression force is developed by direct bearing between the two members. This is frequently a situation involving bearing stress that is at some angle to the grain other than parallel or perpendicular. Common in the past, this form of joint is seldom used today.

For connections, the bearing condition is usually incorporated into the general assessment of the unit value of connecting devices. General considerations for design of bolts are treated in Chapter 10. Design for bearing in beams is discussed in Section 6.4. The situation of stress at an angle to the grain requires the determination of a compromise value somewhere between the allowable values for the two limiting stress conditions for stresses parallel and perpendicular to the wood grain. This situation is discussed in Section 4.4.

4.3 ADJUSTMENT OF DESIGN VALUES

The values given in Table 4.1 are basic references for establishing the allowable values to be used for design. The table values are based on some defined norms, and in many cases the design values will be adjusted for actual use in structural computations. In some cases the form of the modification is a simple increase or decrease achieved by a percentage factor. Table 4.3 lists the various types of adjustment and indicates their applicability to various reference design values. The types of adjustment factors are described in the following discussions.

Load Duration Factor, C_D

The Table 4.1 values are based on so-called *normal* duration loading, which is actually somewhat meaningless. Increases are permitted for very short duration loading, such as wind and earthquakes. A decrease is required when the critical design loading is long time in duration (such as a major dead load). Table 4.4 gives a summary of the *NDS* requirements for modifications for load duration.

TABLE 4.3 Applicability of Adjustment Factors for Sawn Lumber, ASD

	F_b	F_t	F_v	$F_{c\perp}$	F_c	E	E_{min}
ASD only:							
Load Duration	C_D	C_D	C_D	—	C_D	—	—
ASD and LRFD:							
Wet Service	C_M	C_M	C_M	C_M	C_M	C_M	C_M
Temperature	C_t	C_t	C_t	C_t	C_t	C_t	C_t
Beam Stability	C_L	—	—	—	—	—	—
Size	C_F	C_F	—	—	C_F	—	—
Flat Use	C_{fu}	—	—	—	—	—	—
Incising	C_i	C_i	C_i	C_i	C_i	C_i	C_i
Repetitive Member	C_r	—	—	—	—	—	—
Column Stability	—	—	—	—	C_p	—	—
Buckling Stiffness	—	—	—	—	—	—	C_T
Bearing Area	—	—	—	C_b	—	—	—
LRFD only:							
Format Conversion	K_F	K_F	K_F	K_F	K_F	—	K_F
Resistance	ϕ_b	ϕ_t	ϕ_v	ϕ_c	ϕ_c	—	ϕ_s
Time Effect	λ	λ	λ	λ	λ	—	—

TABLE 4.4 Adjustment Factors for Design Values for Structural Lumber Due to Load Duration, C_D[a]

Load Duration	Multiply Design Values by:	Typical Design Loads
Permanent	0.9	Dead load
Ten years	1.0	Occupancy live load
Two months	1.15	Snow load
Seven days	1.25	Construction load
Ten minutes	1.6	Wind or earthquake load
Impact[b]	2.0	Impact load

[a]These factors shall not apply to reference modulus of elasticity E, to reference modulus of elasticity for beam and column stability E_{min}, nor to compression perpendicular to the grain reference design values $F_{c\perp}$ based on a deformation limit.

[b]Load duration factors greater than 1.6 shall not apply to structural members pressure-treated with water-borne preservatives or fire retardant chemicals. The impact load duration factor shall not apply to connections.

Source: Adapted from the *National Design Specification ® for Wood Construction*, 2005 edition (Ref. 1), with permission of the publishers, American Forest & Paper Association.

Wet Service Factor, C_M

The *NDS* document on which Table 4.1 is based defines a specific assumed moisture content on which the table values are based. Increases may be allowed for wood that is specially cured to a lower moisture content. If exposed to weather or other high-moisture conditions, a reduction may be required.

Temperature Factor, C_t

Where prolonged exposure to temperatures over 150° F exists, design values must be reduced. The adjustment factor varies for different reference values and includes consideration for moisture condition and exposure of the wood.

Beam Stability Factor, C_L

Design flexural stress must be adjusted for conditions of potential buckling. The general situation of buckling and the remedies for its prevention are discussed in Section 6.5.

Size Factor, C_F

For dimension lumber, adjustments for size are made for design stresses of bending, shear, and tension as described in Table 4.2. For beams 5 in.

or thicker, with depth exceeding 12 in., adjustment of bending stress is made as described in Section 6.2. Other adjustments may be required for columns and for beams loaded for bending on their wide face.

Flat Use Factor, C_{fu}

When sawn lumber 2 to 4 in. thick is loaded on the wide face (as a plank), adjustments are required as described in the *NDS* references for Table 4.1.

Incising Factor, C_i

Incising refers to small indentation-form cuts made on the surface of lumber that is treated by impregnation of chemicals for enhancement of resistance to fire or rot. A reduction of all reference values is required for this condition.

Repetitive Member Factor, C_r

When wood beams of dimension lumber (mostly joists and rafters) are closely spaced and share a load, they may be eligible for an increase of 15 percent in reference design values; this condition is described as *repetitive member use*. To qualify, the members must be not less than three in number, must support a continuous deck, must be not over 24 in. on center, and must be joined by construction that makes them share deflections (usually bridging or blocking). This increase is also permitted for built-up beams formed by direct attachment of multiple-dimension lumber elements.

The following example illustrates the application of the beam design procedure for the case of a roof rafter.

Example 1. Rafters of Douglas fir-larch, No. 2 grade, are to be used at 16 in. spacing for a span of 20 ft. Solid wood blocking is provided for nailing of the plywood deck panels. Live load without snow is 20 psf and the total dead load, including the rafters, is 15 psf. Find the minimum size for the rafters, based only on bending stress.

Solution: At this spacing the rafters qualify for the increased bending stress described as *repetitive member use*. For the No. 2 grade rafters, the reference value from Table 4.1 for F_b is 900 psi. The loading condition as described (live load without snow) qualifies the situation with regard to load duration for an adjustment factor of $C_D = 1.25$ (see Table 4.4). This live load is usually considered to provide for temporary conditions during roof construction or maintenance and is of short duration. The allowable bending stress for design is thus modified as

$$F'_b = (1.15)(1.25)F_b = (1.15)(1.25)(900) = 1294\,\text{psi}$$

For the rafters at 16 in. spacing, the maximum bending moment is

$$M = \frac{wL^2}{8} = \frac{\left(\frac{16}{12}\right)(20 + 15)(20)^2}{8} = 2333\,\text{ft-lb}$$

and the required section modulus is

$$S = \frac{M}{F'_b} = \frac{2333 \times 12}{1294} = 21.64\,\text{in.}^3$$

From Table A.3, the smallest section with this property is a 2 × 12, with an S of 31.64 in.^{3}. Note that the allowable stress is not changed by Table 4.2 as the table factor is 1.0. Lateral bracing for rafters is discussed in Section 6.5.

Column Stability Factor, C_P

This adjustment is performed in the typical processes of investigation and design of wood columns, which is discussed in Chapter 9. Most often, an adjustment consisting of a reduction of permissible compression stress parallel to the grain is required for relatively slender columns.

Buckling Stiffness Factor, C_T

This adjustment is made only for the modified modulus of elasticity, E_{min}, in certain situations involving wood members subjected to combined compression and bending. Its principal application is in the design of the top chords of wood trusses.

Bearing Area Factor, C_b

This factor is provided for the special case of bearing perpendicular to the grain when the length of bearing is very small. This applies primarily to situations where bearing is transferred from a wood member to a steel plate or washer. The *NDS* provides a formula for determination of an adjusted design stress for these situations.

Adjustments for the LRFD Method

The group of adjustments given at the bottom of Table 4.3 is designated as being for use with the LRFD method only. Use of these adjustments is described in Section 4.5.

Modulus of Elasticity

As discussed in Section 3.8, the modulus of elasticity is a measure of the relative stiffness of a material. For wood, two reference values are used for the modulus of elasticity. The basic reference value is designated E and is the value used for ordinary deformations—primarily the deflection of beams. The other value is designated E_{min} and it is used for stability computations involving the buckling of beams and columns. Values for the stability modulus of elasticity are given in the last column in Table 4.1. Applications for determination of buckling effects in columns are presented in Chapter 9.

4.4 MODIFICATION FOR LOADING WITH RELATION TO GRAIN DIRECTION

Under the condition shown in Figure 4.1, the load from member B exerts a compressive stress on member A on a surface inclined to the grain in member A. The compressive strength of wood is a maximum parallel to the grain and a minimum perpendicular to the grain. The allowable unit compressive stress on an inclined surface is determined

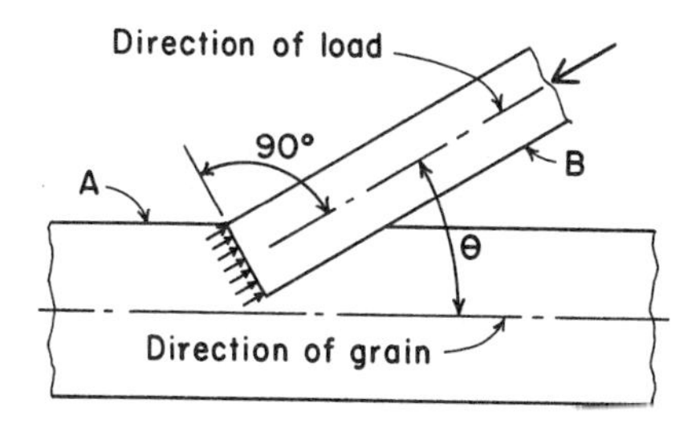

Figure 4.1 Development of stress at an angle to the grain in wood members.

from the following expression, known as the Hankinson formula:

$$F_n = \frac{F_c^* \times F_{c\perp}}{F_c^* \left(\sin^2 \theta\right) + F_{c\perp} \left(\cos^2 \theta\right)}$$

where

F_n = allowable unit stress acting perpendicular to the inclined surface

$F_c{}^*$ = adjusted reference stress in compression parallel to the grain

$F_{c\perp}$ = reference stress in compression perpendicular to the grain

θ = angle between the direction of the load and direction of the grain

When the load is applied parallel to the grain, θ is zero. When the load is applied perpendicular to the grain, θ is 90°. Table 4.5 gives values of $\sin^2$ and $\cos^2$ for various angles from zero to 90 degrees.

Example 2. Two timbers 6 in. wide consisting of Douglas-fir larch, dense No. 1 grade, are framed together as shown in Figure 4.1. The angle between the two pieces is 30°. Compute the allowable unit compressive stress on the inclined surface.

Solution: From Table 4.1, assuming the members to be of "Post and Timbers" classification, the allowable unit stress for bearing parallel to

TABLE 4.5 Values for Use in the Hankinson Formula[a]

Angle (degrees)			Angle (degrees)		
$\sin^2$		$\cos^2$	$\sin^2$		$\cos^2$
0.00000	0	1.00000	0.59682	50	0.41318
0.00760	5	0.99240	0.67101	55	0.32899
0.03015	10	0.96985	0.75000	60	0.25000
0.06698	15	0.93302	0.82140	65	0.17860
0.11698	20	0.88302	0.88302	70	0.11698
0.17860	25	0.82140	0.93302	75	0.06698
0.25000	30	0.75000	0.96985	80	0.03015
0.32899	35	0.67101	0.99240	85	0.00760
0.41318	40	0.58682	1.00000	90	0.00000
0.50000	45	0.50000			

[a] See Figure 4.1.

the grain is 1200 psi, and the allowable stress perpendicular to the grain is 730 psi. Using the appropriate values for the angle from Table 4.5, the allowable bearing stress on the inclined plane is

$$F_n = \frac{1200 \times 730}{1200(0.25) + 730(0.75)} = 1033\,\text{psi}$$

A similar modification is required with the use of some fasteners, such as bolts and split-ring connectors. As an alternative to the computations just illustrated, the functions of the Hankinson formula may be displayed in graphical form, and necessary modifications can be approximated from the graph. The use of such a method is illustrated in Section 10.1.

Problem 4.4.A. Two Douglas fir-larch timbers of ordinary No. 1 grade frame together as shown in Figure 4.1. The angle between the members is 45 degrees. Determine the allowable unit bearing stress on the inclined contact surface.

Problem 4.4.B. Same as Problem 4.4.A, except the wood is dense No. 1 grade and the angle is 35 degrees.

4.5 DESIGN CONTROLS FOR LRFD

For the allowable stress design method (ASD), the concentration—as its name implies—is on maximum developed stress levels as produced by the service loads. The maximum permissible stresses are specified and adjusted for various circumstances. These limiting stress levels are then used to produce a limiting force action (shear, bending, bearing, etc.), and the result is compared to the maximum action produced by the loading.

For the load and resistance design method (LRFD), the same relationships are used, except that the loads are quantified as factored loads (ultimate loads) and the limiting force action is a defined failure limit. The failure limit is defined as the *adjusted resistance* and is designated with the superscript prime, such as *M'N* for adjusted moment resistance. For design purposes, the relationship between the resistance and the load effect is defined as

$$\lambda\, \phi_b\, M' \geq M_u$$

in which

λ = is the time effect factor, see Table 4.8
ϕ_b = is the resistance factor of 0.85 for bending (flexure), see Table 4.7
M' = is the adjusted resisting moment
M_u = is the maximum moment produced by the factored loading

One approach—as defined by the *NDS* (Ref. 1)—is to define the resisting moment (or shear, bearing, etc.) as one produced in the same manner as that derived by the ASD method, except that an adjusted level of stress is used. The adjustment factors are those given in Table 4.6, referring to the defined stress values listed in Table 4.1. It is customary to use stress units of pounds with the ASD method, but

TABLE 4.6 Adjustment Factors for LRFD Using ASD Reference Design Values

Stress	Conversion Factor[a] (ASD to LRFD) K_F
Bending, F_b	$2.16/1000\phi_b$
Tension, F_t	$2.16/1000\phi_t$
Shear, F_v	$2.16/1000\phi_v$
Compression parallel to grain, F_c	$2.16/1000\phi_c$
Compression perpendicular to grain, $F_{c\varsigma}$	$1.875/1000\phi_c$
Connections	$2.16/1000\phi_z$
Modulus of elasticity for stability, E_{min}	$1.5/1000\phi_s$

[a]Produces unit values in kips when ASD values are in pounds.
Source: Adapted from data in the *National Design Specification® for Wood Construction* (Ref. 1), with permission of the publishers, American Forest & Paper Association.

TABLE 4.7 Resistance Factors for Wood Structures, LRFD

ϕ_b	Flexure(Bending)	0.85
ϕ_c	Compression, Bearing	0.90
ϕ_t	Tension	0.80
ϕ_v	Shear, Torsion	0.75
ϕ_S	Stability, E_{min}	0.85
ϕ_z	Connections	0.65

Source: Adapted from data in the *National Design Specification ® for Wood Construction* (Ref. 1), with permission of the publishers, American Forest & Paper Association.

TABLE 4.8 Load Combinations and Time Effect Factors, LRFD

Load Combination	Time Effect Factor, λ
1.4(Dead load)	0.6
1.2(Dead load) + 1.6(Live load):	
When live load is from storage	0.7
When live load is from occupancy	0.8
When live load is from impact	1.25
1.2(Dead load) + 1.6(Wind load) + Live load + 0.5(Roof load)	1.0
1.2(Dead load) +1.6(Roof load) + 0.8(Wind load)	0.8
1.2(Dead load) + Earthquake load + Live load	1.0

Source: Adapted from data in the *National Design Specification ® for Wood Construction* (Ref. 1), with permission of the publishers, American Forest & Paper Association.

to use units of kips with the LRFD method: thus the inclusion of the 1000 factor in the adjustment.

Resistance factors ϕ are as given in Table 4.7. They vary, depending on the type of behavior being considered.

The time effect factor λ is given in Table 4.8. It varies for the different load combinations used to find the required ultimate resistance. Care must be taken when using the adjustment factors to ensure that the ASD reference design value is not modified by the factor for load duration (C_D). For the LRFD method this adjustment is made by use of the λ factor.

Use of this process is discussed in Section 6.7 and its application in design of beams is discussed throughout Chapter 7. Uses of the LRFD procedures for other applications are discussed in later chapters in this book.

5

BEAM FUNCTIONS

This chapter presents considerations that are made in the investigation of the behavior of beams. For basic explanation of relationships, the units used for forces and dimensions are of less significance than their numeric values. For this reason, and for the sake of brevity and simplicity, most numerical computations in the text have been done using only U.S. units. For readers who wish to use metric units, however, the exercise problems have been provided with dual units.

5.1 GENERAL CONSIDERATIONS

A beam is a structural member that resists transverse loads. The supports for beams are usually at or near the ends, and the supporting upward forces are called reactions. The loads acting on a beam tend to *bend* it rather than shorten or lengthen it. *Girder* is the name given to a beam that supports smaller beams; all girders are beams insofar as their structural action is concerned. For construction usage beams carry various names, depending on the form of construction; these include *purlin, joist, rafter, lintel, header*, and *girt*.

There are, in general, five types of beams identified by the number, kind, and position of the supports. Figure 5.1 shows diagrammatically the different types and also the exaggerated shape each beam tends to assume as it bends (deforms) under the loading. In ordinary steel or reinforced concrete beams, these deformations are not usually visible to the eye, but some deformation is always present. In wood beams deformations may be more visible.

A *simple beam* rests on a support at each end, the ends of the beam being free to rotate (Figure 5.1*a*).

A *cantilever beam* is supported at one end only. A beam embedded in a wall and projecting beyond the face of the wall is a typical example (Figure 5.1*b*).

An *overhanging beam* is a beam whose end or ends project beyond its supports. Figure 5.1*c* indicates a beam overhanging one support only.

A *continuous beam* rests on more than two supports (Figure 5.1*d*). Continuous beams are commonly used in reinforced concrete and welded steel construction but are less frequently used in wood structures. A common occurrence is that of decks supported by closely-spaced rafters or joists; deck sheathing ordinarily spans continuously over several supports.

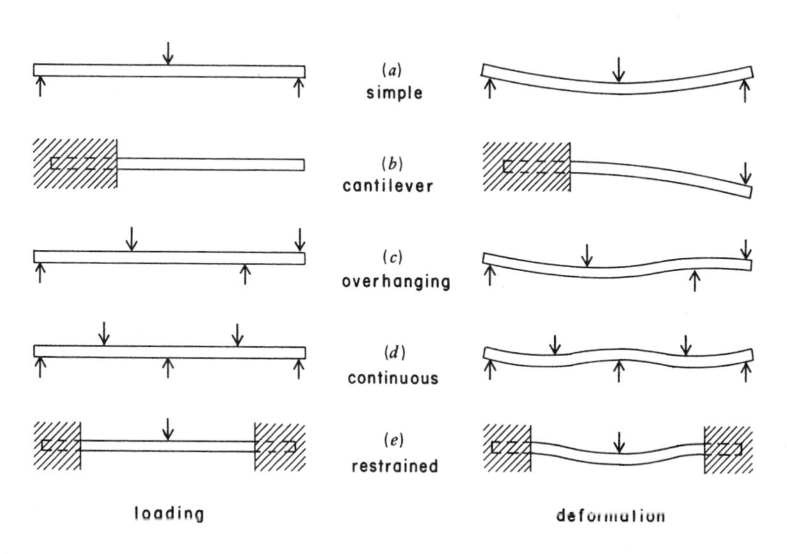

Figure 5.1 Types of beams.

A *restrained beam*, has one or both ends restrained or fixed against rotation (Figure 5.1*e*).

Beams are acted on by external forces that consist of the loads and the reaction forces developed by the beam's supports. The two types of loads that commonly occur on beams are called *concentrated* and *distributed.* A concentrated load is assumed to act at a definite point; such a load is that caused when one beam supports another beam. A distributed load is one that acts over a considerable length of the beam; such a load is one caused by a floor deck supported directly by a beam. If the distributed load exerts a force of equal magnitude for each unit of length of the beam, it is known as a *uniformly distributed load.* The weight of a beam is a uniformly distributed load that extends over the entire length of the beam. However, some uniformly distributed loadings supported by the beam may extend over only a portion of the beam length.

5.2 MOMENTS

The term *moment of a force* is commonly used in engineering problems It is fairly easy to visualize a length of 3 ft, an area of 26 sq in., or a force of 100 lb. A moment, however is less easily understood; it is a force multiplied by a distance. *A moment is the tendency of a force to cause rotation about a given point or axis.* The magnitude of the moment of a force about a given point is the magnitude of the force (pounds, kips, etc.) multiplied by the distance (feet, inches, etc.) from the force to the point of rotation. The point is called the center of moments, and the distance, which is called the *lever arm* or *moment arm*, is measured by a line drawn through the center of moments perpendicular to the line of action of the force. Moments are expressed in compound units such as foot-pounds and inch-pounds or kip-feet and kip-inches. In summary:

$$\text{Moment of force} = \text{magnitude of force} \times \text{moment arm}$$

Consider the horizontal force of 100 lb shown in Figure 5.2*a*. If point *A* is the center of moments, the lever arm of the force is 5 ft. Then the moment of the 100 lb force with respect to point *A* is $100 \times 5 =$ 500 ft-lb. In this illustration the force tends to cause a *clockwise* rotation (shown by the dashed-line arrow) about point *A* and is called a positive moment. If point *B* is the center of moments, the moment arm of the

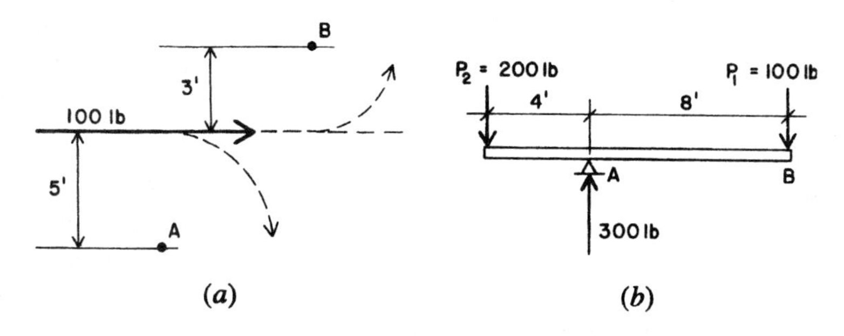

Figure 5.2 Development of moments.

force is 3 ft. Therefore the moment of the 100 lb force about point B is $100 \times 3 = 300$ ft-lb. With respect to point B, the force tends to cause *counterclockwise* rotation; it is called a negative moment. It is important to remember that you can never consider the moment of a force without having in mind the particular point or axis about which it tends to cause rotation.

Figure 5.2*b* represents two forces acting on a bar which is supported at point A. The moment of force P_1 about point A is $100 \times 8 = 800$ ft-lb, and it is clockwise or positive. The moment of force P_2 about point A is $200 \times 4 = 800$ ft-lb. The two moment values are the same, but P_2 tends to produce a counterclockwise, or negative, moment about point A. The positive and negative moments are equal in magnitude and are in *equilibrium*; that is, there is no motion. Another way of stating this is to say that the sum of the positive and negative moments about point A is zero, or

$$\Sigma M_A = 0$$

Stated more generally, *if a system of forces is in equilibrium, the algebraic sum of the moments is zero*. This is one of the laws of equilibrium.

In Figure 5.2*b* point A was taken as the center of the moments, but the fundamental law holds for any point that might be selected. For example, if point B is taken as the center of moments, the moment of the upward supporting force of 300 lb acting at A is clockwise (positive) and that of P_2 is counterclockwise (negative). Then

$$(300 \times 8) - (200 \times 12) = 2400 - 2400 = 0$$

Note that the moment of force P_1 about point B is $100 \times 0 = 0$; it is therefore omitted in writing the equation. The reader should be satisfied that the sum of the moments is zero also when the center of moments is taken at the left end of the bar under the point of application of P_2.

Laws of Equilibrium

When an object is acted on by a number of forces, each force tends to move the object. If the forces are of such magnitude and position that their combined effect produces no motion of the object, the forces are said to be in equilibrium. The three fundamental laws of static equilibrium for a general set of coplanar forces are:

1. The algebraic sum of all the vertical forces equals zero.
2. The algebraic sum of all the horizontal forces equals zero.
3. The algebraic sum of the moments of all the forces about any point equals zero.

These laws, sometimes called the conditions for equilibrium, may be expressed as follows (the symbol Σ indicates a summation, i.e., an algebraic addition of all similar terms involved in the problem):

$$\Sigma V = 0 \quad \Sigma H = 0 \quad \Sigma M = 0$$

The law of moments, $\Sigma M = 0$, was presented in the preceding discussion.

The expression $\Sigma V = 0$ is another way of saying that *the sum of the downward forces equals the sum of the upward forces*. Thus the bar of Figure 5.2*b* satisfies $\Sigma V = 0$ because the upward force of 300 lb equals the sum of P_1 and P_2.

Moments of Forces on a Beam

Figure 5.3*a* shows two downward forces of 100 lb and 200 lb acting on a beam. The beam has a length of 8 ft between the supports; the supporting forces, which are called *reactions*, are 175 lb and 125 lb. The four forces are parallel and for equilibrium to exist, therefore, the two laws, $\Sigma V = 0$ and $\Sigma M = 0$, apply.

First, because the forces are in equilibrium, the sum of the downward forces must equal the sum of the upward forces. The sum of the downward forces, the loads, is $100 + 200 = 300$ lb; and the sum of the upward forces, the reactions, is $175 + 125 = 300$ lb. Thus the force summation is zero.

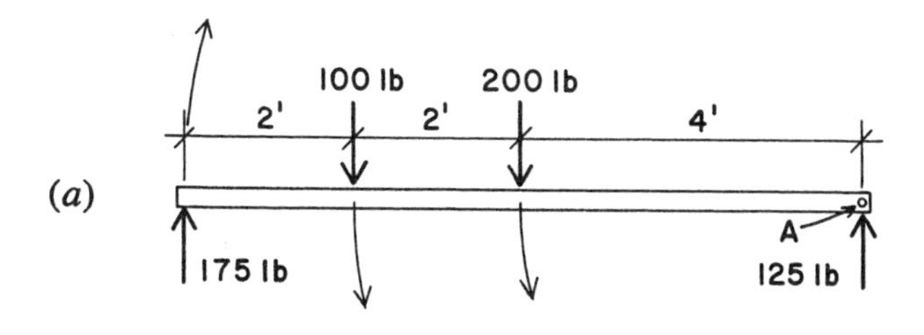

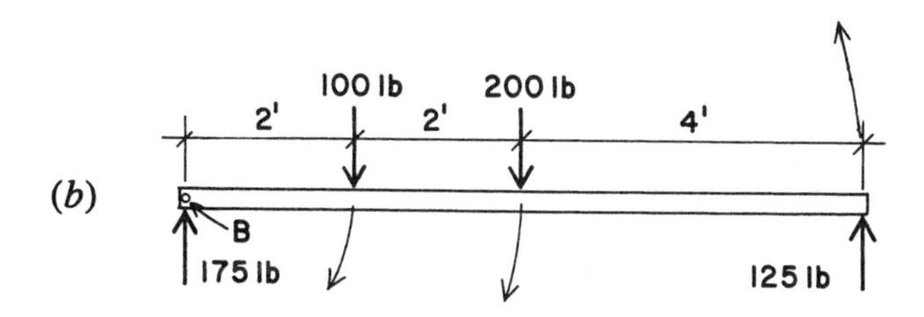

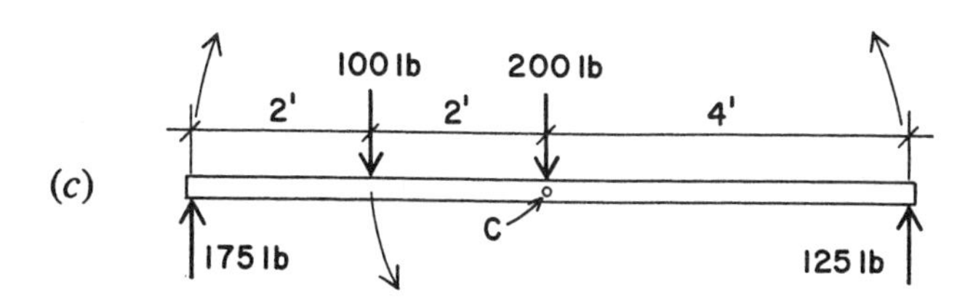

Figure 5.3 Summations of moments about selected points.

Second, because the forces are in equilibrium, the sum of the moments of the forces tending to cause clockwise rotation (positive moments) must equal the sum of the moments of the forces tending to produce counterclockwise rotation (negative moments) about any center of moments. Considering an equation of moments about point A at the right-hand support, the force tending to cause clockwise rotation (shown by the curved arrow) about this point is 175 lb; its moment is $175 \times 8 = 1400$ ft-lb. The forces tending to cause counterclockwise rotation about the *same point* are 100 lb and 200 lb, and their moments are (100×6) and (200×4) ft-lb. Thus, if $\Sigma M_A = 0$, then

$$(175 \times 8) = (100 \times 6) + (200 \times 4)$$
$$1400 = 600 + 800$$
$$1400 \text{ ft-lb} = 1400 \text{ ft-lb}$$

And a condition of equilibrium is demonstrated.

The upward force of 125 lb is omitted from the above moment summation equation because its lever arm about point A is 0 ft, and consequently its moment is zero. A force passing through the center of moments does not cause rotation about that point.

Now select point B at the left support as the center of moments (see Figure 5.3*b*). By the same reasoning, if $\Sigma M_B = 0$, then

$$(100 \times 2) + (200 \times 4) = (125 \times 8)$$
$$200 + 800 = 1000$$
$$1000 \text{ ft-lb} = 1000 \text{ ft-lb}$$

Again the law holds. In this case the force of 175 lb has a lever arm of 0 ft about the center of moments and its moment is zero.

The reader should verify this case by selecting any other point, such as point C in Figure 5.3*c* as the center of moments, and confirming that the sum of the moments is zero for this point.

Problem 5.2.A. Figure 5.4 represents a beam in equilibrium with three loads and two reactions. Select five different centers of moments and write the equation of moments for each, showing that the sum of the clockwise moments equals the sum of the counterclockwise moments.

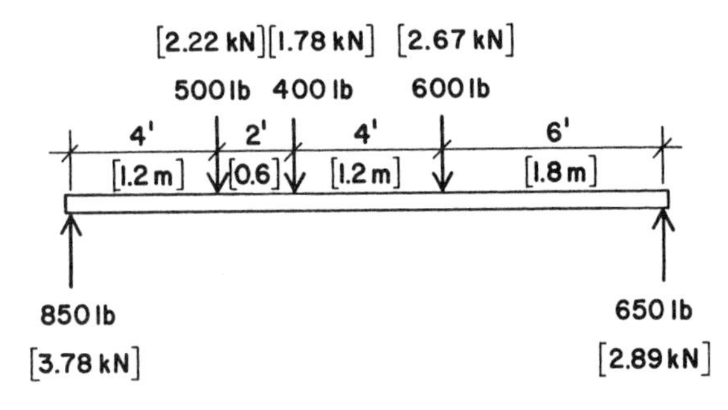

Figure 5.4 Reference for Problem 5.2.A.

5.3 BEAM LOADS AND REACTION FORCES

Reactions are the forces acting at the supports that hold in equilibrium the forces, or loads, that are applied to a beam. The left and right reactions of a simple beam as viewed are usually called R_1 and R_2, respectively. Determination of reactions for simple beams is achieved with the use of equilibrium conditions for parallel force systems.

If a beam 18 ft long has a concentrated load of 9000 lb located 9 ft from the supports, it is readily seen that each upward force at the supports will be equal and will be one half the load in magnitude, or 4500 lb. But consider, for instance, the 9000 lb load placed ten ft from one end, as shown in Figure 5.5. What will the upward supporting forces be? This is where the principle of moments can be used. Consider a summation of moments about the right-hand support R_2.
Thus,

$$\Sigma M = 0 = +(R_1 \times 18) - (9000 \times 8)$$

$$R_1 = \frac{72000}{18} = 4000 \text{ lb}$$

Then, considering the equilibrium of vertical forces,

$$\Sigma V = 0 = +R_1 + R_2 - 9000 \quad \text{or} \quad R_2 = 9000 - 4000 = 5000 \text{ lb}$$

The accuracy of this solution can be verified by taking moments about the left-hand support.
Thus,

$$\Sigma M = 0 = -(R_2 \times 18) + (9000 \times 10), \quad R_2 = \frac{90{,}000}{18} = 5000 \text{ lb}$$

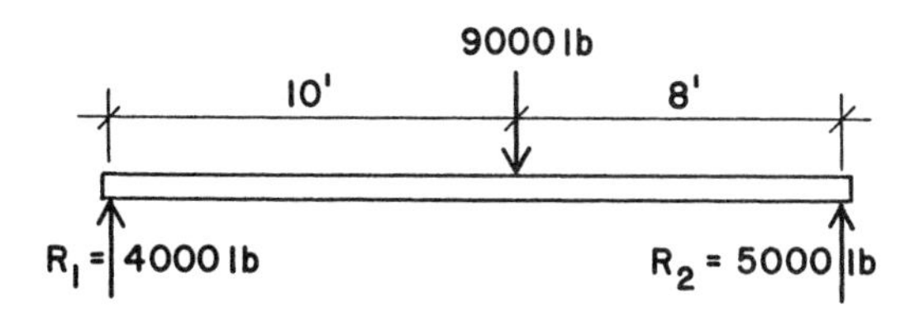

Figure 5.5 Beam reactions for a single load.

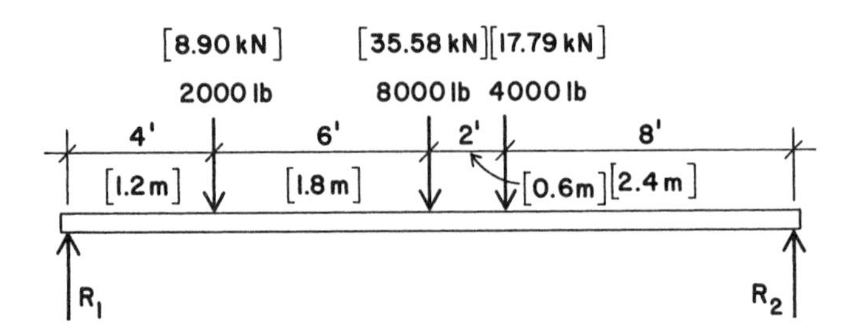

Figure 5.6 Reference for Example 1.

Example 1. A simple beam 20 ft long has three concentrated loads, as indicated in Figure 5.6. Find the magnitudes of the reactions.

Solution: Using the right-hand support as the center of moments,

$$\Sigma M = +(R_1 \times 20) - (2000 \times 16) - (8000 \times 10) - (4000 \times 8)$$

from which

$$R_1 = \frac{(32{,}000 + 80{,}000 + 32{,}000)}{20} = 7200 \text{ lb}$$

From a summation of the vertical forces,

$$\Sigma V = 0 = +R_2 + 7200 - 2000 - 8000 - 4000, \quad R_2 = 6800 \text{ lb}$$

With all forces determined, a summation about the left-hand support, or any point except the right-hand support, will verify the accuracy of the work.

The following example demonstrates a solution with uniformly distributed loading on a beam. A convenience in this work is to consider the total uniformly distributed load as a concentrated force placed at the center of the distributed load.

Example 2. A simple beam 16 ft long carries the loading shown in Figure 5.7*a*. Find the reactions.

Solution: The total uniformly distributed load may be considered as a single concentrated load placed at 5 ft from the right-hand support;

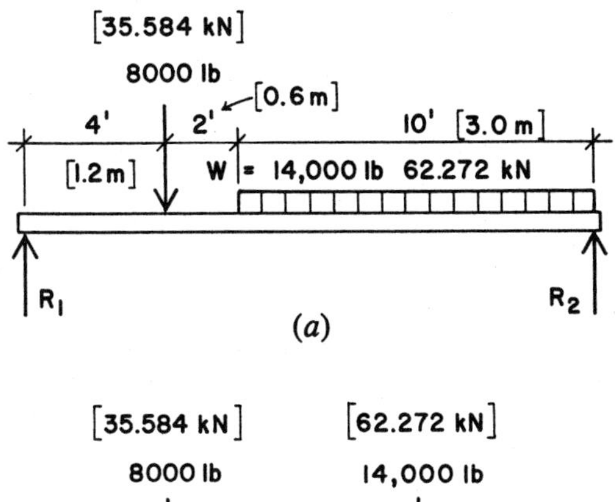

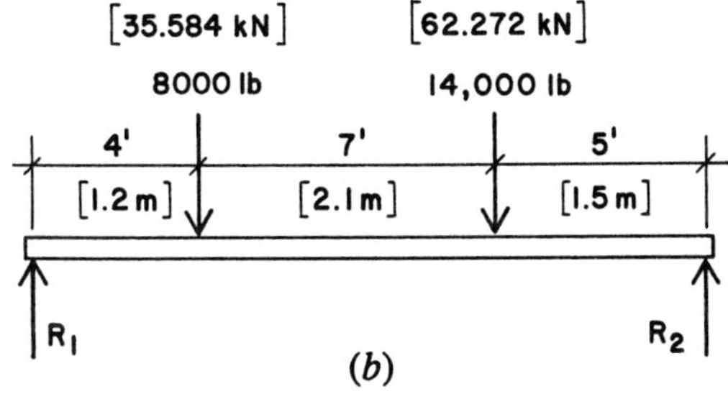

Figure 5.7 Reference for Example 2.

this loading is shown in Figure 5.7*b*. Considering moments about the right-hand support,

$$\Sigma M = 0 = +(R_1 \times 16) - (8000 \times 12) - (14{,}000 \times 5)$$

from which,

$$R_1 = \frac{166{,}000}{16} = 10{,}375 \text{ lb}$$

And from a summation of vertical forces,

$$R_2 = (8000 + 14{,}000) - 10{,}375 = 11{,}625 \text{ lb}$$

Again, a summation of moments about the left-hand support will verify the accuracy of the work.

In general, any beam with only two supports, where the supports develop only vertical reaction forces, will be statically determinate. This includes the simple span beams in the preceding examples as well as beams with overhanging ends.

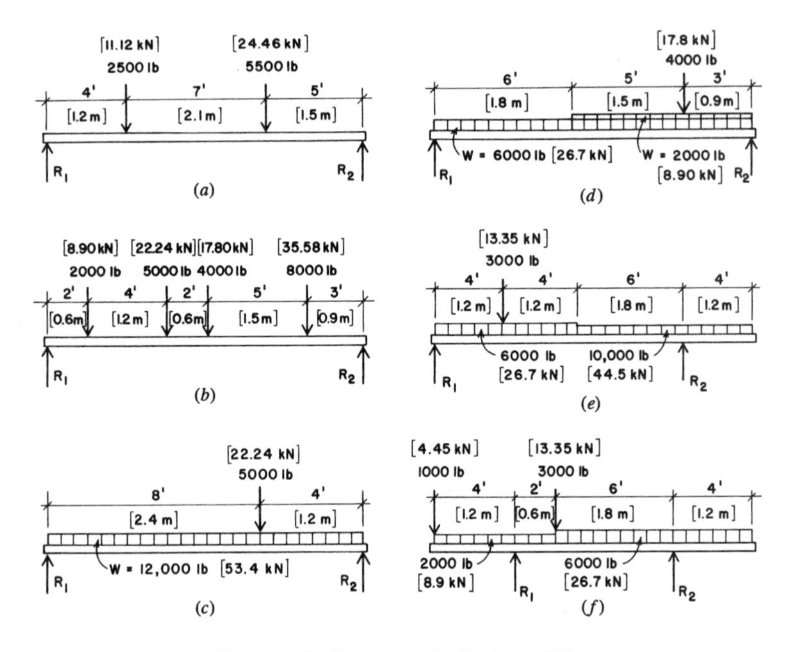

Figure 5.8 Reference for Problem 5.3.

Problem 5.3.A–F. Find the reactions for the beams shown in Figure 5.8.

5.4 BEAM SHEAR

Figure 5.9*a* represents a simple beam with a uniformly distributed load over its entire length. Examination of an actual beam so loaded probably would not reveal any effects of the loading on the beam. However, there are three distinct major tendencies for the beam to fail. Figure 5.9*b–d* illustrates the three phenomena.

First, there is a tendency for the beam to fail by dropping between the supports (Figure 5.9*b*). This is called *vertical shear.* Second, the beam may fail by bending (Figure 5.9*c*). Third, there is a tendency in wood beams for the fibers of the beam to slide past each other in a horizontal direction (Figure 5.9*d*), an action described as horizontal shear. Naturally, a beam properly designed does not fail in any of the ways just mentioned, but these tendencies to fail are always present and must be considered in structural design.

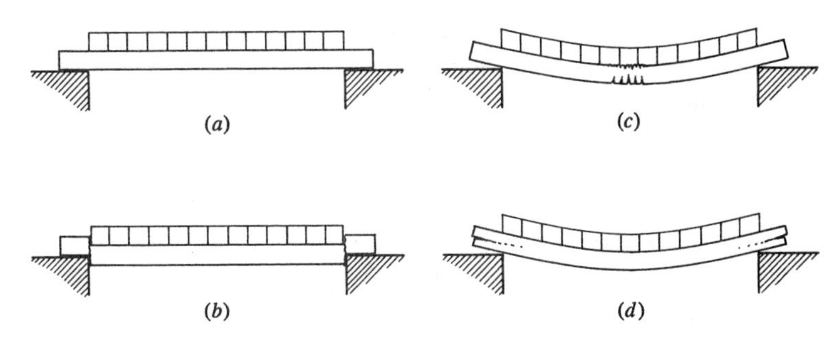

Figure 5.9 Stress failures in beams: (*b*) bending, (*c*) vertical shear, (*d*) horizontal shear.

Vertical Shear

Vertical shear is the tendency for one part of a beam to move vertically with respect to an adjacent part. The magnitude of the shear force at any section in the length of a beam is equal to the algebraic sum of the vertical forces on either side of the section. Vertical shear is usually represented by the letter V. In computing its values in the examples and problems, consider the forces to the left of the section, but keep in mind that the same resulting force magnitude will be obtained with the forces on the right. To find the magnitude of the vertical shear at any section in the length of a beam, simply add up the forces to the right or the left of the section. It follows from this procedure that the maximum value of the shear for simple beams is equal to the greater reaction.

Example 3. Figure 5.10*a* illustrates a simple beam with concentrated loads of 600 lb and 1000 lb. The problem is to find the value of the

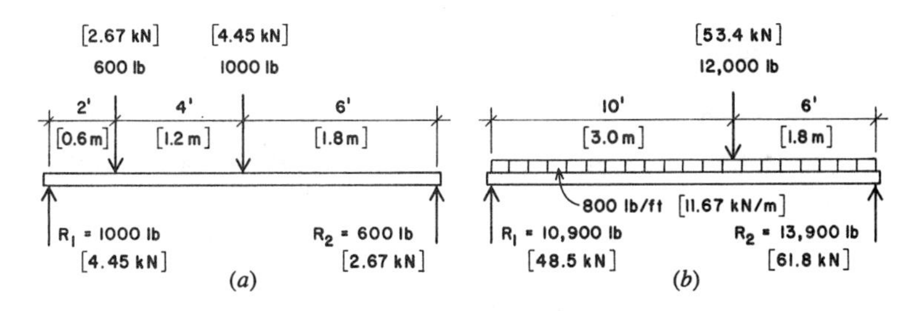

Figure 5.10 Reference for Examples 3 and 4.

vertical shear at various points along the length of the beam. Although the weight of the beam constitutes a uniformly distributed load, it is neglected in this example.

Solution: The reactions are computed as previously described, and are found to be $R_1 = 1000$ lb and $R_2 = 600$ lb.

Consider next the value of the vertical shear V at an infinitely short distance to the right of R_1. Applying the rule that the shear is equal to the reaction minus the loads to the left of the section, we write

$$V = R_1 - 0, \quad \text{or} \quad V = 1000 \text{ lb}$$

The zero represents the value of the loads to the left of the section, which of course, is zero. Now take a section 1 ft to the right of R_1; again

$$V_{(x=1)} = R_1 - 0, \quad \text{or} \quad V_{(x=1)} = 1000 \text{ lb}$$

The subscript $(x = 1)$ indicates the position of the section at which the shear is taken, the distance of the section from R_1. At this section the shear is still 1000 lb and has the same magnitude up to the 600 lb load.

The next section to consider is a very short distance to the right of the 600 lb load. At this section,

$$V_{(x=2+)} = 1000 - 600 = 400 \text{ lb}$$

Because there are no loads intervening, the shear continues to be the same magnitude up to the 1000 lb load. At a section a short distance to the right of the 1000 lb load,

$$V_{(x=6+)} = 1000 - (600 + 1000) = -600 \text{ lb}$$

This magnitude continues up to the right-hand reaction R_2.

Example 4. The beam shown in Figure 5.10*b* supports a concentrated load of 12,000 lb located 6 ft from R_2 and a uniformly distributed load of 800 pounds per linear foot (lb/ft) over its entire length. Compute the value of vertical shear at various sections along the span.

Solution: By use of the equations of equilibrium, the reactions are determined to be $R_1 = 10{,}900$ lb and $R_2 = 13{,}900$ lb. Note that the total distributed load is 800 × 16 = 12,800 lb. Now consider the vertical shear force at the following sections at a distance measured from the left support.

$$V_{(x=0)} = 10{,}900 - 0 = 10{,}900 \text{ lb}$$

$$V_{(x=1)} = 10{,}900 - (800 \times 1) = 10{,}100 \text{ lb}$$

$$V_{(x=5)} = 10{,}900 - (800 \times 5) = 6{,}900 \text{ lb}$$

$$V_{(x=10-)} = 10{,}900 - (800 \times 10) = 2{,}900 \text{ lb}$$

$$V_{(x=10+)} = 10{,}900 - \{(800 \times 10) + 12{,}000)\} = -9{,}100 \text{ lb}$$

$$V_{(x=16)} = 10{,}900 - \{(800 \times 16) + 12{,}000)\} = -13{,}900 \text{ lb}$$

Shear Diagrams

In the two preceding examples the value of the shear at several sections along the length of the beams was computed. In order to visualize the results it is common practice to plot these values on a diagram, called the *shear diagram* which is constructed as explained below.

To make such a diagram, first draw the beam to scale and locate the loads. This has been done in Figures 5.11*a* and *b* by repeating the load diagrams of Figures 5.10*a* and *b*, respectively. Beneath the beam draw a horizontal base line representing zero shear. Above and below this line, plot at any convenient scale the values of the shear at the various sections; the positive, or plus, values are placed above the line and the

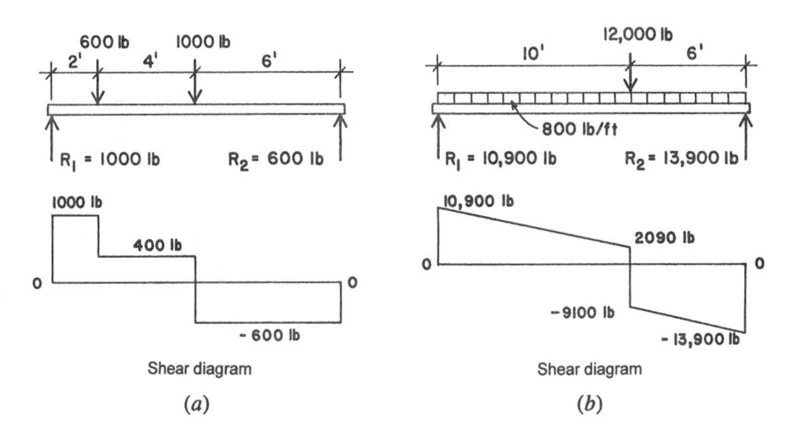

Figure 5.11 Construction of shear diagrams.

negative, or minus, values below. In Figure 5.11*a*, for instance, the value of the shear at R_1 is + 1000 lb. The shear continues to have the same value up to the load of 600 lb, at which point it drops to 400 lb. The same value continues up to the next load, 1000 lb, where it drops to −600 lb and continues to the right-hand reaction. Obviously, to draw a shear diagram it is necessary to compute the values at significant points only. Having made the diagram, we may readily find the value of the shear at any section of the beam by scaling the vertical distance in the diagram. The shear diagram for the beam in Figure 5.11*b* is made in the same manner.

There are two important facts to note concerning the vertical shear. The first is the maximum value. The diagrams in each case confirm the earlier observation that the maximum shear is at the reaction having the greater value, and its magnitude is equal to that of the greater reaction. In Figure 5.11*a* the maximum shear is 1000 lb, and in Figure 5.11*b* it is 13,900 lb. We disregard the positive or negative signs in reading the maximum values of the shear, for the diagrams are merely conventional

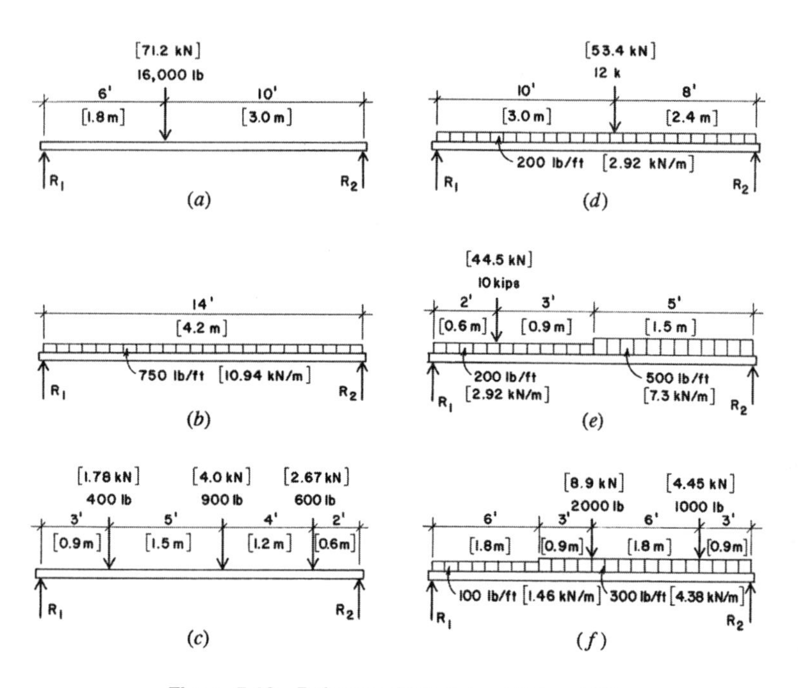

Figure 5.12 Reference for Problems 5.4 and 5.5.

methods of representing the absolute numerical values. Another important fact to note is the point at which the shear changes from a plus to a minus quantity. We call this the point at which the shear passes through zero. In Figure 5.11*a* it is under the 1000 lb load, 6 ft from R_1. In Figure 5.11*b* it is under the 12,000 lb load, 10 ft from R_1. A major concern for noting this point is that it indicates the location of the maximum value of bending moment in the beam, as discussed in the next section.

Problems 5.4.A–F. For the beams shown in Figure 5.12 draw the shear diagrams and note all critical values for shear. Note particularly the maximum value for shear and the point at which the shear passes through zero.

5.5 BENDING MOMENT

The forces that tend to cause bending in a beam are the reactions and the loads. Consider the section $X - X$, 6 ft from R_1 (Figure 5.13). The force R_1, or 2000 lb, tends to cause a clockwise rotation about this point. Because the force is 2000 lb and the lever arm is 6 ft, the moment of the force is $2000 \times 6 = 12{,}000$ ft-lb. This same value may be found by considering the forces to the right of the section $X - X$: R_2, which is 6000 lb, and the load 8000 lb, with lever arms of 10 and 6 ft, respectively. The moment of the reaction is $6000 \times 10 = 60{,}000$ ft-lb, and its direction is counterclockwise with respect to the section $X - X$. The moment of force 8000 lb is $8000 \times 6 = 48{,}000$ ft-lb, and its direction is clockwise. Then 60,000 ft-lb − 48,000 ft-lb = 12,000 ft-lb, the resultant moment tending to cause conterclockwise rotation about the section $X - X$. This is the same magnitude as the moment of the forces on the left which tend to cause a clockwise rotation.

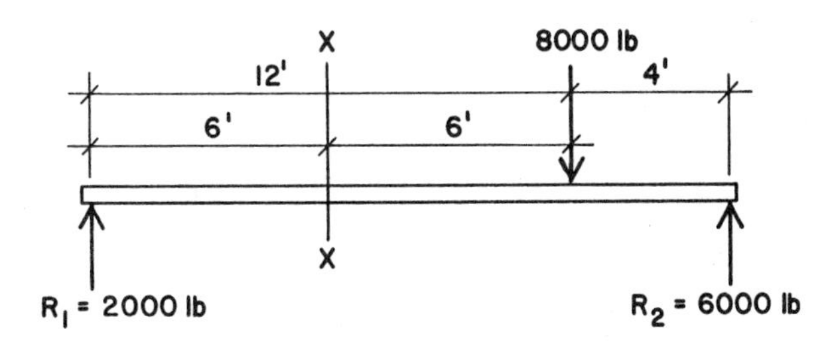

Figure 5.13 Internal bending at a selected beam cross section.

Thus it makes no difference whether use is made of the forces to the right of the section or the left, the magnitude of the moment is the same. It is called the *bending moment* (or the *internal bending moment*) because it is the moment of the forces that cause bending stresses in the beam. Its magnitude varies throughout the length of the beam. For instance, at 4 ft from R_1 it is only 2000×4, or 8000 ft-lb. The bending moment is the algebraic sum of the moments of the forces on either side of the section. For simplicity, take the forces on the left; then the bending moment at any section of a beam is equal to the moments of the reactions minus the moments of the loads to the left of the section. Because the bending moment is the result of multiplying forces by distances, the denominations are foot-pounds or kip-feet.

Bending Moment Diagrams

The construction of bending moment diagrams follows the procedure used for shear diagrams. The beam span is drawn to scale showing the locations of the loads. Below this, and usually below the shear diagram, a horizontal base line is drawn representing zero bending moment. Then the bending moments are computed at various sections along the beam span, and the values are plotted vertically to any convenient scale. In simple beams all bending moments are positive and therefore are plotted above the base line. In overhanging or continuous beams there are also negative moments, and these are plotted below the base line.

Example 5. The load diagram in Figure 5.14 shows a simple beam with two concentrated loads. Draw the shear and bending moment diagrams.

Solution: R_1 and R_2 are first computed and are found to be 16,000 lb and 14,000 lb, respectively. These values are recorded on the load diagram.

The shear diagram is drawn as described in Section 5.4. Note that in this instance it is necessary to compute the shear at only one section (between the concentrated loads) because there is no distributed load, and we know that the shear at the supports is equal in magnitude to the reactions.

Because the value of the bending moment at any section of the beam is equal to the moments of the reactions minus the moments of the loads to the left of the section, the moment at R_1 must be zero, for there are no forces to the left. Other values in the length of the beam are computed

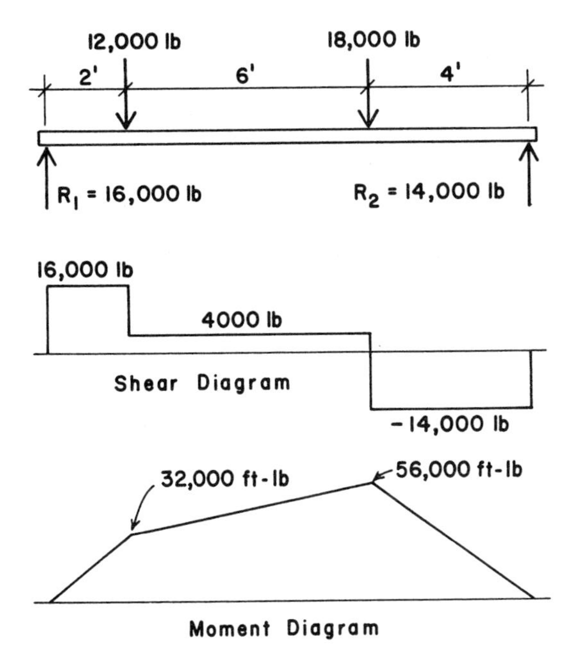

Figure 5.14 Reference for Example 5.

as follows. The subscripts ($x = 1$, etc.) show the distance from R_1 at which the bending moment is computed.

$$M_{(x=1)} = (16{,}000 \times 1) = 16{,}000 \text{ ft-lb}$$

$$M_{(x=2)} = (16{,}000 \times 2) = 32{,}000 \text{ ft-lb}$$

$$M_{(x=5)} = (16{,}000 \times 5) - (12{,}000 \times 3) = 44{,}000 \text{ ft-lb}$$

$$M_{(x=8)} = (16{,}000 \times 8) - (12{,}000 \times 6) = 56{,}000 \text{ ft-lb}$$

$$M_{(x=10)} = (16{,}000 \times 10) - \{(12{,}000 \times 8) + (18{,}000 \times 2)\}$$

$$= 28{,}000 \text{ ft-lb}$$

$$M_{(x=12)} = (16{,}000 \times 12) - \{(12{,}000 \times 10) + (18{,}000 \times 4)\} = 0$$

The result of plotting these values is shown in the bending moment diagram of Figure 5.14. More moments were computed than were necessary. We know that the bending moments at the supports of simple beams are zero, and in this instance only the bending moments directly under the loads were needed in order to construct the moment diagram.

Relations between Shear and Moment

In simple beams the shear diagram passes through zero at some point between the supports. As stated earlier, an important principle in this respect is that the bending moment has a maximum magnitude wherever the shear passes through zero. In Figure 5.14 the shear passes through zero under the 18,000-lb load, that is, at $x = 8$ ft. Note that the bending moment has its greatest value at this same point, 56,000 ft-lb.

Example 6. Draw the shear and bending moment diagrams for the beam shown in Figure 5.15, which carries a uniformly distributed load of 400 lb per lin ft and a concentrated load of 21,000 lb located 4 ft from R_1.

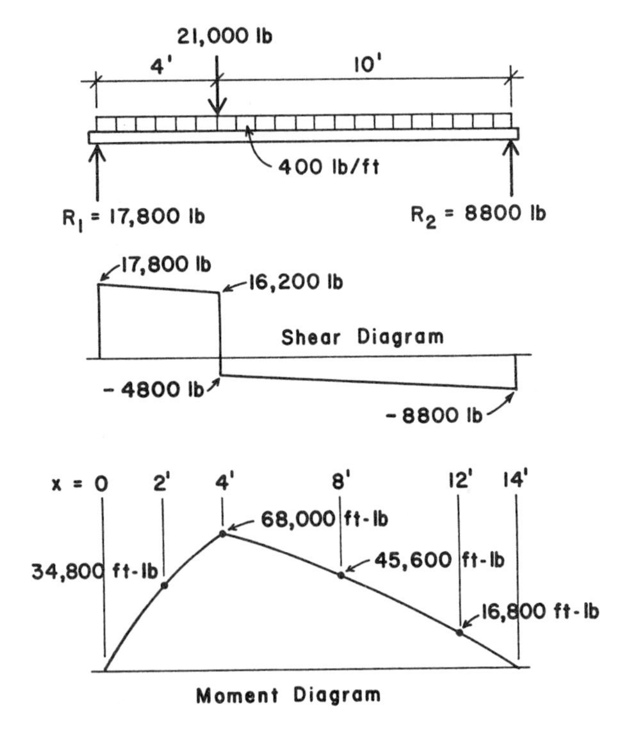

Figure 5.15 Reference for Example 6.

Solution: Computing the reactions, we find $R_1 = 17{,}800$ lb and $R_2 = 8800$ lb. By use of the process described in Section 5.4, the critical shear values are determined and the shear diagram is drawn as shown in the figure.

Although the only value of bending moment that must be computed for design use is that where the shear passes through zero, some additional values are determined in order to plot the true form of the moment diagram. Thus,

$$M_{(x=2)} = (17{,}800 \times 2) - (400 \times 2 \times 1) = 34{,}800 \text{ ft-lb}$$

$$M_{(x=4)} = (17{,}800 \times 4) - (400 \times 4 \times 2) = 68{,}000 \text{ ft-lb}$$

$$M_{(x=8)} = (17{,}800 \times 8) - \{(400 \times 8 \times 4) + (21{,}000 \times 4)\}$$
$$= 45{,}600 \text{ ft-lb}$$

$$M_{(x=12)} = (17{,}800 \times 12) - \{(400 \times 12 \times 6) + (21{,}000 \times 8)\}$$
$$= 16{,}800 \text{ ft-lb}$$

From the two preceding examples (Figures 5.14 and 5.15), it will be observed that the shear diagram for the parts of the beam on which no loads occur is represented by horizontal lines. For the parts of the beam on which a uniformly distributed load occurs, the shear diagram consists of straight inclined lines. The bending moment diagram is represented by straight inclined lines when only concentrated loads occur and by a curved line if the load is distributed. Occasionally, when a beam has both concentrated and uniformly distributed loads, the shear does not pass through zero under one of the concentrated loads. This frequently occurs when the distributed load is relatively large compared with the concentrated loads. Because it is necessary in designing beams to find the maximum bending moment, we must know the point at which it occurs. This, of course, is the point where the shear passes through zero, and its location is readily determined by the procedure illustrated in the following example.

Example 7. The load diagram in Figure 5.16 shows a beam with a concentrated load of 7000 lb, applied 4 ft from the left reaction, and a uniformly distributed load of 800 lb per lin ft extending over the full span. Compute the maximum bending moment on the beam.

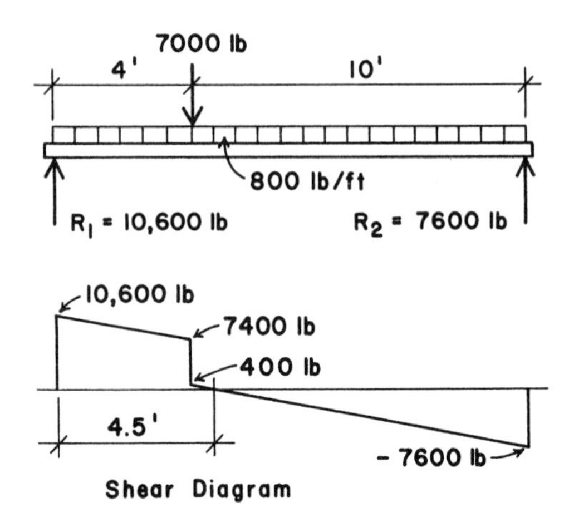

Figure 5.16 Reference for Example 7.

Solution: The values of the reactions are found to be $R_1 = 10{,}600$ lb and $R_2 = 7600$ lb and are recorded on the load diagram.

The shear diagram is constructed and it is observed that the shear passes through zero at some point between the concentrated load of 7000 lb and the right reaction. Call this distance x ft from R_2. The value of the shear at this section is zero; therefore an expression for the shear for this point, using the reaction and loads, is equal to zero. This equation contains the distance x:

$$V_{(\text{at}\,x)} = -7600 + 800x = 0, \quad x = \frac{7600}{800} = 9.5 \text{ ft}$$

The zero shear point is thus at 9.5 ft from the right support and (as shown in the diagram) at 4.5 ft from the left support. This location can also be determined by writing an equation for the summation of shear from the left of the point, which should produce the answer of 4.5 ft.

Following the convention of summing up the moments from the left of the section, the maximum moment is determined as

$$M_{(x=4.5)} = +(10{,}600 \times 4.5) - (7000 \times 0.5) - \left[800 \times 4.5 \times \left(\frac{4.5}{2}\right)\right]$$

$$= 36{,}100 \text{ ft}$$

Problem 5.5.A–F. Draw the shear and bending moment diagrams for the beams in Figure 5.12, indicating all critical values for shear and moment and all significant dimensions. (*Note:* These are the beams for Problem 5.4, for which the shear diagrams were constructed.)

Sense of Bending in Beams

When a simple beam bends, it has a tendency to assume the shape shown in Figure 5.17*a*. In this case the fibers in the upper part of the beam are in compression. For this condition the bending moment is considered as positive. Another way to describe a positive bending moment is to say that it is positive when the curve assumed by the bent beam is concave upward. When a beam projects beyond a support (Figure 5.17*b*), this portion of the beam has tensile stresses in the upper part. The bending moment for this condition is called negative; the beam is bent concave downward. When constructing moment diagrams, following the method previously described, the positive and negative moments are shown graphically.

Example 8. Draw the shear and bending moment diagrams for the overhanging beam shown in Figure 5.18.

Solution: Computing the reactions

$$\text{from } \Sigma M \text{ about } R_1: R_2 \times 12 = 600 \times 16 \times 8, \quad R_2 = 6400 \text{ lb}$$

$$\text{from } \Sigma M \text{ about } R_2: R_1 \times 12 = 600 \times 16 \times 4, \quad R_1 = 3200 \text{ lb}$$

With the reactions determined, the construction of the shear diagram is quite evident. For the location of the point of zero shear, considering its distance from the left support as x

$$3200 - 600x = 0, \quad x = 5.33 \text{ ft}$$

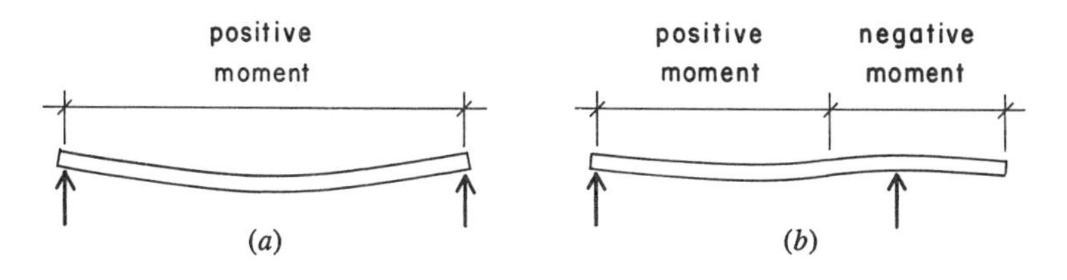

Figure 5.17 Sense of bending moment in beams.

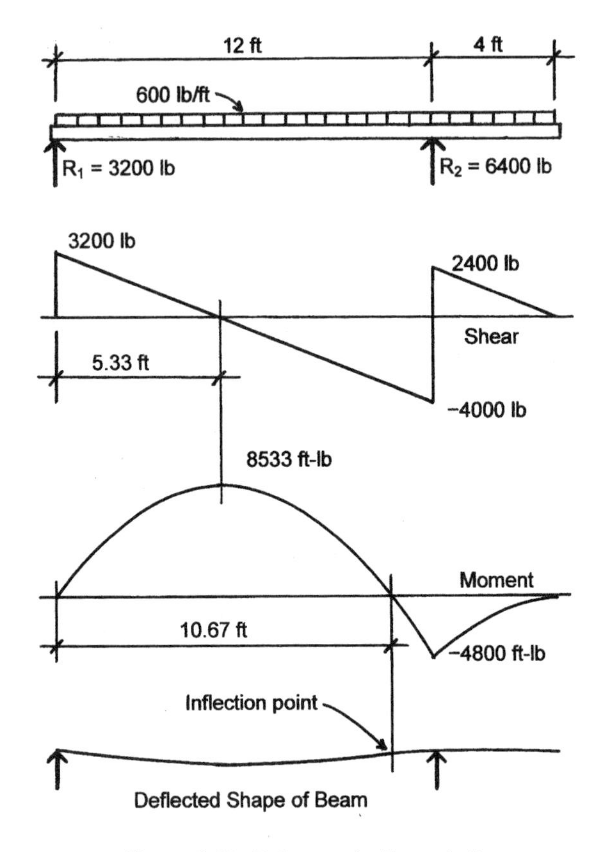

Figure 5.18 Reference for Example 8.

For the critical values needed to plot the moment diagram:

$$M_{(x=5.33)} = +(3200 \times 5.33) - \left[600 \times 5.33 \times \left(\frac{5.33}{2}\right)\right]$$

$$= 8533 \quad \text{ft-lb}$$

$$M_{(x=12)} = (3200 \times 12) - (600 \times 12 \times 6) = -4800 \text{ ft-lb}$$

The form of the moment diagram for the distributed loading is a curve (parabolic), which may be verified by plotting some additional points on the graph. For this case the shear diagram passes through zero twice, both of which points indicate peaks of the moment diagram—one positive and one negative. As the peak in the positive portion of the moment diagram is actually the apex of the parabola, the location of the zero moment value is simply twice the value previously determined as x. This point corresponds to the change in the form of curvature on the elastic curve (deflected shape) of the beam; this point is described as the *inflection point* for the deflected shape. The location of the point of zero moment can also be determined by writing an equation for the sum of moments at the unknown location. In this case, calling the new unknown point x

$$M = 0 = +(3200 \times x) - \left[600 \times x \times \left(\frac{x}{2}\right)\right] = 3200\ x - 300\ x^2$$

Solution of this quadratic equation should produce the value of $x = 10.67$ ft.

Example 9. Compute the maximum bending moment for the overhanging beam shown in Figure 5.19.

Solution: Computing the reactions, $R_1 = 3200$ lb and $R_2 = 2800$ lb. As usual, the shear diagram can now be plotted as the graph of the loads and reactions, proceeding from left to right. Note that the shear passes through zero at the location of the 4000 lb load and at both supports. As usual, these are clues to the form of the moment diagram. With the usual moment summations, values for the moment diagram can now be found at the locations of the supports and all of the concentrated loads. From this plot it will be noted that there are two inflection points (locations of zero moment). As the moment diagram is composed of straight-line segments in this case, the locations of these points may be found by writing simple linear equations for their locations. However, use can also be made of some relationships between the shear and moment graphs. One of these has already been used, relating to the correlation of zero shear and maximum moment. Another relationship is that the change of the value of moment between any two points along the beam is equal to the total area of the shear diagram between the two points. If the value of moment is known at some point, it is thus a simple matter to find values at other points. For example, starting from the left end, the value of moment is known to be zero at the left end of the beam; then the value of the moment at the support is the area of the rectangle on the shear diagram with base of 4 ft and height of 800 lb—the area being $4 \times 800 = 3200$ ft-lb.

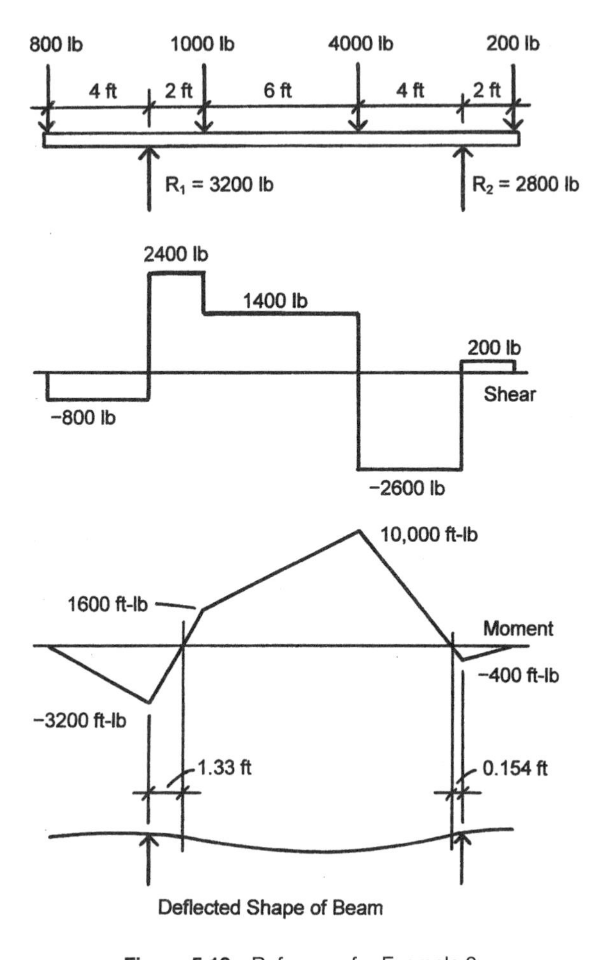

Figure 5.19 Reference for Example 9.

Now, proceeding along the beam to the point of zero moment (call it x distance from the support), the change is again 3200, which relates to an area of the shear diagram that is $x \times 2400$.
Thus,

$$2400x = 3200, \quad x = \frac{3200}{2400} = 1.33 \text{ ft}$$

And, now calling the distance from the right support to the point of zero moment x,

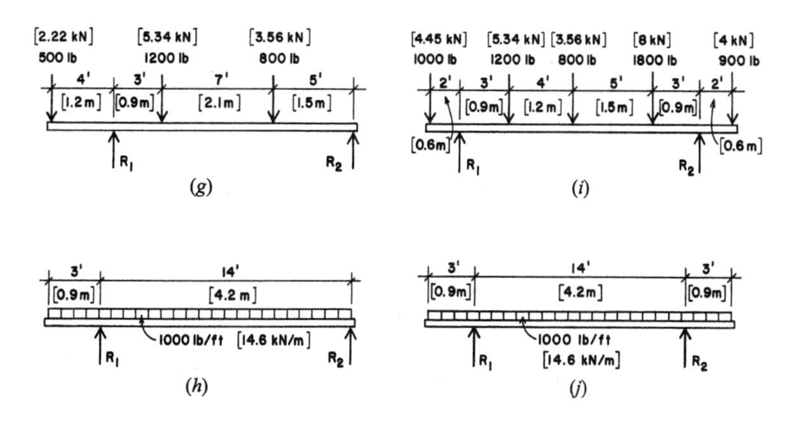

Figure 5.20 Reference for Problem 5.5.G–J.

$$2600x = 400, \quad x = \frac{400}{2600} = 0.154 \text{ ft}$$

Problems 5.5.G–J. Draw the shear and bending moment diagrams for the beams in Figure 5.20, indicating all critical values for shear and moment and all significant dimensions.

Cantilever Beams

In order to keep the signs for shear and moment consistent with those for other beams, it is convenient to draw a cantilever beam with its fixed end to the right, as shown in Figure 5.21. We then plot the values for the shear and moment on the diagrams as before, proceeding from the left end.

Example 10. The cantilever beam shown in Figure 5.21*a* projects 12 ft from the face of the wall and has a concentrated load of 800 lb at the unsupported end. Draw the shear and moment diagrams. What are the values of the maximum shear and maximum bending moment?

Solution: The value of the shear is −800 lb throughout the entire length of the beam. The bending moment is maximum at the wall; its value is 800 × 12 = −9600 ft-lb. The shear and moment diagrams are shown in Figure 5.21*a*. Note that the moment is negative for the entire length of the cantilever beam, corresponding to its concave downward shape throughout its length.

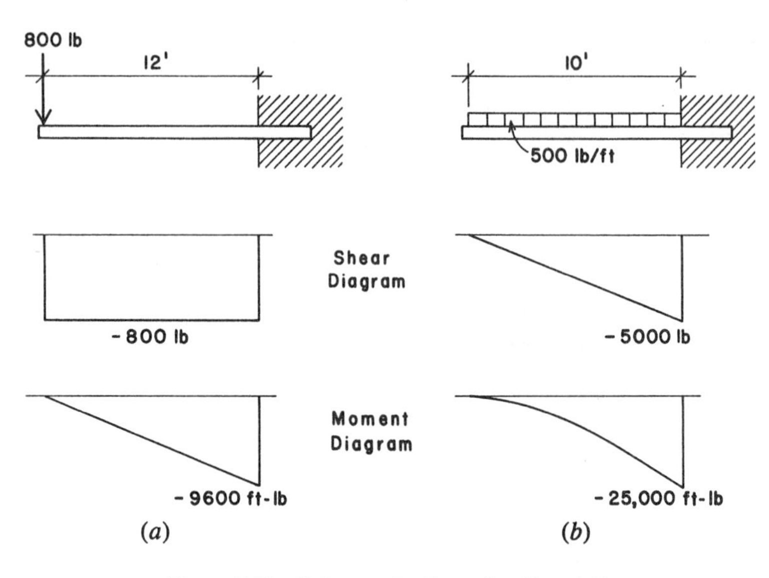

Figure 5.21 Reference for Examples 10 and 11.

Although they are not shown in the figure, the reactions in this case are a combination of an upward force of 800 lb and a clockwise resisting moment of 9600 ft-lb.

Example 11. Draw shear and bending moment diagrams for the beam in Figure 5.21*b*, which carries a uniformly distributed load of 500 lb/ft over its full length.

Solution: The total load is $500 \times 10 = 5000$ lb. The reactions are an upward force of 5000 lb and a moment determined as

$$M = -500 \times 10 \times \left(\frac{10}{2}\right) = -25{,}000 \text{ ft-lb}$$

which, it may be noted, is also the total area of the shear diagram between the outer end and the support.

Example 12. The cantilever beam indicated in Figure 5.22 has a concentrated load of 2000 lb and a uniformly distributed load of 600 lb per lin ft at the positions shown. Draw the shear and bending moment

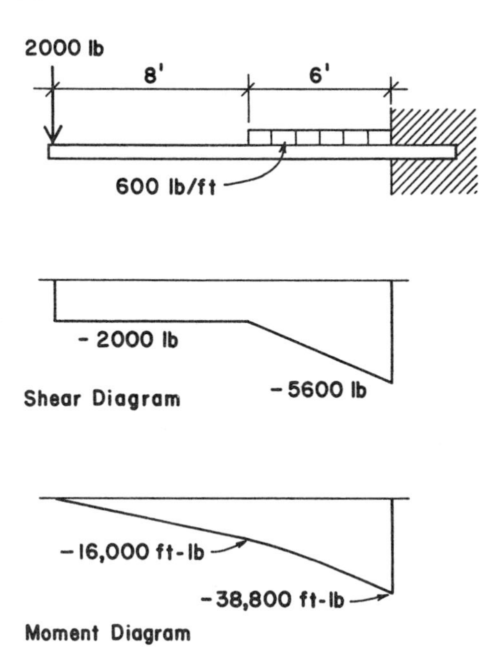

Figure 5.22 Reference for Example 12.

diagrams. What are the magnitudes of the maximum shear and maximum bending moment?

Solution: The reactions are actually *equal* to the maximum shear and bending moment. Determined directly from the forces, they are

$$V = 2000 + (600 \times 6) = 5600 \text{ lb}$$

$$M = -(2000 \times 14) - \left[600 \times 6 \times \left(\frac{6}{2}\right)\right] = -38{,}800 \text{ ft-lb}$$

The diagrams are quite easily determined. The other moment value needed for the moment diagram can be obtained from the moment of the concentrated load or from the simple rectangle of the shear diagram: $2000 \times 8 = 16{,}000$ ft-lb.

Note that the moment diagram has a straight-line shape from the outer end to the beginning of the distributed load, and becomes a curve from this point to the support.

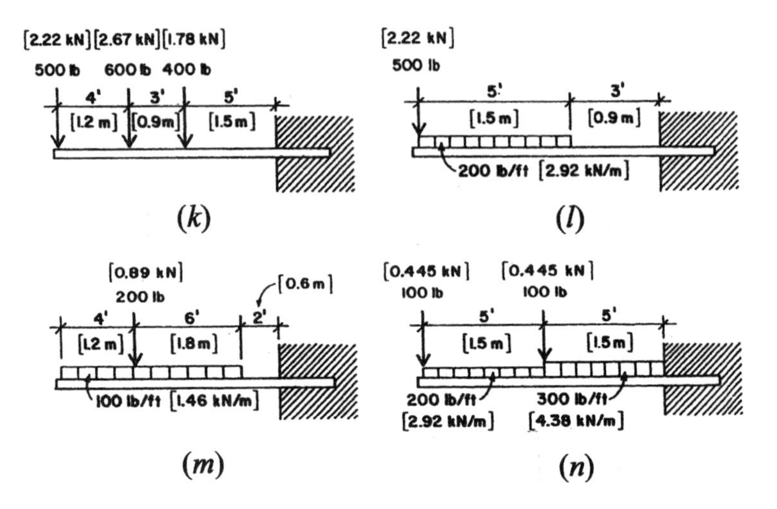

Figure 5.23 Reference for Problem 5.5.K–N.

It is suggested that Example 10 be reworked with Figure 5.22 reversed, left for right. All numerical results will be the same, but the shear diagram will be positive over its full length.

Problems 5.5.K–N. Draw the shear and bending moment diagrams for the beams in Figure 5.23, indicating all critical values for shear and moment and all significant dimensions.

5.6 TABULATED VALUES FOR BEAM BEHAVIOR

Bending Moment Formulas

The methods of computing beam reactions, shears, and bending moments presented thus far in this chapter make it possible to find critical values for design under a wide variety of loading conditions. However, certain conditions occur so frequently that it is convenient to use formulas that give the maximum values directly. Structural design handbooks contain many such formulas; two of the most commonly used formulas are derived in the following examples.

Simple Beam, Concentrated Load at Center of Span

A simple beam with a concentrated load at the center of the span occurs very frequently in practice. Call the load P and the span length between supports L, as indicated in the load diagram of Figure 5.24*a*. For this symmetrical loading each reaction is $P/2$, and it is readily apparent that the shear will pass through zero at distance $x = L/2$ from R_1. Therefore the maximum bending moment occurs at the center of the span, under the load. Computing the value of the bending moment at this section,

$$M_{\max} = \frac{P}{2} \times \frac{L}{2} = \frac{PL}{4}$$

Example 13. A simple beam 20 ft in length has a concentrated load of 8000 lb at the center of the span. Compute the maximum bending moment.

Solution: As just derived, the formula giving the value of the maximum bending moment for this condition is $M = PL/4$. Therefore,

$$M = \frac{PL}{4} = \frac{8000 \times 20}{4} = 40{,}000 \text{ ft-lb}$$

Simple Beam, Uniformly Distributed Load

This is probably the most common beam loading; it occurs time and again. For any beam, its own dead weight as a load to be carried is

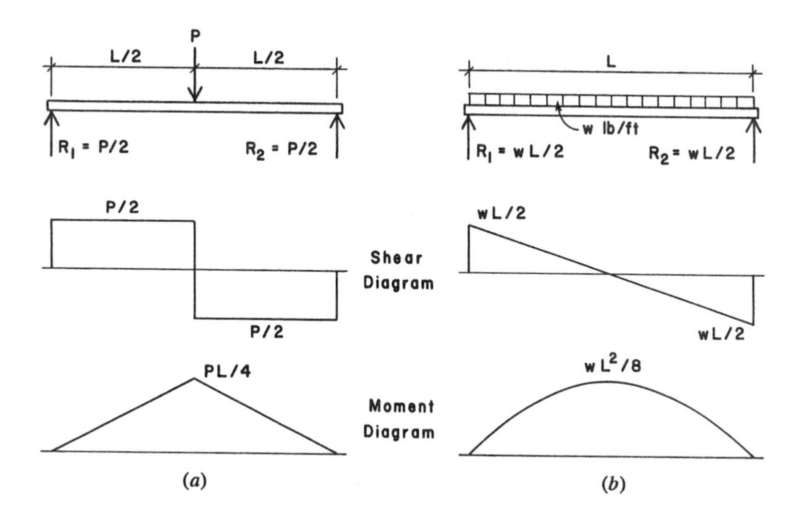

Figure 5.24 Values for simple beam loadings.

usually of this form. Calling the span L and the unit load w, as indicated in Figure 5.24b, the total load on the beam is $W = wL$; hence each reaction is $W/2$ or $wL/2$. The maximum bending moment occurs at the center of the span at distance $L/2$ from R_1. Writing the value of M for this section:

$$M = +\left[\frac{wL}{2} \times \frac{L}{2}\right] - \left[w \times \left(\frac{L}{2}\right) \times \left(\frac{L}{4}\right)\right] = \frac{wL^2}{8}, \quad \text{or} \frac{WL}{8}$$

Note the alternative use of the unit load w or the total load W in this formula. Both forms will be seen in various references. It is important to carefully identify the use of one or the other.

Example 14. A simple beam 14 ft long has a uniformly distributed load of 800 lb/ft. Compute the maximum bending moment.

Solution: As just derived, the formula that gives the maximum bending moment for a simple beam with uniformly distributed load is $M = wL^2/8$. Substituting these values:

$$M = \frac{wL^2}{8} = \frac{800 \times 14^2}{8} = 19{,}600 \text{ ft-lb}$$

or, using the total load of $800 \times 14 = 11{,}200$ lb,

$$M = \frac{wL}{8} = \frac{11{,}200 \times 14}{8} = 19{,}600 \text{ ft-lb}$$

Use of Tabulated Values for Beams

Some of the most common beam loadings are shown in Figure 5.25. In addition to the formulas for the reactions R, for maximum shear V, and for maximum bending moment M, expressions for maximum deflection Δ are also given. Discussion of deflections formulas will be deferred for the time being but will be considered in Section 6.3.

In Figure 5.25 if the loads P and W are in pounds or kips, the vertical shear V will also be in the same units of pounds or kips. When the loads are given in pounds or kips and the span is given in feet, the bending moment M will be in units of foot-pounds or kip-feet.

Also given in Figure 5.25 are values designated ETL, which stands for *equivalent tabular load.* These may be used to derive a hypothetical uniformly distributed load that when applied to the beam will produce

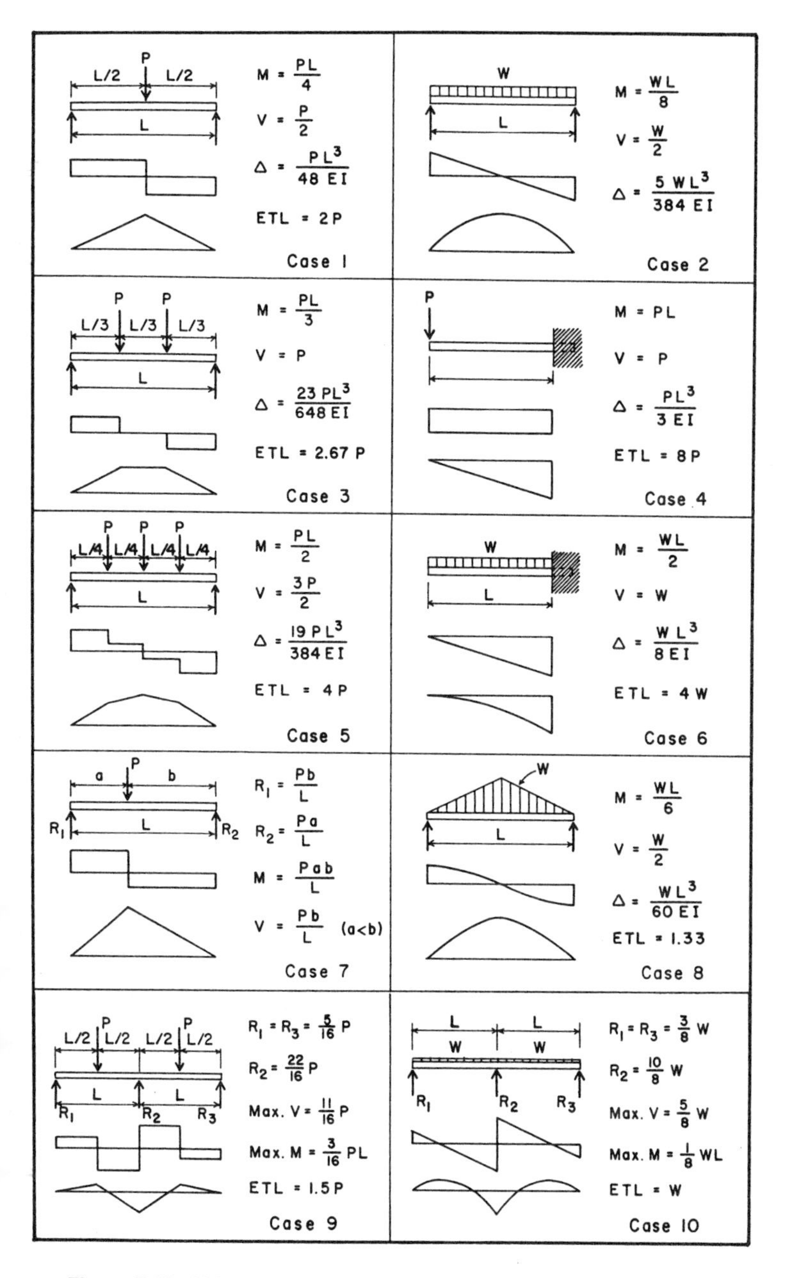

Figure 5.25 Values for typical beam loadings and support conditions.

the same magnitude of maximum bending moment as that for the given case of loading. Use of these factors is illustrated in later parts of the book.

Problem 5.6.A. A simple span beam has two concentrated loads of 4 kips [17.8 kN] each placed at the third points of the 24-ft [7.32-m] span. Find the value for the maximum bending moment in the beam.

Problem 5.6.B. A simple span beam has a uniformly distributed load of 2.5 kips/ft [36.5 kN/m] on a span of 18 ft [5.49 m]. Find the value for the maximum bending moment in the beam.

Problem 5.6.C. A simple beam with a span of 32 ft [9.745 m] has a concentrated load of 12 kips [53.4 kN] at 12 ft [3.66 m] from one end. Find the value for the maximum bending moment in the beam.

Problem 5.6.D. A simple beam with a span of 36 ft [11 m] has a distributed load that varies from a value of 0 at its ends to a maximum of 1000 lb/ft [14.59 kN/m] at its center (Case 8 in Figure 5.25). Find the value for the maximum bending moment in the beam.

5.7 MULTIPLE-SPAN BEAMS

As noted in Section 5.1, a continuous beam is a beam that rests on three or more supports. The magnitude of the reactions cannot be determined solely by the principle of moments used for beams with only two supports. For this reason, continuous beams are said to be "statically indeterminate," and the discussion of the necessary computations is beyond the scope of this book. Continuous beams are common in reinforced concrete structures and in welded steel construction but occur infrequently in wood structures. An exception is the situation of wood decks, for which units are ordinarily continuous over several closely-spaced rafters or joists.

Two continuous beam situations that sometimes occur in wood construction are illustrated in Cases 9 and 10 in Figure 5.25. Case 10 shows a beam continuous over two equal spans with equal uniformly distributed loads W on each span. Case 9 has equal concentrated loads P at the centers of two equal continuous spans. In both cases formulas are given for finding the reactions, the maximum vertical shear, and the maximum bending moment. For these two cases the maximum bending moment occurs over the center support and is negative (tension in the top of the beam). The maximum positive moments for Cases 9 and 10 (values not shown in Figure 5.25) are $M = 5PL/32$ and $WL/14.2$, respectively. In most cases the maximum bending moment is critical for design, regardless of whether it is positive or negative.

Example 15. A continuous beam has two equal spans of 10 ft each and a total uniformly distributed load of 1000 lbs/ft extending over each span. Determine the reactions, construct the shear and moment diagrams, and note the magnitudes of the maximum shear and maximum bending moment. The beam diagram is shown in Figure 5.26*a*. Note that there are two inflection points (points of zero moment).

Solution: For the reactions, refer to Case 10 in Figure 5.25 and note that $R_1 = R_3 = 3/8(W)$ and $R_2 = 10/8(W)$. Then

$$R_1 = R_3 = \frac{3W}{8} = \frac{3(10{,}000)}{8} = 3750 \text{ lb}$$

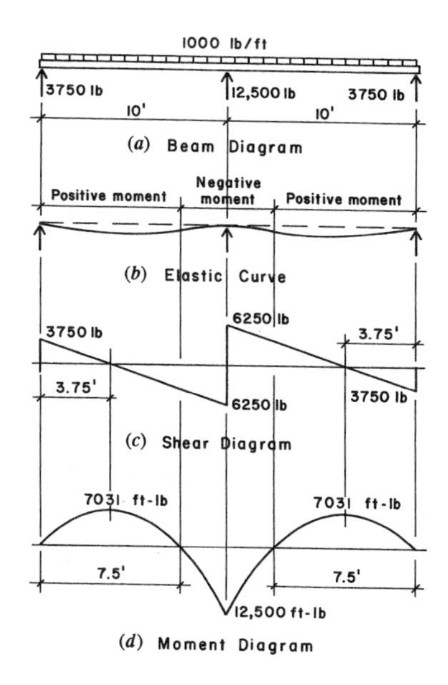

Figure 5.26 Diagrams for the two-span beam.

and

$$R_2 = \frac{10W}{8} = \frac{10(10{,}000)}{8} = 12{,}500 \text{ lb}$$

Values of the vertical shear are computed for various points on the beam, as described in Section 5.4, and the shear diagram is plotted as shown in Figure 5.26*c*. For the conditions of this example, Case 10 of Figure 5.25 shows that the maximum shear is 5/8(W). Thus

$$V = \frac{5W}{8} = \frac{5(10{,}000)}{8} = 6250 \text{ lb}$$

This value occurs on both sides of the center support. The shear passes through zero at the center support and at two other places within the spans. In the left span zero shear occurs at

$$x = \frac{R_1}{w} = \frac{3750}{1000} = 3.75 \text{ ft}$$

From observation of Case 10 in Figure 5.25, the maximum moment in the beam occurs as a negative moment over the center support. Its value is determined as

$$M = \frac{WL}{8} = \frac{10{,}000(10)}{8} = 12{,}500 \text{ ft-lb}$$

The maximum positive moment occurs at the point of zero shear, or 3.75 ft from the end support. Using the left span, a summation of moments to the left of this point about the zero shear point is as follows:

$$\begin{aligned} M &= R_1(3.75) - \frac{W}{L}(3.75)\left(\frac{3.75}{2}\right) \\ &= 3750(3.75) - 1000(3.75)(1.875) \\ &= 14{,}062.5 - 7031.25 = 7031.25 \text{ ft-lb} \end{aligned}$$

As the moment passes through zero, changing sign, the beam curvature inflects; this occurs at the internal points of zero moment on the diagram in Figure 5.26*d*. Since the positive portion of the moment diagram is a symmetrical parabola, this point occurs at $2(3.75) = 7.5$ ft from the end supports.

Problem 5.7.A. The two spans of a continuous beam are each 12 ft in length. Each span carries a concentrated load of 4 kips at its center. Find the reactions and construct the shear and moment diagrams for the beam showing all critical values. Find the locations of the points of zero shear and zero moment.

Problem 5.7.B. A two-span continuous beam has 16 ft spans and a uniformly distributed load of 0.8 kips/ft. Find the reactions and construct the shear and moment diagrams showing all critical values. Find the locations of the points of zero shear and zero moment.

6

BEHAVIOR OF BEAMS

The preceding chapter dealt with the resolution of force actions in terms of beam loads, support forces, the development of the internal forces of shear and bending, and the form of the primary deformation of deflection. This chapter treats the various considerations for a beam that relate to its development of the resisting force actions as it relates to quantified stresses and actual dimensions of deflection. Since the treatment of beam investigation by the LRFD method as used here is a modification of the ASD method, our procedure here will be to first treat the problems in general using the ASD method, followed by a treatment of the LRFD method.

6.1 SHEAR IN BEAMS

The development of internal shear force in beams was discussed in the preceding chapter in Section 5.4. We now consider the effect of this shear force on the beam, resulting in the development of shear stresses. This is a general occurrence in any beam, regardless of its material or the form of its cross section. However, some aspects of the stress behavior are special for each type of beam. In the situation considered here, the concern is primarily with the development of shear in beams consisting of solid-sawn wood and having rectangular cross sections. A more general case for development of shear is considered later.

Resistance to Vertical Shear

Shear stresses in beams were discussed briefly in Section 3.4. The tendency for a beam to fail by vertical shear is illustrated in Figure 3.3*b*. In a wood beam this type of shear is cross-grain shear, and failures of the kind shown in the figure seldom occur. However, the vertical shear force generates stresses of horizontal shear, diagonal tension, and diagonal compression, all of the same magnitude as the vertical shear stress. Diagonal tension causes cracking in tension-weak concrete; diagonal compression causes buckling of the thin webs of steel beams; and horizontal shear causes horizontal splitting of wood beams. Thus, although horizontal shear is the culprit for wood beams, we generally visualize the effect of vertical shear, understanding that the unit stress is the same for both responses.

The formula used to compute the maximum horizontal unit shear in a wood beam is

$$f_v = \frac{3}{2} \times \frac{V}{bd}$$

In which f_v = maximum unit shear stress in psi
V = total vertical shear force in the beam in lb at the section being investigated
b = width (breadth) of the beam cross section in in.
d = depth of the beam cross section in in.

This formula is derived later. It applies only to rectangular cross sections.

If the end shear force—that is, the maximum shear at the beam support—is used, this formula yields conservative values for stress. The stress computed occurs at the neutral axis, and shearing failure here will split the beam into two parts—upper and lower. Although probably undesirable, this split will actually produce two beams with the same width, and their combined shear resistance will actually be greater than that of the nonsplit beam. Consequently, it is permitted to use the vertical shear force at some point from the support, usually assumed as a distance equal to the beam depth. The usual procedure for investigation is to use the end shear force, but, if an overstress is found, to recompute the stress at the actual critical section.

Example 1. A simple beam with a span of 14 ft supports a uniformly distributed dead load of 300 lb/ft and a uniformly distributed live load of 500 lb/ft. A 10 × 14 beam of Douglas fir of select structural grade is used. Is the beam safe with respect to shear stress?

Solution: From Table 4.1 the allowable shear stress f_v is 170 psi. Table A.3 indicates dimensions of 9.5 × 13.5 in. The total load on the beam is 800 × 14 = 11,200 lb. Thus each reaction is 11,200/2 = 5600 lb. The maximum unit shear stress is

$$f_v = \frac{3V}{2bd} = \frac{3(5600)}{2(9.5 \times 13.5)} = 66 \text{ psi}$$

As this is well below the limiting stress of 170 psi, the beam is quite safe for shear. Had this investigation indicated a critical condition, it might have been of interest to investigate for the stress at the true critical section at a d distance from the support. The following computations illustrate this procedure.

The beam depth of 13.5 in. is equal to 13.5/12 = 1.125 ft. At this distance from the support the shear force is

$$V = 5600 - (800 \times 1.125) = 5600 - 900 = 4700 \text{ lb}$$

and the shear stress at this section is

$$f_v = \frac{4700}{5600} \times 66 = 55 \text{ psi}$$

Of course, there is no reason why the shear stress in a beam cannot be computed by using the modified end shear in the first place, if it appears that horizontal shear may be critical for a given situation.

Example 2. An 8 × 14 beam is used for a simple span of 16 ft and supports three concentrated live loads of 4000 lbs each at the quarter points of the span (see Figure 5.25, Case 5). The lumber used has an allowable shear stress of 170 psi. Investigate the beam for shear.

Solution: There are two loadings on this beam—the concentrated live loads and the dead load consisting of the uniformly distributed weight of the beam. A simplified procedure is to investigate first only for the live load. Then if the result indicates a stress condition close to the limit, the investigation for the beam weight may be performed and the two stresses added together. For the live load

$$V = \frac{3P}{2} = \frac{3(4000)}{2} = 6000\,\text{lb}$$

and the shear stress is:

$$f_v = \frac{3V}{2bd} = \frac{3(6000)}{2(7.5 \times 13.5)} = 89\,\text{psi}$$

Since this stress is considerably below the limit, it is probably evident that the beam weight will not be a critical concern. However, it is quite easy to investigate this using the procedure in the previous example.

Shear Investigation for Design

In the design of beams it is customary to first determine the size of the beam required for bending moment and then to investigate that chosen size for shear and deflection. This process is demonstrated in Section 7.1. An exception to this case is a beam that carries a large load on a short span, where shear may be a controlling critical concern.

Problem 6.1.A. A 10 × 14 in. [241 × 343 mm] beam has a span of 15 ft [4.5 m] with a single concentrated live load of 9 kips [40 kN] located at 5 ft [1.5 m] from the left end. Compute the maximum shear stress.

Problem 6.1.B. A 6 × 12 in. [152 × 241 mm] beam carries a total uniformly distributed live load of 3600 lb [16.0 kN] and a dead load of 3000 lb [13.3 kN] on a span of 10 ft [3 m]. If the maximum allowable shear stress is 170 psi, is the beam adequate with respect to shear?

Problem 6.1.C. A 10 × 12 in. [241 × 292 mm] beam of Douglas fir-larch, No. 1 grade, is used on a simple span of 12 ft [3.6 m]. It supports a uniformly distributed live load of 800 lb/ft [11.67 kN/m] and a total uniformly distributed dead load of 400 lb/ft [5.84 kN/m]. Is the beam size adequate for shear?

General Formula for Shear Stress

The formula used previously to determine the maximum horizontal shear stress in beams with rectangular cross section is:

$$f_v = \frac{3V}{2bd}$$

Because wood beams ordinarily have rectangular cross sections, this formula is appropriate for most beams. In recent years, however, important advances have been made in the fabrication of wood beams of various sizes and shapes. Glued-laminated beams are often employed and I-shapes and box sections are built up with wood elements. To find the horizontal shear stress for beams in which the cross sections are not rectangular, use is made of the general formula for shear stress

$$f_v = \frac{VQ}{Ib}$$

where f_v = unit horizontal shear stress at any point in the beam cross section (psi)

V = total vertical shear at the section selected (lb)

Q = statical moment with respect to the neutral axis of the beam cross section of the area of the cross section above or below the point at which stress is to be determined (in.^{3})

I = moment of inertia of the beam cross section with respect to its neutral axis (in.^{4})

b = width of the beam at the point where stress is computed (in.)

A *statical moment* is an area multiplied by the distance of its centroid from a reference axis.

Consider the rectangular cross section of which b and d are the width and depth, respectively, as shown in Figure 6.1*a*. Using the general formula, find the maximum shear stress at the neutral axis X - X. The area above the neutral axis is $b \times d/2$ and its centroid is at a distance of $d/4$ from the netral axis. Then the statical moment is

$$Q = \left(b \times \frac{d}{2}\right)\left(\frac{d}{4}\right) = \frac{bd^2}{8}$$

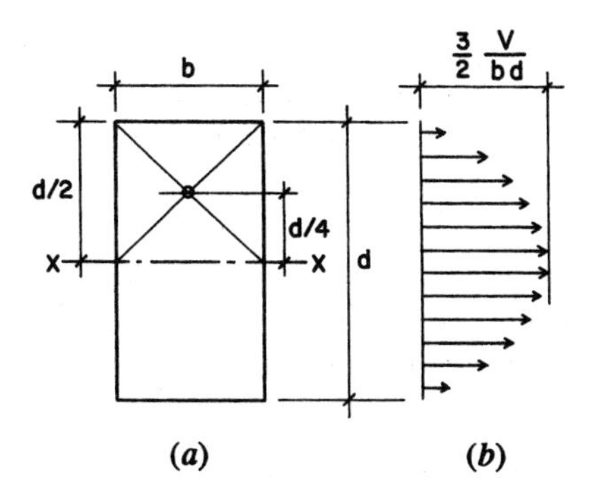

Figure 6.1 Development of beam shear stress.

From Appendix A we know that the moment of inertia of a rectangle is $I = bd^3/12$. Substituting in the general formula,

$$f_v = \frac{VQ}{Ib} = \frac{V\left(\frac{bd^2}{8}\right)}{\left(\frac{bd^3}{12}\right)b} = \frac{3V}{2bd}$$

Which is the formula given previously for the maximum shear stress for a rectangular section. The shear stress in beams is not distributed evenly over the cross section. The form of distribution for a rectangular section is shown in Figure 6.1*b*, with the maximum shear occurring at the neutral axis.

The general shear stress formula can be used for any shape of cross section.

Example 3. Determine the horizontal shear stress at the horizontal glue line of the box beam shown in Figure 6.2 for a vertical shear force of 4000 lb.

Solution: The point at which the shear stress is desired is the joint between the top and bottom 2 × 6 pieces and the three middle pieces. The statical moment for this computation is that of one of the 2 × 6 pieces, thus

$$Q = (1.5 \times 5.5)\left(\frac{11.25}{2} + \frac{1.5}{2}\right) = 52.6 \text{ in.}^3$$

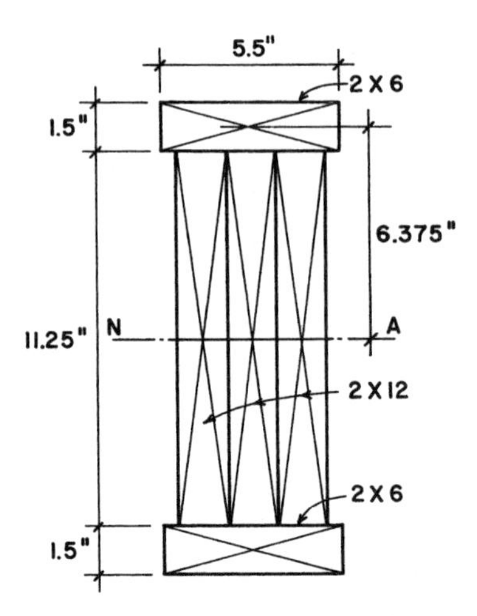

Figure 6.2 Reference for Example 3.

The moment of inertia of the section is determined by using the process described in Section A.3 of the Appendix. For the central members, the moment of inertia is simply 3 times the listed property for the member; thus $I = 3(178) = 534\,\text{in.}^4$. For the top and bottom pieces, the moment of inertia about the neutral axis of the combined section for one piece is determined as

$$I = I_0 + Az^2$$

In this equation, I_0 is the moment of inertia of the 2×6 about its minor axis (I_y), which is $1.547\,\text{in.}^4$. The area of one piece is $8.25\,\text{in.}^2$, and the distance z is 6.375 in., as shown in Figure 6.2.Thus the total moment of inertia of the top and bottom pieces is

$$I = 2[1.547 + 8.25(6.375)^2] = 674\,\text{in.}^4$$

The total moment of inertia of the 5-piece section is thus

$$I_{NA} = 674 + 534 = 1208\,\text{in.}^4$$

The stress at the glue line between the outer top and bottom pieces and the center pieces is now determined as

$$f_v = \frac{VQ}{Ib} = \frac{4000 \times 52.6}{1208 \times 4.5} = 38.7\,\text{psi}$$

Problem 6.1.D. Five pieces of lumber are glued together to form a beam similar to that shown in Figure 6.2. The top and bottom pieces are 2 × 8s and the vertical center portion consists of two 3 × 10s and one 2 × 10. Find the shear stress at the glue line between the top and bottom pieces and the central portion for a shear force of 6600 lb.

Problem 6.1.E. Same as Problem 6.1.D, except the top and bottom pieces are 3 × 8s, the center portion consists of five 2 × 12s, and the shear force is 11,400 lb.

6.2 BENDING IN BEAMS

Resisting Moment and the Flexure Formula

Figure 6.3*a* shows a simple beam of rectangular cross section supporting a single concentrated load *P*. An enlarged view of the portion of the beam to the left of the cut section *X* - *X* is shown in Figure 6.3*b*. In resisting the bending moment at this section, compression stresses are developed in the upper portion of the section and tension stresses in the lower portion. At some point in the depth of the beam there is a

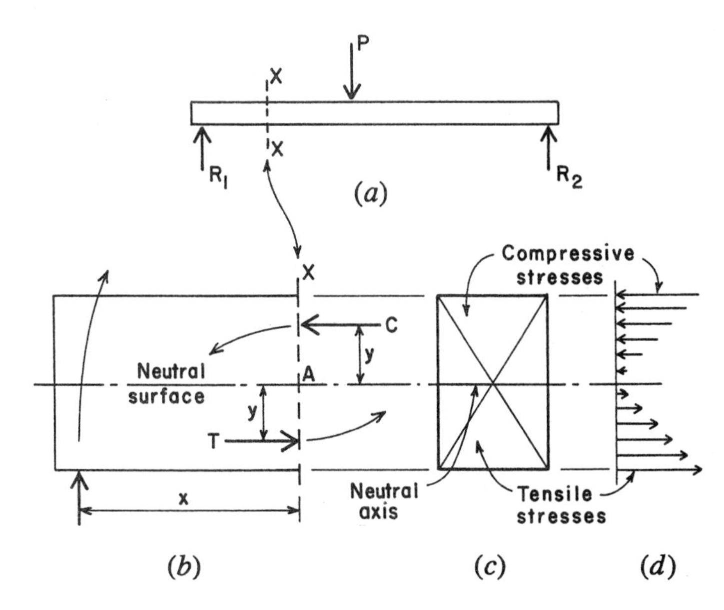

Figure 6.3 Development of bending stress in a beam.

horizontal plane called the *neutral surface*, on which bending stress is zero, as the stress development switches from compression to tension. With regard to the cut section, as shown in Figure 6.3*c*, the neutral surface appears as an axis line; this is called the *neutral axis*.

The compressive stresses on the upper portion of the section combine to produce a resultant compressive force, and similarly, the tensile stresses in the bottom produce a tension force. As shown in Figure 6.3*b*, equilibrium of forces on the left portion of the beam requires that the active moment effect of the reaction R times x be resisted by the combined effects of the internal compression and tension forces, or

$$R_1 \times x = (C \times y) + (T \times y)$$

The combined effect of the internal forces of tension and compression is called the *resisting moment*. Development of the resisting moment may be expressed in terms of the stresses at the cross section, and the stresses can then be related directly to the external bending moment at the section. The following discussion explains this process for the simple rectangular section.

In Figure 6.3*b* the compressive and tensile stresses acting on the rectangular cross section are represented by their resultants C and T, respectively. These stresses are not developed as uniformly distributed stresses, but rather they vary from values of zero at the neutral axis to a maximum at the top and bottom of the section, as shown in Figure 6.3*d*. This pattern of stress distribution is repeated in Figure 6.4, where the form of development of the compression force is indicated. The crosshatched area indicates the area on which the compressive stresses work. It is assumed that the stresses vary proportionally from zero at the neutral axis to a maximum at the upper and lower edges of the section. The average compressive stress is thus $f/2$, and the centroid of the triangular stress is at $d/3$ from the neutral axis. This permits a definition of the portion of the resisting moment developed by compression as

$$M_C = b \times \frac{d}{2} \times \frac{f}{2} \times \frac{d}{3} = f\left(\frac{bd^2}{12}\right)$$

And, since the effect of the tensile stresses is similar, the total resisting moment can be expressed as

$$M_R = M_C + M_T = 2\left[f\left(\frac{bd^2}{12}\right)\right] = f\left(\frac{bd^2}{6}\right)$$

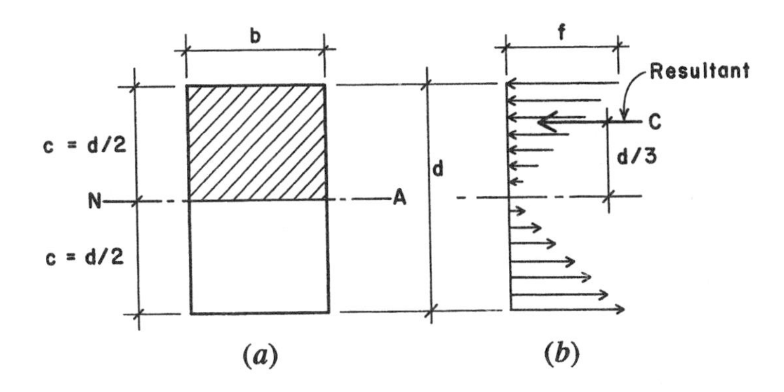

Figure 6.4 Moment resistance developed by bending stress.

As discussed in Appendix A, the expression $bd^2/6$ is the section modulus S of the rectangular shape. Then, since the resisting moment must be equal to the external bending moment, a relationship between bending moment and bending stress is simply stated as

$$M = fS$$

This simple expression is called the *flexure formula.*

Application of the Flexure Formula

The expression $M = fS$ may be stated in three different forms, depending on the information desired. These are given here with a notation that makes a distinction between allowable bending stress (F_b) and computed maximum bending stress (f_b). Thus

$$(1)\quad M = F_b S \qquad (2)\quad f_b = \frac{M}{S} \qquad (3)\quad S = \frac{M}{F_b}$$

Form (1) yields the maximum potential resisting moment when the section modulus of the beam and the maximum allowable bending stress are known. Form (2) yields the computed maximum bending stress when the beam section modulus and the moment acting on the beam are known. These two forms are used to investigate the adequacy of a given beam.

Form (3) is used for design work. It yields the required section modulus when the maximum bending moment and the limiting allowable bending stress are known. When the required section modulus has been determined, a beam with an S equal to or larger than the computed value is selected.

When using the flexure formula, care must be exercised with respect to the units in which terms are expressed. Stress values are ordinarily given in pounds per square inch (psi) or kips per square inch (ksi) [kPa or MPa in metric units]. S is usually given in in.^{3} or mm^3. Bending moments are usually computed in ft-lb or kip-ft, thus requiring a conversion to in.-lb or kip-in. [kN-m in metric units with attention to the powers of 10].

Example 4. The maximum bending moment on a beam is 13,100 ft-lb [17.8 kN-m]. If the allowable bending stress for the wood is 1400 psi [9.65 MPa], determine the required beam section.

Solution: The required section modulus is

$$S = \frac{M}{F_b} = \frac{13{,}100 \times 12}{1400} = 112.3 \text{ in.}^3 \; [1840 \times 10^3 \text{ mm}^3]$$

A possible choice from Table A.3 is a 6 × 12 beam with $S = 121.2$ in.^{3}.

Problem 6.2.A. If the maximum bending moment on an 8 × 10 beam is 16 kip-ft [22 kN-m], what is the value of the maximum bending stress?

Problem 6.2.B. Find the required standard lumber shape with the least cross-sectional area from Table A.3 for a beam in which the maximum bending moment is 20 kip-ft [27 kN-m] if the allowable stress is 1600 psi [11.0 MPa].

Problem 6.2.C. Find the maximum resisting moment for a beam consisting of a wood 6 × 10 if the maximum allowable bending stress is 1400 psi [9.65 MPa].

Size Factors for Rectangular Beams

For solid-sawn beams deeper than 12 in. and thicker than 4 in., the allowable bending stress must be reduced. This is accomplished by multiplying the base allowable stress by the appropriate *size factor*

TABLE 6.1 Actual Beam Depth

(in.)	(mm)	C_F
13.5	343	0.987
15.5	394	0.972
17.5	445	0.959
19.5	495	0.947
21.5	546	0.937
23.5	597	0.928

determined from the formula:

$$C_F = \left(\frac{12}{d}\right)^{1/9}$$

where C_F = the size factor
d = the depth of the beam in in.

Values of C_F for a few depths of beams greater than 12 in. are given in Table 6.1. For design work, where the depth of the beam is not known in advance, the procedure is to find the section required and then to determine its capacity with the allowable bending stress reduced as necessary. Knowing this situation in advance, designers know to select deep sections for a slightly higher S value than that determined by the unmodified allowable bending stress.

6.3 DEFLECTION

Deflections in wood structures tend to be most critical for rafters and joists, where span-to-depth ratios are often pushed to the limit. However, long-term high levels of bending stress can also produce sag, which may be visually objectionable or cause problems with the construction. In most situations it is the maximum value of deflection that is of concern; for a simple beam, as shown in Figure 6.5, this

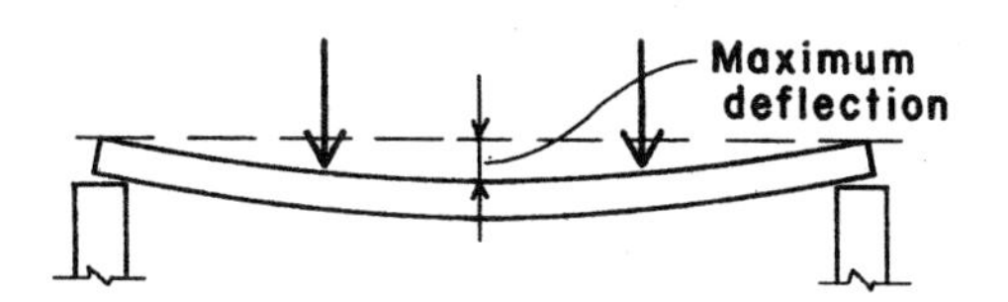

Figure 6.5 Beam deflection.

deflection occurs at midspan. In general it is wise to be conservative with deflections of wood structures. Push the limits and you will surely get sagging floors and roofs and possibly very bouncy floors. This may in some cases make a strong argument for use of glued-laminated beams, fabricated joists, or even steel beams.

For the common uniformly loaded simple beam, the deflection takes the form of the equation

$$\Delta = \frac{5WL^3}{384EI}$$

Substitutions of relations between W, M, and flexural stress in this equation can result in the form

$$\Delta = \frac{5L^2 f_b}{24Ed}$$

Using average values of 1500 psi for f_b and 1,500,000 psi for E, the expression reduces to

$$\Delta = \frac{0.03L^2}{d}$$

where Δ = deflection in inches
L = span in feet
d = beam depth in inches

Figure 6.6 is a plot of this expression with curves for nominal dimensions of depth for standard lumber. For reference the lines on the graph corresponding to ratios of deflection of *L*/180, *L*/240, and *L*/360 are shown. *L*/180 is sometimes used as a limit for total load for roofs. *L*/240 is more commonly used for total load deflection. *L*/360 is commonly used for live load deflection. Also shown for reference is the limiting span-to-depth ratio of 25 to 1, which is commonly considered to be a practical span limit for general purposes. For beams with other values for bending stress and modulus of elasticity, true deflections can be obtained as follows:

$$\text{true } \Delta = \frac{\text{true } f_b}{1500} \times \frac{1{,}500{,}000}{\text{true } E} \times \Delta \text{ from graph}$$

The following examples illustrate problems involving deflection. Douglas fir-larch is used for these examples and for the problems that follow them.

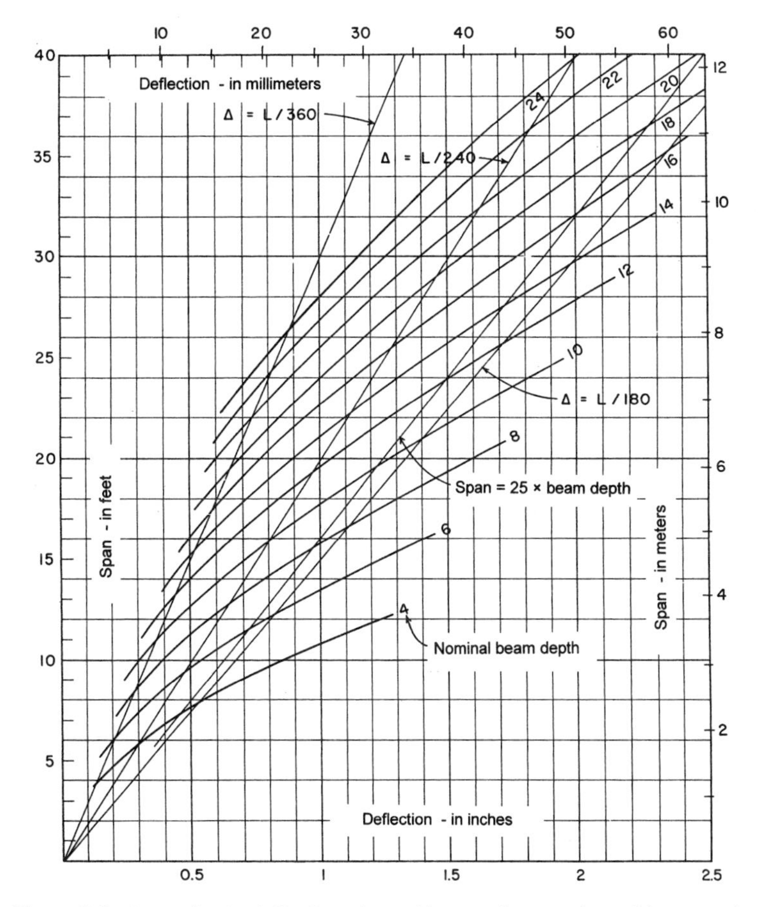

Figure 6.6 Approximate deflection of wood beams. Assumed conditions: maximum bending stress of 1500 psi and modulus of elasticity of 1,500,000 psi.

Example 5. An 8 × 12 wood beam with $E = 1{,}600{,}000$ psi is used to carry a total uniformly distributed load of 10 kips on a simple span of 16 ft. Find the maximum deflection of the beam.

Solution: From Table A.3 find the value of $I = 950\,\text{in.}^4$ for the 8 × 12 section. Then, using the deflection formula for this loading,

$$\Delta = \frac{5WL^3}{384EI} = \frac{5 \times 10{,}000 \times (16 \times 12)^3}{384 \times 1{,}600{,}000 \times 950} = 0.61\ \text{in.}$$

Or, using the graph in Figure 6.6,

$$M = \frac{WL}{8} = \frac{10{,}000 \times 16}{8} = 20{,}000 \text{ ft-lb}$$

From Table A.3, $S = 165 \text{ in.}^3$, and the beam maximum stress is

$$f_b = \frac{M}{S} = \frac{20{,}000 \times 12}{165} = 1455 \text{ psi}$$

From Figure 6.6, $\Delta =$ approximately 0.66 in. Then

$$\Delta = \frac{1455}{1500} \times \frac{1{,}500{,}000}{1{,}600{,}000} \times 0.66 = 0.60 \text{ in.}$$

which shows reasonable agreement with the value computed using the deflection formula.

Example 6. A beam consisting of a 6×10 section with $E = 1{,}400{,}000$ psi spans 18 ft and carries two concentrated loads. One load is 1800 lb and is placed at 3 ft from one end of the beam, and the other load is 1200 lb, placed at 6 ft from the opposite end of the beam. Find the maximum deflection due only to the concentrated loads.

Solution: For an approximate computation, use the equivalent uniform load method, consisting of finding the hypothetical total uniform load that will produce a moment equal to the actual maximum moment in the beam. Then the deflection for uniformly distributed load may be used with this hypothetical (equivalent uniform) load. Thus

$$\text{if:} \quad M = \frac{WL}{8}, \qquad \text{then:} \quad W = \frac{8M}{L}$$

For this loading the maximum bending moment is 6600 ft-lb (the reader should verify this by the usual procedures) and the equivalent uniform load is thus

$$W = \frac{8M}{L} = \frac{8 \times 6600}{18} = 2933 \text{ lb}$$

From Table A.3, $I = 393 \text{ in.}^4$, and the approximate deflection is

$$\Delta = \frac{5WL^3}{384EI} = \frac{5 \times 2933 \times (18 \times 12)^3}{384 \times 1{,}400{,}000 \times 393} = 0.70 \text{ in.}$$

As in the previous example, the deflection could also be found by using Figure 6.6, with adjustments made for the true maximum bending stress and the true modulus of elasticity.

Note: For the following problems, neglect the beam weight and consider deflection to be limited to 1/240 of the beam span. Wood is Douglas fir-larch.

Problem 6.3.A. A 6 × 14 beam of No. 1 grade is 16 ft [4.88 m] long and supports a total uniformly distributed load of 6000 lb [26.7 kN]. Investigate the deflection.

Problem 6.3.B. An 8 × 12 beam of dense No. 1 grade is 12 ft [3.66 m] in length and has a concentrated load of 5 kips [22.2 kN] at the center of the span. Investigate the deflection.

Problem 6.3.C. Two concentrated loads of 3500 lb [15.6 kN] each are located at the third points of a 15 ft [4.57 m] beam. The 10 × 14 beam is of select structural grade. Investigate the deflection.

Problem 6.3.D. An 8 × 14 beam of select structural grade has a span of 16 ft [4.88 m] and a total uniformly distributed load of 8 kips [35.6 kN]. Investigate the deflection.

Problem 6.3.E. Find the least weight section that can be used for a simple span of 18 ft [5.49 m] with a total uniformly distributed load of 10 kips [44.5 kN] based on deflection. Wood is No. 1 grade.

6.4 BEARING

Bearing occurs at beam ends when a beam sits on a support, or when a concentrated load is placed on top of a beam within the span. The stress developed at the bearing contact area is compression perpendicular to the grain, for which an allowable value ($F_{c\perp}$) is given in Table 4.1.

Although the design values given in the table may be safely used, when the bearing length is quite short the maximum permitted level of stress may produce some indentation in the edge of the wood member. If the appearance of such a condition is objectionable, a reduced stress is recommended. Excessive deformation may also produce some significant vertical movement, which may be a problem for the general building construction.

Example 7. An 8 × 14 beam of Douglas fir-larch, No. 1 grade, has an end bearing length of 6 in. [152 mm]. If the end reaction is 7400 lb [32.9 kN], is the beam safe for bearing?

Solution: The developed bearing stress is equal to the end reaction divided by the product of the beam width and the length of bearing. Thus

$$F_{c\perp} = \frac{\text{bearing force}}{\text{contact area}} = \frac{7400}{7.5 \times 6} = 164\,\text{psi} \quad [1.13\,\text{MPa}]$$

This is compared to the allowable stress of 625 psi from Table 4.1, which shows the beam to be quite safe.

Example 8. A 2 × 10 rafter cantilevers over and is supported by the 2 × 4 top plate of a stud wall. The load from the rafter is 800 lb [3.56 kN]. If both the rafter and the plate are Douglas fir-larch, No. 2 grade, is the situation adequate for bearing?

Solution: The bearing stress is determined as

$$F_{c\perp} = \frac{800}{1.5 \times 3.5} = 152\,\text{psi} \quad [1.05\,\text{MPa}]$$

This is considerably less than the allowable stress of 625 psi, so the bearing is safe.

Example 9. A two-span 3 × 12 beam of Douglas fir-larch, No. 1 grade, bears on a 3 × 14 beam at its center support. If the reaction force is 4200 lb [18.7 kN], is this safe for bearing?

Solution: Assuming the bearing to be at right angles, the stress is

$$F_{c\perp} = \frac{4200}{2.5 \times 2.5} = 672\,\text{psi} \quad [4.63\,\text{MPa}]$$

This is slightly in excess of the allowable stress of 625 psi.

Problem 6.4.A. A 6 × 12 beam of Douglas fir-larch, No. 1 grade, has 3 in. of end bearing to develop a reaction force of 5000 lb [22.2 kN]. Is the situation adequate for bearing?

Problem 6.4.B. A 3 × 16 rafter cantilevers over a 3 × 16 support beam. If both members are of Douglas fir-larch, No. 1 grade, is the situation adequate for bearing? The rafter load on the support beam is 3000 lb [13.3 kN].

6.5 BUCKLING OF BEAMS

Buckling of beams—in one form or another—is mostly a problem with beams that are relatively weak on their transverse axes, that is, the axis of the beam cross section at right angles to the axis of bending. This is not a frequent condition in concrete beams, but is a common one with beams of wood or steel or with trusses that perform beam functions.

When buckling is a problem one solution is to redesign the beam for more resistance to lateral movement. Another possibility is to analyze for the lateral buckling effect and reduce the usable bending capacity as appropriate. However, the solution most often used is to brace the beam against the movement developed by the buckling effect. To visualize where and how such bracing should be done it is first necessary to consider the various possibilities for buckling. The three main forms of beam buckling are shown in Figure 6.7.

Figure 6.7*b* shows the response described as *lateral* (that is *sideways*) *buckling*. This action is caused by the compressive stresses in the beam that make it act like a long column, which is thus subject to a sideways buckling movement as with any slender column. Bracing the beam for this action means simply preventing its sideways movement at the beam edge where compression exists. For simple span beams this edge is the top of the beam. For beams, joists, rafters, or trusses that directly support roof or floor decks, the supported deck may provide this bracing if it is adequately attached to the supporting members. For beams that support other beams in a framing system, the supported beams at right angles to the supporting member may provide lateral bracing. In the latter case, the unsupported length of the buckling member becomes the distance between the supported beams, rather than its entire span length.

Another form of buckling for beams is that described as *torsional buckling*, as shown in Figure 6.7*d*. This action may be caused by tension stress, resulting in a rotational, or twisting, effect. This action can occur even when the top of the beam is braced against lateral movement, and is often due to a lack of alignment of the plane of the loading and the vertical axis of the beam. Thus a beam that is slightly tilted

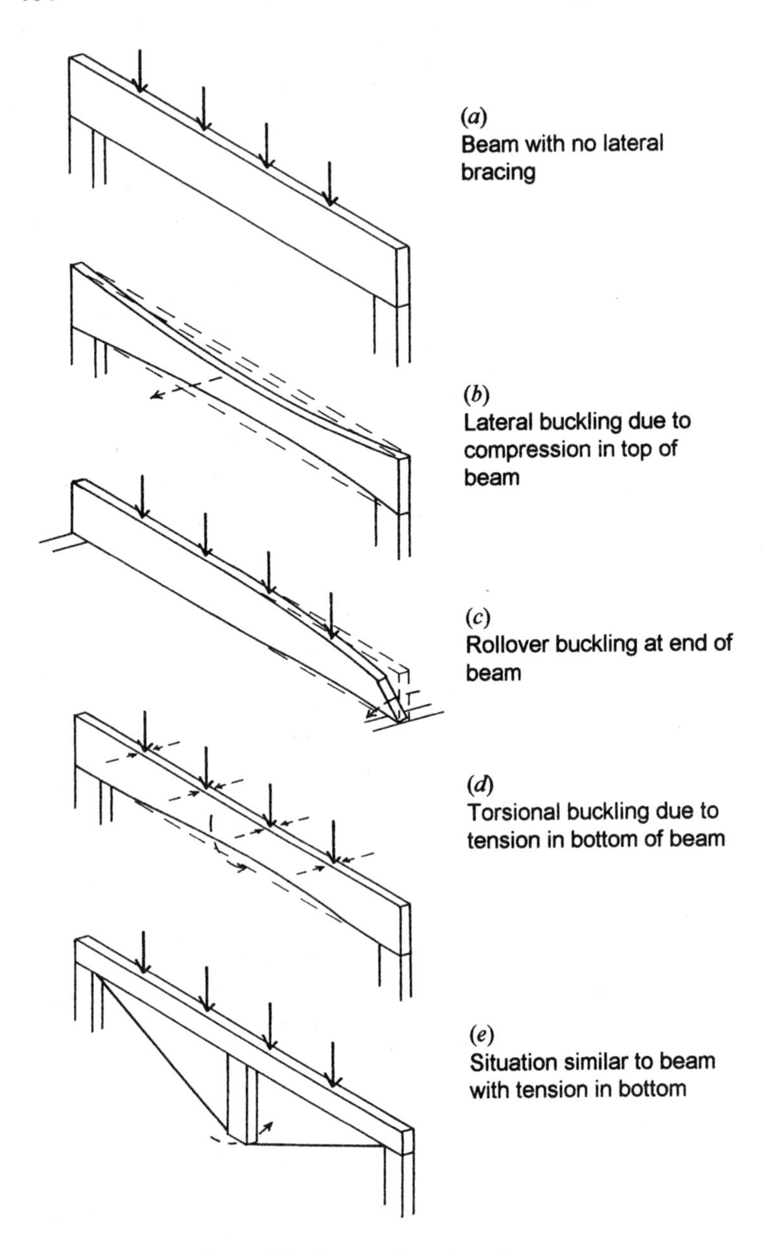

Figure 6.7 Forms of buckling of beams.

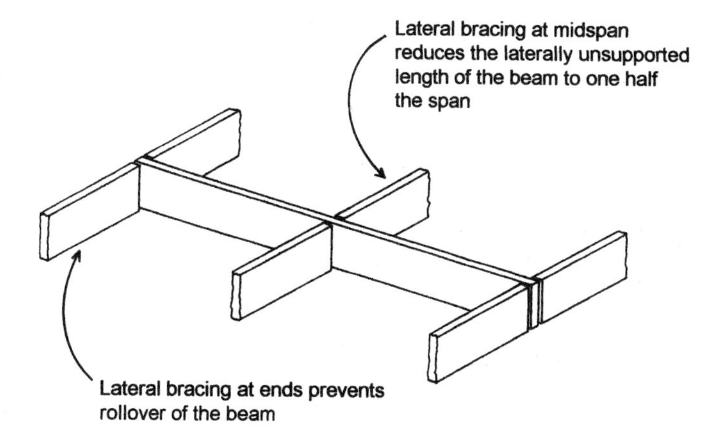

Figure 6.8 Lateral bracing for beams.

is predisposed to a torsional response. An analogy for this is shown in Figure 6.7*e*, which shows a trussed beam with a vertical post at the center of the span. Unless this post is perfectly vertical, a sideways motion at the bottom end of the post is highly likely. Another situation where torsional buckling occurs is at beam supports, as shown in Figure 6.7*c*.

To prevent both lateral and torsional buckling, it is necessary to brace the beam sideways at both its top and bottom. If the roof or floor deck is capable of bracing the top of the beam, the only extra bracing required is that for the bottom. For closely spaced trusses this bracing is usually provided by simple horizontal ties between adjacent trusses. For beams in wood or steel framing systems, lateral bracing may be provided as shown in Figure 6.8. The beam shown is braced for both lateral and torsional buckling.

Design specifications provide for the adjustment of bending capacity or allowable stress when a beam is vulnerable to buckling failure. To eliminate this adjustment, the NDS establishes requirements for lateral bracing; these are summarized in Table 6.2. This situation applies mostly to relatively thin wood joists and rafters, and is discussed more thoroughly in Section 7.3.

6.6 UNSYMMETRICAL BENDING

Beams in flat-spanning structures are ordinarily positioned so the loads and the plane of bending are perpendicular to one of the principal axes

TABLE 6.2 Lateral Support Requirements for Rectangular Sawn Wood Beams[a]

Ratio of Depth To Breadth $(d/b)^b$	Required Conditions to Avoid Reduction of Bending Stress
$d/b \leq 2$	No support required.
$2 < d/b \leq 4$	Ends held in position to prevent rotation or lateral displacement.
$4 < d/b \leq 5$	Compression edge held in position for entire span, and ends held in position to prevent rotation or lateral displacement.
$5 < d/b \leq 6$	Compression edge held in position for entire span, ends held in position to prevent rotation or lateral displacement, and bridging or blocking at intervals not exceeding 8 ft.
$6 < d/b \leq 7$	Both edges held in position for entire span, and ends held in position to prevent rotation or lateral displacement.

[a] Adapted from data in *National Design Specification ® for Wood Construction* (Ref. 1), with permission of the publishers, American Forest & Paper Association.
[b] Ratio of nominal dimensions for standard sections.

of the beam section. In this case, the bending stresses are symmetrically distributed, and the principal axis about which bending occurs is also the neutral stress axis.

There are various situations in which a beam is subjected to bending in a manner that results in simultaneous bending about both principal axes of the beam cross section. If the beam is adequately braced against torsion and buckling, the result may simply be a case of what is called *biaxial bending* or *unsymmetrical bending*. Figure 6.9 shows a beam in a sloping roof structure, in which the beam spans between the beams or trusses that define the roof slope. With respect to wind loading on the sloped surface, the beam will be bent about a major axis. However, with respect to gravity loads this is a case of unsymmetrical bending. Considered separately, the bending stresses produced by the components of the bending with respect to the beam's axes are

$$f_{bx} = \frac{M_x}{S_x} \quad \text{and} \quad f_{by} = \frac{M_y}{S_y}$$

These are maximum stresses that occur at the edges of the section. Their distribution is of a form such as that shown in Figure 6.9*c*, which can be described by determining the values for the stresses at the four corners of the section. Noting the senses of the moments with respect to the two axes, and using plus for tensile stress and minus for compressive

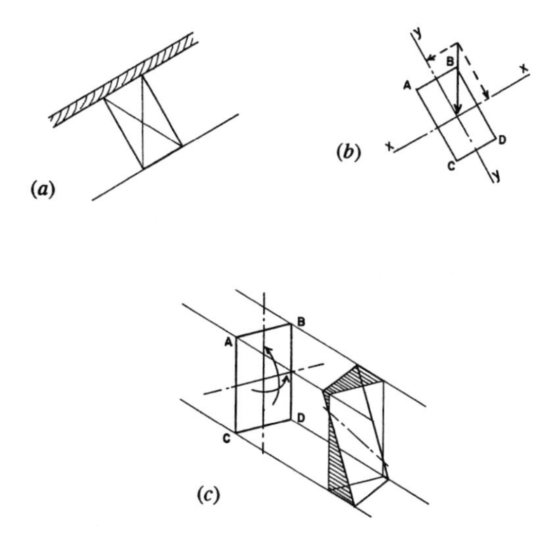

Figure 6.9 Development of unsymmetrical bending.

stress, the net stresses at the corners are:

$$\text{at } A\text{: } - f_x + f_y$$

$$\text{at } B\text{: } - f_x - f_y$$

$$\text{at } C\text{: } + f_x + f_y$$

$$\text{at } D\text{: } + f_x - f_y$$

This is a somewhat idealized condition that ignores problems of torsion and buckling, and is valid only for sections with biaxial symmetry (such as the rectangle or I-shape). If a member is used in the situation shown in Figure 6.9, it should preferably be one with low susceptibility to these actions, such as the almost square section shown, or the construction should be developed so as to provide bracing to prevent actions other than the simple bending illustrated in Figure 6.9.

Example 10. Figure 6.10*a* shows the use of a beam in a sloping roof. The beam section is rotated to a position that corresponds to the roof slope of 30 degrees, and consists of an 8 × 10 wood member (actual dimensions from Table A.3 of 7.5 × 9.5 in.). Find the net bending stress condition for the beam if the gravity load generates a moment of 10 kip-ft [14 kN-m] in a vertical plane.

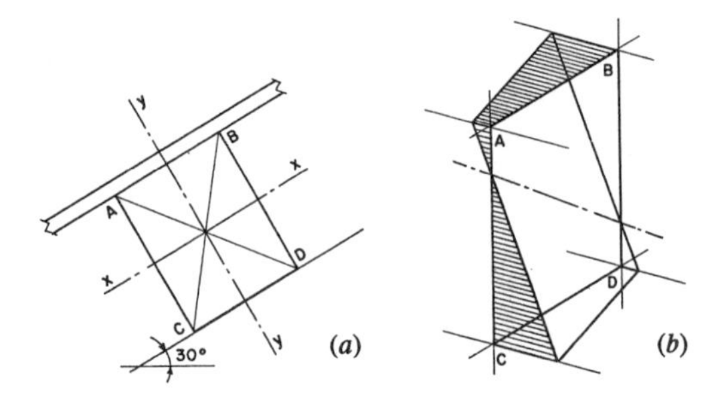

Figure 6.10 Reference for Example 10.

Solution: From Table A.3, the properties of the beam are $S_x = 112.8\,\text{in.}^3$ [$1.85 \times 10^6\ \text{mm}^3$] and $S_y = 89.1\,\text{in.}^3$ [$1.46 \times 10^6\ \text{mm}^3$]. The components of the moment with respect to the x and y axes are:

$$M_x = 10\cos 30 = 10(0.866) = 8.66\,\text{kip-ft}\ [12.12\ \text{kN-m}]$$

$$M_x = 10\sin 30 = 10(0.5) = 5\,\text{kip-ft}\ [7\ \text{kN-m}]$$

And the corresponding maximum stresses are

$$f_x = \frac{M_x}{S_x} = \frac{8.66(12)}{112.8} = 0.921\,\text{ksi}\ [6.55\,\text{MPa}]$$

$$f_y = \frac{M_y}{S_y} = \frac{5(12)}{89.1} = 0.673\,\text{ksi}\ [4.79\,\text{MPa}]$$

The net stress conditions at the four corners of the section—lettered A through D in Figure 6.10*a*—are then as follows, using plus for tensile stress and minus for compressive stress:

$$\text{at } A: -0.921 + 0.673 = -0.248\,\text{ksi}\ [1.76\ \text{MPa}]$$

$$\text{at } B: -0.921 - 0.673 = -1.594\,\text{ksi}\ [11.34\ \text{MPa}]$$

$$\text{at } C: +0.921 + 0.673 = +1.594\,\text{ksi}\ [11.34\ \text{MPa}]$$

$$\text{at } D: +0.921 - 0.673 = +0.248\,\text{ksi}\ [1.76\ \text{MPa}]$$

The form of the stress distribution is as shown in Figure 6.10*b*.

Problem 6.6.A. A wood beam consisting of a 6 × 10 of Douglas fir-larch, dense No. 1 grade, is used as a beam in a sloping roof, as shown in Figure 6.9. Assuming no stress modifications apply, and that torsion and buckling are prevented, is the beam adequate for a bending moment of 8 kip-ft [10.8 kN-m] in a vertical plane with a roof slope of 25 degrees?

Problem 6.6.B. Same as Problem 6.6.A, except the beam is a 10 × 12, the wood is Douglas fir, select structural grade, the bending moment is 18 kip-ft, and the roof slope is 20 degrees.

6.7 BEHAVIOR CONSIDERATIONS FOR LRFD

Investigation of beams in the LRFD method uses many of the relationships derived for the ASD method. One difference, of course, is the use of ultimate load for design, rather than service load. The other main difference has to do with the basic form of expression for the resistance of the structural member to force effects (bending, shear, bearing). The ASD method uses a limiting resistance expressed in terms of a safe limiting stress condition. Thus the limiting resistance is directly related to service loads. In the LRFD method, resistance is derived in total force effect form: based on some stress analysis, but expressed in pounds, ft-lb, etc.

One method for handling the LRFD analysis is to use the basic relationships of the ASD method with adjusted values for stress and some additional modification factors. Thus the process is basically an altered version of the ASD method, but the answers are expressed in LRFD terms.

Because the numbers tend to get larger in LRFD analysis, it is customary to use force in kips and stress in ksi, rather than the pounds and psi used in the ASD method. This is somewhat arbitrary, but references are developed in this form, so the reader should become accustomed to the practice.

The following discussions treat the topics in the earlier sections of this chapter, illustrating the applications in the LRFD method.

Shear in Beams

As described in Section 6.1, the total usable shear in a beam, at a given stress level, may be expressed as

$$V = \frac{2 f_v A}{3}$$

where f_v is the defined value for shear stress and A is the area of the beam cross section for a rectangular beam. For a limiting condition in the ASD method, the stress used is that obtained from Table 4.1, designated F_v. This value for stress is actually only the so-called reference value, and is frequently modified for size of the member, moisture content, and so on. The modified stress is designated F'_v, and thus the true usable shear force is expressed as

$$V = \frac{2F'_v A}{3}$$

For the LRFD method, this expression is modified in two ways. First the usable stress is adjusted, as explained in Section 4.5, by using the adjustment factor from Table 4.6. For shear this factor is $2.16/1000\phi_v$, for which ϕ_v is 0.75 (see Table 4.7). The fully adjusted and modified shear resistance is then expressed as

$$\lambda\phi_v V' = \lambda\phi_v \left\{ \frac{2}{3} \left(\frac{2.16}{1000\phi_v} \right) (F'_v)(A) \right\}$$

In this formula the term λ (lambda) is the *time effect factor* described in Section 4.5. This factor depends on the load combination for which the shear is being determined, and its values are given in Table 4.8 for five common load combinations.

The formula for shear is typically used in one of two ways. The first way involves a beam of determined size for which the usable ultimate shear resistance is to be established, usually for comparison to the actual shear in a beam. For this problem the formula is used in the form above, with λ chosen for the appropriate load combination. In the design of beams the beam size is often first established on the basis of flexure, and consideration for shear involves a check on the beam's adequacy.

The second way the formula is used is in a design process if it is desired to find the appropriate size beam based on shear. For this problem the formula is transformed into an expression of the required area. This process may be used when a high shear force makes it possible that shear, and not flexure, may be critical for the beam.

The following two examples illustrate these two problems. The first example is the same situation as that described for Example 1 in Section 6.1.

Example 11. A simple beam with a span of 14 ft supports a uniformly distributed dead load of 300 lb/ft and a uniformly distributed live load of 500 lb/ft. A 10 × 14 beam of Douglas fir of select structural grade is used. Is the beam safe with respect to shear stress?

Solution: The first step is to determine the required ultimate shear force. For the combined dead load and live load the ultimate combination is

$$w_u = 1.2(300) + 1.6(500) = 1160\,\text{lb/ft} \quad \text{or} \quad 1.16\,\text{kip/ft}$$

and the total load on the beam is 1.16(14) = 16.24 kip.

The maximum shear V_u is one half the load, or 8.12 kip. The test for adequacy involves the satisfaction of the basic LRFD formula

$$\lambda\phi_v V' \geq V_u$$

Which is to say, the factored shear must be equal to or greater than the ultimate shear. To find the factored shear we first determine the adjusted shear stress. Using the formula given above, we find the reference shear stress from Table 4.1 to be 170 psi, or 0.170 ksi. For λ, with the combined dead and live loading, Table 4.8 yields a value of 0.8. The factored shear is thus found as

$$\begin{aligned}
\lambda\phi_v V' &= \lambda\phi_v \left\{\frac{2}{3}\left(\frac{2.16}{1000\phi_v}\right)(F_v')(A)\right\} \\
&= (0.8)(0.75)\left\{\frac{2}{3}\left(\frac{2.16}{1000(0.75)}\right)(170)(128.25)\right\} \\
&= 25.1\,\text{kip}
\end{aligned}$$

As this is greater than the ultimate shear of 8.12 kip, the section is adequate (or safe).

As with the ASD method, a reduced value of shear may be used with uniformly loaded beams. This permits the exclusion of the shear within a distance from the support equal to the beam depth. This will result in a design value for ultimate shear force of less than 8.12 kip. Since the section has been demonstrated to be far from critical, this investigation is not indicated in this situation.

The following example illustrates the second type of shear problem, that in which it is desired to determine the size of beam required for a given shear performance.

Example 12. A 20-ft-span beam carries a dead load of 160 lb/ft and a live load of 240 lb/ft and is to consist of Douglas fir-larch, select structural grade. Find the minimum size for the beam, based on shear resistance.

Solution: The first step involves the determination of the required ultimate shear force at the end of the beam. The loads are thus adjusted to factored ones as follows:

$$w_u = 1.2(160) + 1.6(240) = 576\,\text{lb/ft or } 576/1000 = 0.576\,\text{kip/ft}$$

The maximum shear for the simple beam is determined as

$$V_u = \frac{wL}{2} = \frac{0.576(20)}{2} = 5.76\,\text{kip}$$

Assuming a beam depth of 12 in., the reduced shear at beam depth distance from the support becomes

$$V_u = 5.76 - 0.576 = 5.184\,\text{kip}$$

Using the adjustments for load time and resistance reduction, the required shear resistance becomes

$$V' = \frac{V_u}{\lambda\phi} = \frac{5.184}{(0.8)(0.75)} = 8.64\,\text{kip}$$

Using this value for the required shear, we can transform the shear equation into one for finding A, as follows

$$V' = \frac{2}{3}\left\{\frac{2.16}{1000\phi_v}(F'_v)(A)\right\}$$

from which

$$A = \frac{2}{3}\left\{\frac{V'}{\frac{2.16}{1000\phi_v}(F'_v)}\right\} = \frac{3}{2}\left\{\frac{8.64}{\frac{2.16}{1000(0.75)}(170)}\right\} = 26.5\,\text{in.}^2$$

From Table A.3, sections of 3 × 12, 4 × 10, or 6 × 6 may be used to obtain this area. The 3 × and 4 × members are in the class of *Dimension* Lumber in Table 4.1, and thus a slightly larger shear stress (180 psi) is permitted—still not a critical issue. However, as in many situations, considerations for bending and deflection require larger (mostly deeper) sections. Thus, initial design for shear is only justified when an exceptionally high load is carried on a relatively short span. In this example, if the load is doubled and the span is cut in half, the critical shear is approximately the same but bending and deflection are much less critical.

Problem 6.7.A. A simple beam with a span of 18 ft supports a uniformly distributed dead load of 240 lb/ft and a uniformly distributed live load of 480 lb/ft. An 8 × 16 beam of Douglas fir-larch of No. 1 grade is used. Is the beam safe with regard to shear force?

Problem 6.7.B. Same as Problem 6.7.A, except span is 24 ft, dead load is 360 lb/ft, live load is 560 lb/ft, and the section is a 10 × 20.

Problem 6.7.C. A 10 × 18 in. beam of Douglas fir-larch, dense No. 1 grade, is used for a 16 ft span. The beam supports a uniformly distributed dead load of 160 lb/ft and three concentrated live loads of 4 kips each at the quarter points of the span (4 ft on center). Find the minimum size for the beam, based on shear resistance.

Problem 6.7.D. Same as Problem 6.7.C, except the dead load is 240 lb/ft, the live load is 6 kips, the span is 20 ft, and the section is a 12 by 20.

Bending in Beams

For bending the procedures are essentially the same as described for shear. The relationship between beam resistance and load-generated moment is expressed as

$$\lambda \phi_b M' \geq M_u$$

M' is expressed as the product of a limiting bending stress and the beam's section modulus: $M' = F_b'S$. F_b' is the reference value stress for the ASD method, F_b, multiplied by any applicable adjustment factors. With these substitutions, the moment equation becomes

$$\lambda \phi_b \left(\frac{2.16}{1000\phi_b} \right) (F_b')(S) \geq M_u$$

As with shear, there are two common types of problems. The first is an investigation, or a *design check*, which consists of determining the moment resistance capacity of a given member and comparing it to a required, load-generated, ultimate moment. The second problem is a basic design situation, where the member required for a given ultimate moment is to be determined. The following examples demonstrate these problems.

Example 13. A simple beam with a span of 14 ft supports a uniformly distributed dead load of 300 lb/ft and a uniformly distributed live load of 500 lb/ft. A 10 × 14 beam of Douglas fir of Select Structural grade is used. Is the beam safe with respect to bending?

Solution: First, the required ultimate moment is determined. Thus

$$w_u = 1.2(300) + 1.6(500) = 1160\,\text{lb/ft} \quad \text{or} \quad 1.16\,\text{kip/ft}$$

$$M_u = \frac{wL^2}{8} = \frac{1.16(14)^2}{8} = 28.42\,\text{kip-ft}$$

For the resistance of the beam, the value of S from Table A.3 is 288.563 in.^{3}. From Table 4.1 the design value of F_b is 1600 psi. This must be modified by any appropriate factors, but it is assumed for this example that no modification is required. For the situation of dead load plus live load, λ from Table 4.8 is 0.8. The resistance factor for bending is 0.85 from Table 4.7. The moment capacity is thus

$$\begin{aligned} \lambda\phi_b M' &= \lambda\phi_b \left(\frac{2.16}{1000\phi_b} F'_b\right)(S) \\ &= (0.8)(0.85)\left(\frac{2.16}{1000(0.85)}(1600)\right)(288.563) \\ &= 798\,\text{kip-in.} \quad \text{or} \quad \frac{798}{12} = 66.5\,\text{kip-ft} \end{aligned}$$

Since this is considerably greater than the required moment, the beam is more than adequate.

Example 14. A 20-ft-span beam carries a dead load of 160 lb/ft and a live load of 240 lb/ft and is to consist of Douglas fir-larch, select structural grade. Find the minimum size for the beam, based on bending resistance.

Solution: For the ultimate moment:

$$w_u = 1.2(160) + 1.6(240) = 576\,\text{lb/ft} \quad \text{or} \quad 576/1000 = 0.576\,\text{kip/ft}$$

and the ultimate moment is

$$M_u = \frac{w_u L^2}{8} = \frac{0.576(20)^2}{8} = 28.8\,\text{kip-ft}$$

From Table 4.1 $F_b = 1600\,\text{psi}$, and

$$\begin{aligned}\lambda\phi_b M' &= \lambda\phi_b \left(\frac{2.16}{1000\phi_b} F_b\right)(S)\\ &= (0.8)(0.85)\left(\frac{2.16}{1000(0.85)}(1600)\right)(S)\\ &= 2.7648S \text{ (in kip-in.)}\end{aligned}$$

Equating this to M_u

$$\begin{aligned}M_u &= 28.8\,\text{kip-ft} = 28.8(12) = 345.6\,\text{kip-in.} = 2.7648S\\ S &= \frac{345.6}{2.7648} = 125\,\text{in.}^3\end{aligned}$$

From Table A.3, possible choices are 6 × 14, 8 × 12, 10 × 10.

Problem 6.7.E. A simple beam with a span of 18 ft supports a uniformly distributed dead load of 240 lb/ft and a uniformly distributed live load of 480 lb/ft. An 8 × 16 beam of Douglas fir-larch of No. 1 grade is used. Is the beam safe with regard to bending?

Problem 6.7.F. Same as Problem 6.7.E, except span is 24 ft, dead load is 360 lb/ft, the live load is 560 lb/ft., and the section is a 10 × 20.

Problem 6.7.G. A 10 × 18 in. beam of Douglas fir-larch, dense No. 1 grade, is used for a 16 ft span. The beam supports a uniformly distributed dead load of 160 lb/ft and three concentrated live loads of 4 kips each at the quarter points of the span (4 ft on center). Is the beam safe with regard to bending?

Problem 6.7.H. Same as Problem 6.7.G, except the dead load is 240 lb/ft, the live load is 6 kips, the span is 20 ft, and the section is a 12 × 20.

Problem 6.7.I. A simple beam with a span of 22 ft supports a uniformly distributed dead load of 200 lb/ft and a uniformly distributed live load of 600 lb/ft. Douglas fir-larch of No. 1 grade is to be used. Design the beam for bending ignoring the weight of the beam.

Problem 6.7.J. Same as Problem 6.7.I, except the span is 26 ft, dead load is 300 lb/ft, and live load is 400 lb/ft.

Deflection

In both the ASD and LRFD methods deflections are considered to occur under the service loads. Computations are thus the same for both methods.

Bearing

Bearing is treated in the same manner as shear and bending. The bearing load is a factored load, adjusted stress is used, and the process is similar in form to that used for shear and bending.

Unsymmetrical (Biaxial) Bending

Investigation of biaxial bending is treated in the NDS LRFD Specification. The result is similar in form to that described in Section 6.6, but the process should be used as described in the specification. This is not a frequently encountered problem, so it will not be illustrated here. An understanding of the basic concerns can be obtained from the work in Section 6.6.

Design Process

The general beam design process is discussed in Chapter 7. Use of the LRFD method for beam design is also discussed. The broadest context for beam design is illustrated in the building design examples in Chapter 15.

7

DESIGN OF BEAMS

Design of beams uses the information gained from the analytical work treated in Chapters 5 and 6. However, the design process begins with the final form of the beam unknown. This does not affect the determination of reactions, shear forces, or bending moments due to loads, but it does relate to investigations for internal forces and their stress effects, and to the determination of conditions regarding deflection. The material in this chapter treats the issues of the design process.

7.1 DESIGN PROCEDURE

The design of a sawn wood beam is accomplished in five steps. With the beam not chosen, the initial work must be done without consideration for the beam weight; or an approximate weight may be assumed. Weights of standard sawn lumber sizes are given in Table A.3, assuming a density of 35 lb/ft^3, which is that given for Douglas fir-larch.

Step 1: Determine the loads the beam must support and find the reaction forces.

Step 2: Determine the maximum bending moment and the required section modulus for the beam. A beam section with an adequate section modulus can then be selected from Table A.3. If the beam depth exceeds two times the width, consideration should be given to possible lateral or torsional buckling.

Step 3: Investigate the beam selected in Step 1 for shear. If necessary, increase the dimensions of the beam.

Step 4: Investigate the beam for deflection, using any established limits for the situation.

Step 5: If the support for the beam involves a bearing condition, investigate for the necessary provisions for the bearing details.

7.2 BEAM DESIGN EXAMPLES

The following examples illustrate application of the foregoing procedure to the design of sawn wood beams under different loading conditions.

ASD Method

Example 1. A simple beam has a span of 14 ft [4.2 m] and carries a load of 7200 lb [32 kN] uniformly distributed over the beam length. Design the beam using Douglas fir-larch, No. 1 grade, sawn lumber. Deflection under total load is limited to 0.5 in. [13 mm].

Solution: From Table 4.1 reference values are $F_b = 1350$ psi [9.3 MPa], $F_v = 170$ psi [1.17 MPa], $E = 1{,}600{,}000$ psi [11 GPa]. Since no special conditions have been stated for this example, the reference values will be used without modification. For the symmetrical beam and its loading, the reactions are each equal to one half the total load, or 3600 lb [16 kN]. From Figure 5.25, Case 2, the maximum bending moment is

$$M = \frac{WL}{8} = \frac{(7200)(14)}{8} = 12{,}600 \text{ ft-lb } [16.8 \text{ kN-m}]$$

and the required section modulus is

$$S = \frac{M}{F_b} = \frac{(12{,}600)(12)}{1350} = 112 \text{ in.}^3 \ [1836 \times 10^3 \text{ mm}^3]$$

From Table A.3, a 6 × 12 has a section modulus of 121.3 in.^{3} and weighs 15.4 lb/ft, making a total beam weight of 15.4 × 14 = 216 lb. The total load is thus 7200 + 216 = 7416 lb, which increases the required section modulus to (7416/7200)(112) = 115.4 in.^{3}. As this is still less than the section modulus of the chosen beam, the chosen shape is adequate.

The maximum shear is equal to one half the total load, or 3708 lb. As described in Section 6.1, the critical design value for shear may be that which occurs at a distance from the support equal to the beam depth. Using the beam depth as 12 in. or one ft, this shear will be

$$V = 3708 - \frac{7416}{14} = 3708 - 530 = 3178\,\text{lb}$$

Using the area of the beam from Table A.3, the shear stress is found as

$$f_v = \frac{3}{2} \times \frac{3178}{63.25} = 75.4\,\text{psi}$$

As this value is considerably less than the limiting stress, the beam is not critical for shear.

For deflection, using the formula given for Case 2 in Figure 5.25, the computation is

$$\Delta = \frac{5WL^3}{384EI} = \frac{5(7416)(14 \times 12)^3}{384(1{,}600{,}000)(697)} = 0.411\,\text{in.}$$

as this is less than the limit of 0.5 in, the beam is adequate for deflection.

Stability of the beam with respect to torsional rotation and lateral buckling should be considered. Referring to Table 6.2, it may be observed that the depth-to-breadth ratio for this beam—6/12 = 2—means that no stability considerations must be made. If this ratio is higher, the provisions for bracing given in Table 6.2 must be considered, or else a reduced bending capacity can be determined using stability analysis equations from the NDS.

LRFD Method

Example 2. A simple beam has a span of 14 ft [4.2 m] and carries a total dead load of 3000 lb [13.3 kN] and a total live load of 4200 lb [18.3 kN], both of which are distributed uniformly over the beam length. Design the beam using Douglas fir-larch, No. 1 grade, sawn lumber.

Deflection under total load is limited to 0.5 in. (This is the same problem as in Example 1.)

Solution: For the LRFD method the work begins with the determination of the ultimate load. Using the factored combination for dead load plus live load:

$$W_u = 1.2(DL) + 1.6(LL) = 1.2(3000) + 1.6(4200) = 3600 + 6720 = 10{,}320\,\text{lb}$$

and the maximum moment is

$$M_u = \frac{WL}{8} = \frac{(10{,}320)(14)}{8} = 18{,}060\,\text{lb-ft [24.5 kN-m]}$$

$$= \frac{18{,}060 \times 12}{1000} = 216.7\,\text{kip-in.}$$

The required section modulus for this moment can be determined by equating this moment to the factored resisting moment, thus

$$\lambda\phi_b M' = \lambda\phi_b\left(\frac{2.16}{1000\phi_b}\right)(F_b)(S) = M_u$$

$$(0.8)(0.85)\left(\frac{2.16}{1000(0.85)}\right)(1350)(S) = 216.7$$

$$S = \frac{216.7}{2.33} = 93.0\,\text{in.}^3$$

From Table A.3, possible choices are 6 × 12, 8 × 10. Try the 6 × 12, for which Table A.3 yields the following data: $A = 63.25\,\text{in.}^2$, $S = 121.3\,\text{in.}^3$, $I = 697\,\text{in.}^4$, and the weight per ft is 15.4 lb/ft. The additional dead load of the beam weight increases the total load as follows:

$$W = 10{,}320 + 1.2(15.4 \times 14) = 10{,}320 + 258 = 10{,}579\,\text{lb}$$

This slightly increases the maximum moment and the required section modulus, but the minor change is not critical for the member choice in this case.

For shear the maximum value is that at the beam end, equal to one half the beam load, or 10,579/2 = 5290 lb. This value may be compared to the adjusted shear resistance of the member, determined as

$$\lambda\phi_v V' = \lambda\phi_v \left\{\frac{2}{3}\left(\frac{2.16}{1000\phi_v}\right)(F'_v)(A)\right\}$$
$$= (0.8)(0.75)\left\{\frac{2}{3}\left(\frac{2.16}{1000(0.75)}\right)(170)(63.25)\right\}$$
$$= 12.387\,\text{kips or } 12{,}387\text{ lb}$$

which considerably exceeds the required shear. Of course, the actual critical ultimate shear may be reduced to that at a distance from the support equal to the beam depth. This procedure is demonstrated for the ASD method in Example 1. Having demonstrated that the beam is adequate for a larger value of shear, this adjusted shear need not be considered in this case.

For deflection the computation is exactly the same as that for the ASD method, as shown in Example 1, since deflection is always caused by service loads, not by factored loads.

Problem 7.2.A. A simple beam with a span of 15 ft carries a total dead load of 4000 lb [17.8 kN] and a total live load of 5000 lb [22.2 kN] in addition to the beam weight. These loads are all uniformly distributed on the beam span. Deflection is limited to 0.625 in. under the total load. Design the beam using Douglas fir-larch, Select Structural grade, by the ASD method.

Problem 7.2.B. Repeat Problem 7.2.A using the LRFD method.

Problem 7.2.C. A simple beam with a span of 20 ft carries a total dead load of 2000 lb [8.9 kN] and a total live load of 3000 lb [13.3 kN] in addition to the beam weight. These loads are all uniformly distributed on the beam span. Deflection is limited to 1.25 in. under the total load. Design the beam using Douglas fir-larch, Dense No. 1 grade, by the ASD method.

Problem 7.2.D. Repeat Problem 7.2.C using the LRFD method.

7.3 JOISTS AND RAFTERS

Floor joists and roof rafters are closely spaced beams that support floor or roof decks. They are common elements of the structural system described as the *light wood frame*. These may consist of sawn lumber, light trusses, laminated pieces, or composite elements achieved with combinations of sawn lumber, laminated pieces, plywood, or particleboard. The discussion in this chapter deals only with sawn lumber, typically in the class called *dimension lumber* having nominal thickness of 2 to 4 in. Although the strength of the structural deck is a factor, spacing of joists and rafters is typically related to the dimensions of the panels used for decking. The most used panel size is 48 by 96 in., from which are derived spacings of 12, 16, 19.2, 24, and 32 in.

Floor Joists

A common form of floor construction is shown in Figure 7.1. The structural deck shown in the figure is plywood, which produces a top surface not generally usable as a finished surface. Thus, some finish must be used, such as the hardwood flooring shown here. More common now for most interiors is carpet or thin tile, both of which require some smoother surface than the structural plywood panels, resulting in the use of *underlayment* typically consisting of wood fiber panels.

A drywall panel finish (paper-faced gypsum plaster board) is shown here for the ceiling directly attached to the underside of the joists. Since the floor surface above is usually required to be horizontally flat, the same surface can thus be developed for the ceiling. With rafter construction for roofs that must normally be sloped, or when more space is required in the floor/ceiling construction, a separate *suspended* structure must be provided for the ceiling.

Lateral bracing for joists is usually provided by the attached deck. If additional bracing is required (see Table 6.2), it may consist of bridging

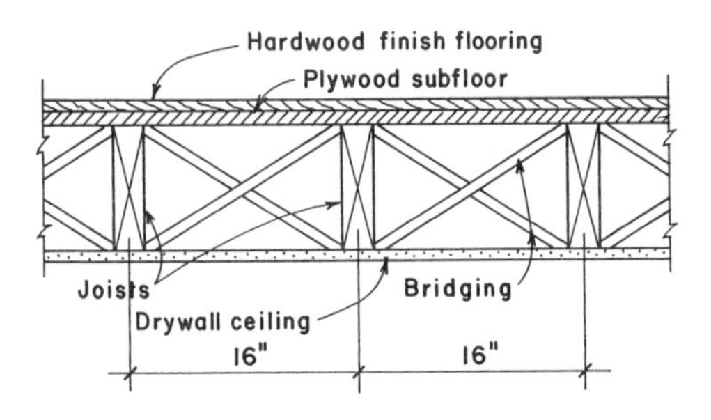

Figure 7.1 Typical joist floor construction.

as shown in Figure 7.1, or of solid blocking consisting of short pieces of the joist elements aligned in rows between the joists. If blocking is used, it will normally be located so as to provide for edge nailing of the deck panels. This nailing of all the edges of panels is especially critical when the joist and deck construction is required to serve as a horizontal diaphragm for wind or seismic forces. (See discussion of diaphragms in Chapter 14.)

Solid blocking is also used under any supported walls perpendicular to the joists, or under walls parallel to the joists but not directly above a joist. Any loading on the joist construction, other than the usual assumed dispersed load on the deck, should be considered for reinforcement of the regular joist system. A simple way to give extra local strength to the system is to double up joists. Doubling of joists is a common practice at the edges of large openings in the floor.

With the continuous, multiple-span effect of decking, and possible inclusion of bridging or blocking, there is typically a potential for load sharing by adjacent joists. This is the basis for classification as a *repetitive member*, permitting an increase of 15 percent in the allowable bending stress.

Floor joists may be designed as beams by the procedure illustrated in Section 7.2. However, the most frequent use of joists in light wood framing systems is in situations that are well defined in a short range of conditions. Spans are usually quite short, both dead and live loads are predictable, and a relatively few wood species and grades are most commonly used. This allows for the development of tabulated lists of joist sizes from which appropriate choices can be made. Table 7.1 is an abbreviated sample of such a table

Example 3. Using Table 7.1, select joists to carry a live load of 40 psf and a dead load of 10 psf on a span of 15 ft 6 in. Wood is Douglas fir-larch, No. 2 grade.

Solution: From Table 7.1, possible choices are 2 × 10 at 12 in., 2 × 12 at 16 in., or 2 × 12 at 19.2 in.

Note that the values in Table 7.1 are based on a maximum deflection of 1/360 of the span under live load. In using the table for the example, it is also assumed that there is no modification of the reference stress values. If true conditions are significantly different from those assumed for Table 7.1, the full design procedure from Section 7.2 is required.

TABLE 7.1 Maximum Spans for Floor Joists (ft-in.)[a]

	Joist Size			
Spacing (in.)	2 × 6	2 × 8	2 × 10	2 × 12
	Live load = 40 psf, Dead load = 10 psf, Maximum deflection = $L/360$			
12	10–9	14–2	17–9	20–7
16	9–9	12–7	15–5	17–10
19.2	9–1	11–6	14–1	16–3
24	8–1	10–3	12–7	14–7
	Live load = 40 psf, Dead load = 20 psf, Maximum deflection = $L/360$			
12	10–6	13–3	16–3	18–10
16	9–1	11–6	14–1	16–3
19.2	8–3	10–6	12–10	14–10
24	7–5	9–50	11–6	13–4

[a] Joists are Douglas fir-larch, No. 2 grade. Assumed maximum available length of single piece is 26 ft.

Source: Compiled from data in the *International Building Code* (Ref. 4), with permission of the publishers, International Code Council.

Rafters

Rafters are used for roof decks in a manner similar to floor joists. While floor joists are typically installed dead flat, rafters are commonly sloped to achieve roof drainage. For structural design it is common to consider the rafter span to be the horizontal projection, as indicated in Figure 7.2.

As with floor joists, rafter design is frequently accomplished with the use of safe load tables. Table 7.2 is representative of such tables and has been reproduced from the *IBC* (Ref. 4). Organization of the table is similar to that for Table 7.1. The following example illustrates the use of the data in Table 7.2.

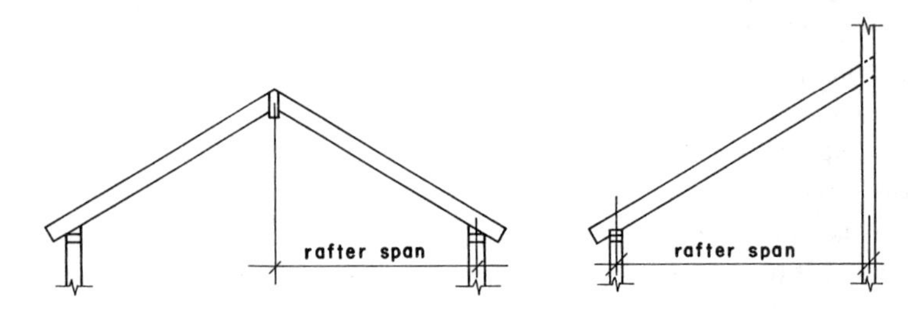

Figure 7.2 Span of sloping rafters.

TABLE 7.2 Maximum Spans for Rafters (ft-in.)[a]

	Rafter Size				
Spacing (in.):	2 × 4	2 × 6	2 × 8	2 × 10	2 × 12
	Live load = 20 psf, Dead load = 10 psf, Maximum deflection = L/240				
12	9–10	15–6	20–5	25–8	26–0
16	8–11	14–1	18–2	22–3	25–9
19.2	8–5	13–1	16–7	20–3	23–6
24	7–10	11–9	14–10	18–2	21–0
	Live load = 20 psf, Dead load = 20 psf, Maximum deflection = L/240				
12	9–10	14–4	18–2	22–3	25–9
16	8–6	12–5	15–9	19–3	22–4
19.2	7–9	11–4	14–4	17–7	20–4
24	6–11	10–2	12–10	15–8	18–3

[a]Rafters are Douglas fir-larch, No. 2 grade. Ceiling is not attached to rafters. Assumed maximum available length of single piece is 26 ft.

Source: Compiled from data in the *International Building Code* (Ref. 4), with permission of the publishers, International Code Council.

Example 4. Rafters are to be used for a roof span of 16 ft. Live load is 20 psf; total dead load is 10 psf; live load deflection is limited to 1/240 of the span. Find the rafter size required for Douglas fir-larch of No. 2 grade.

Solution: From Table 7.2, possible choices are for 2 × 8 at 16 in., 2 × 8 at 19.2 in., or 2 × 10 at 24 in.

Problems 7.3.A,B,C,D. Using Douglas fir-larch, No. 2 grade, pick the joist size required from Table 7.1 for the stated conditions. Live load is 40 psf, dead load is 10 psf, deflection is limited to L/360 under live load only.

	Joist Spacing (in.)	Joist Span (ft)
A	16	14
B	12	14
C	16	16
D	12	20

Problems 7.3.E,F,G,H. Using Douglas fir-larch, No. 2 grade, pick the rafter size required from Table 7.2 for the stated conditions. Live load is 20 psf, dead load is 20 psf, deflection is limited to *L*/240 under live load only.

	Rafter Spacing (in.)	Rafter Span (ft)
E	16	12
F	24	12
G	16	18
H	24	18

7.4 ALTERNATIVE SPANNING ELEMENTS

Various types of structural components can be produced with assembled combinations of plywood, fiber panels, laminated products, and solid-sawn lumber. Figure 7.3 shows some commonly used elements that can serve as structural components for buildings.

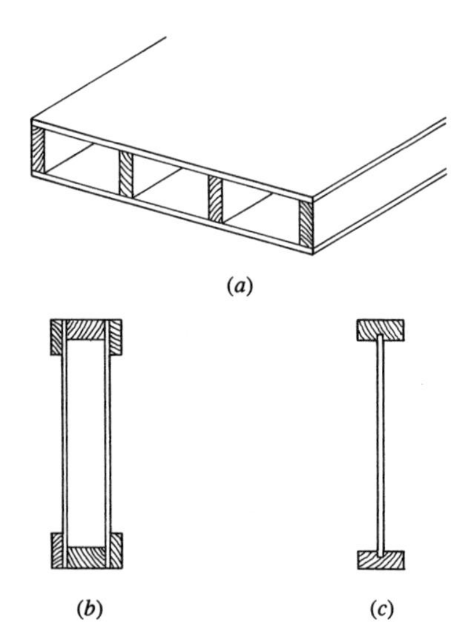

Figure 7.3 Composite, built-up components with elements of solid-sawn lumber and panels of plywood or wood fiber.

The unit shown in Figure 7.3*a* consists of two panels of plywood attached to a frame of solid-sawn lumber elements. This is generally described as a *sandwich panel*; however, when used for structural purposes it is called a *stressed-skin panel*. For spanning actions, the plywood panels serve as bending stress resisting flanges and the lumber elements as beam webs for shear development.

Another common type of product takes the form of the box-beam (Figure 7.3*b*) or the built-up I-beam (Figure 7.3*c*). In this case the roles defined for the sandwich panel are reversed, with the solid-sawn elements serving as flanges and the panel material as the web. These elements are highly variable, using both plywood and fiber products for the panels, and solid-sawn lumber or glued-laminated products for the flange elements. It is also possible to produce various profiles, such as one with a horizontally flat bottom and a sloped or curved top.

The box-beam shown in Figure 7.3*b* can be assembled with attachments of ordinary nails or screws. The I-beam uses glued joints to attach the web and flanges. Box-beams may be custom-assembled at the building site, but the I-beams are produced in highly controlled factory conditions.

I-beam products have become highly popular for use in the range of spans just beyond the feasibility for solid-sawn lumber joists and rafters; that is, over about 20 ft for joists and about 25 ft for rafters. See discussion in Section 12.5.

Two types of light wood trusses are widely used. The W-truss, shown in Figure 7.4*a*, is widely used for short-span gable-form roofs.

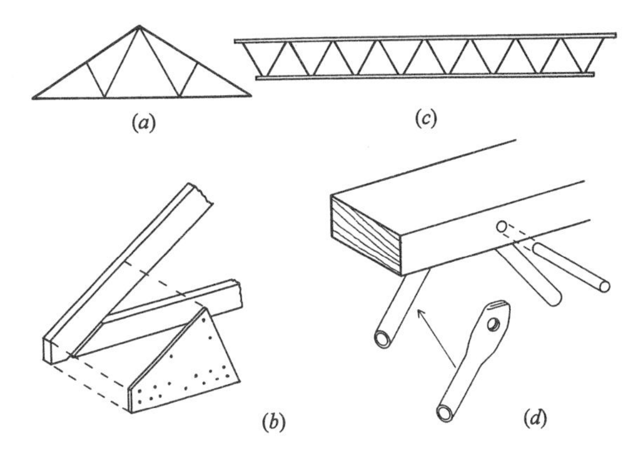

Figure 7.4 Light wood trusses.

Achieved with a single layer of 2× lumber members, and with simple gusset-plated joints (Figure 7.4*b*), this has been a common form of the roof structure for small wood-framed buildings for many years. Gussets may consist of pieces of plywood, attached with nails, but are now mostly factory-assembled with metal connector plates.

For flat spanning structures—both roofs and floors—the truss shown in Figure 7.4*c* is used, mostly for spans just beyond the spanning length feasible for solid-sawn wood rafters or joists. One possible assembly is shown in Figure 7.4*d*, using steel tubes with flattened ends, connected to the chords with pins driven through drilled holes. Chords may be simple solid-sawn lumber elements, but are also made of proprietary laminated elements that permit virtually unlimited length for single-piece members. See discussion in Section 11.8.

8

WOOD DECKS

Materials used to produce structural decks with wood frames include the following:

1. Boards of nominal 1-in.-thick sawn wood, typically with tongue-and-groove edges.
2. Sawn wood elements thicker than 1 inch nominal dimensions, usually called planks or planking. These have tongue-and-groove edges or some other form of edge development to prevent slipping between adjacent units.
3. Plywood of appropriate thickness for the span and the construction.
4. Other panel materials, including those composed of compressed wood fibers or particles.
5. Proprietary products, usually used for roof decks.

8.1 BOARD DECKS

Before plywood established its dominance as a decking material, most roof and floor decks were made with $^{3}/_{4}$ in. (nominal 1 in.) boards, as

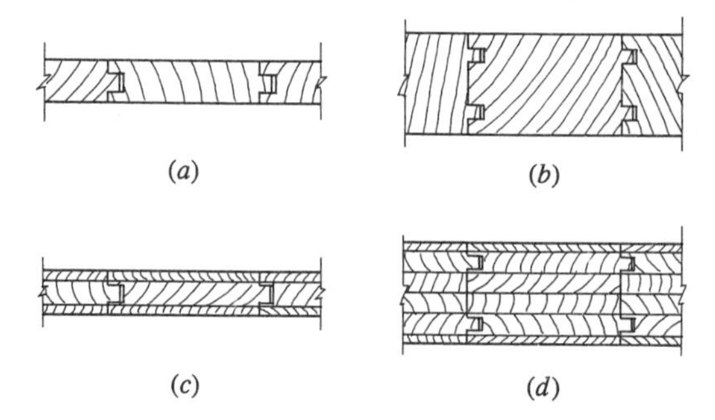

Figure 8.1 Units for board and plank decks.

shown in Figure 8.1*a*. Today this type of deck is seldom used, except in some areas where sawn wood products are readily available and the cost of construction with the boards is competitive with plywood or other decking.

Decks of nominal 1 in. thickness are usually adequate for roofs or floors where supporting members are not more than 24 in. on center. However, the type of roofing or the type of finish floor treatments must be considered. As with most structural decks, it is common to use some filler material on top of the structural deck to achieve a smoother surface. This is especially necessary for flat roofs and for floors with carpet or thin tiles.

Plank Decks

The most widely used form of plank deck is that made with 2 in. nominal thickness units (approximately 1.5 in. in actual thickness). This form of deck is usually quite expensive, so there are usually specific reasons for selecting such a deck, including one or more of the following:

1. The deck is to be exposed to view on the underside, and the appearance of the planks is considerably better than that of plywood or other panel units.
2. Exposed to view or not, the deck may require a fire rating, and the thicker decks are much better in this regard. The 2 in. nominal thickness is usually the minimum required with construction qualified by building codes as *heavy timber* for fire rating.
3. It may be desired to have supporting members with spacing exceeding that which is feasible with panel deck units.

4. Concentrated loads from vehicles or heavy equipment may be too high for thinner decks.

Nominal 2 in. plank deck may be of the same form as board deck (Figure 8.1*a*) but is often made with laminated units, as shown in Figures 8.1*c* and *d*. Plank deck is also available in thickness greater than 2 in. nominal. When thickness exceeds 2.5 in. or so, the units usually have a double tongue-and-groove on each face, as shown in Figure 8.1*b* for a solid-sawn unit and in Figure 8.1*d* for a laminated unit.

One problem with board and plank deck is the low capacity of the deck for diaphragm action in resisting lateral forces. This is generally the case when the boards or planks are installed at right angles to supporting members. Placing the deck units on a diagonal is one way of increasing diaphragm capability, although a more practical means is to simply add a thin panel of plywood to the top of the deck.

Four types of spans for plank floors are recognized: simple, two-span continuous, combination simple and two-span continuous, and controlled random. The last three types are all stiffer, in varying degree, than the simple span because of the continuity introduced by the different arrangements of the pieces of planking. The four span types are identified in Figure 8.2.

Examination of the figure shows that all planks are the same length in the simple span with end joints over each beam. The plank lengths are also equal for the two-span continuous arrangement with end joints over

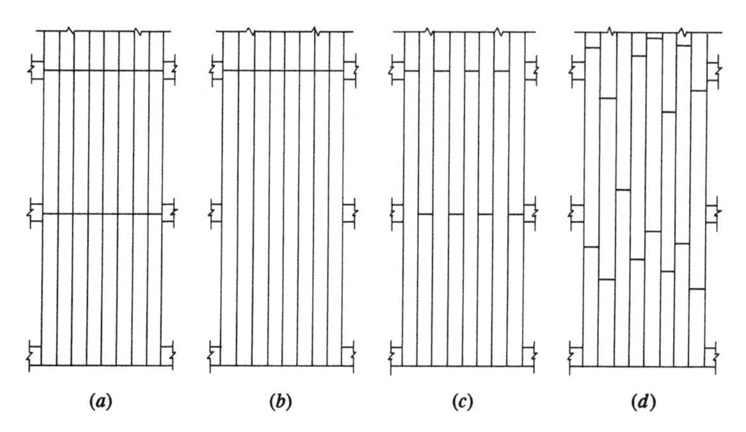

Figure 8.2 Plank deck installation and span conditions: (a) simple span, (b) two-span continuous, (c) combination simple and two-span continuous, (d) controlled random.

every other beam. For the combination span all pieces are two spans in length except for every other piece in the end span, but the end joints over intermediate beams are staggered in adjacent lines of decking. The random arrangement permits the use of economical random lengths of decking with less waste required to trim lengths to fit spans. The principal controls for the random type span are that end joints be well scattered and that each piece of plank bear on at least one beam.

Plank decks are commercially fabricated products, and information about them should be obtained from the suppliers or manufacturers of the products. Type of units available, finishes of exposed undersides, installation specifications, and structural capacities vary greatly, and the products available regionally should be selected for design work.

8.2 WOOD FIBER DECKS

Various products are produced from wood that is reduced to fiber, particle, or stand form. Major considerations are those for the size and shape of the reduced elements and their arrangement in the finished products. For paper, cardboard, and some fine hardboard products, the wood is reduced to very fine particles and is generally randomly placed in the mass of the products. This results in little orientation of the sheet-form materials, other than that resulting from the manufacturing process of the particular products.

For some structural products, use is made of somewhat larger pieces of wood, and some orientation of the end product is obtained from the form of placement of the particles in the mass of the product. Products of this type are discussed in Chapter 12, Section 12.3. Standards for these products are generally available from industry-wide recognized organizations, and are often now included in building codes.

This is definitely a growth area, as plywood and solid-sawn elements become more expensive and the resource for large logs becomes less available. Acceptable fibers and particles can be obtained from smaller, fast-growth trees, and with the large markets for paper products well established, the industry for production of fiber products is firmly in place.

8.3 PLYWOOD DECKS

Structural plywood consists of panels of multiple plies of softwood—usually Douglas fir—with the grain direction in adjacent plies at right angles to each other, and with the face plies having their grain in the same direction. There are thus always an uneven number of plies and a greater strength for bending with the span in the direction of the face grain. Thinner panels may have only three plies, but thicker panels have more plies, and the difference in strength becomes less pronounced. Panels are rated on the basis of the span capability in the face ply direction but always have some spanning capability in the weaker direction as well. In the United States the standard panel size is 48 by 96 in. and the least waste is obtained when panel supports are spaced with whole number divisions of 96 in.: 12, 16, 19.2, 24, 32, or 48 in.

In addition to the span capability, panel thickness depends on two other considerations. The first consideration is the panel strength for diaphragm action in shear, as discussed in Chapter 13. The second consideration has to do with the finish materials that must be attached to the panels. The latter concern may well establish a minimum thickness for anchorage of fasteners, which is mostly an issue with wall sheathing and roof decks, since floors are seldom made with deck materials of less than a $^3/_4$ in. thickness.

Structural plywood is discussed at length in Chapter 12.

8.4 SPANNING CAPABILITY OF DECKS

Spanning capabilities for decks may be a basis for establishment of required spacing of supports when a particular deck is preferred. On the other hand, a desired support spacing will establish the particular related deck that can be used. Other design considerations may also affect both deck choice and support spacing.

With a given support spacing, a range of choices for deck may be considered. Of course, loading and use (roof or floor) will also affect this choice. Table 8.1 presents the possible span capability for various decking materials, assuming a minimum roof live load of 20 psf and a dead load of 10 psf. Because most decking is installed with multiple spans, deflection is less critical. In a similar form, Table 8.2 presents the span capability of floor decking, using a live load of 40 psf and a total dead load of 15 psf. For floors deflection is more critical, mainly as related to bouncing, with thickness established primarily by experience and not by some arbitrary ratio of span to thickness. Plank deck units in these tables represent an average of values given for commercial products.

TABLE 8.1 Span Capability for Various Roof Decking Materials

Decking Material	Range of possible span (in.)								
	12	16	19.2	24	32	48	60	72	84
Board deck, 0.75 in.	......	......	......	......					
Plank deck, 1.375 in	......	......	......	......	......	......			
Plank deck, 2.25 in.	......	......	......	......	......	......	......	......	
Plank deck, 3.25 in.	......	......	......	......	......	......	......	......	......
Plywood, 0.375 in.	......	......	......	......					
Plywood, 15/32 in.	......	......	......	......	......	......			
Particleboard (OSB), 0.75 in.	......	......	......	......					

TABLE 8.2 Span Capability for Various Floor Decking Materials

Decking Material	Range of possible span (in.)							
Decking Material	12	16	19.2	24	32	48	60	72
Board deck, $3/4$ in.	......	......	......					
Plank deck, 1.375 in.	......	......	......	......	......			
Plank deck, 2.25 in.	......	......	......	......	......	......		
Plank deck, 3.25 in.	......	......	......	......	......	......	......	......
Plywood, 3/4 in. Face grain parallel to supports	......	......	......	......				
Plywood, 3/4 in. Face grain perpendicular to supports	......	......	......	......				

All deck materials are commercial products. Those widely used may have span recommendations in building codes. Products manufactured by several companies may have properties, span capabilities, and installation requirements included in industry-wide standards. In the end, use of a specific proprietary product should first be considered with information supplied by the product's manufacturer and distributors.

9

WOOD COLUMNS

A column is a compression member, the length of which is several times greater than its least lateral dimension. The term *column* is generally applied to relatively heavy vertical members, and the term *strut* is given to smaller compression members not necessarily in a vertical position. The type of wood column used most frequently is the *simple solid column*, which consists of a single sawn piece of wood that is square or oblong in cross section. Solid columns of circular cross section are also considered simple solid columns and typically consist of trimmed, but not sawn, tree trunks which are called *poles*. A *spaced column* is an assembly of two or more sawn pieces with their longitudinal axes parallel and separated at their ends and at middle points of their length by blocking. Two other types are *built-up columns*, consisting of multiple sawn pieces bound by mechanical fasteners, and *glued-laminated columns*. The *studs* in light wood framing are also columns. All of these types of columns are treated in this chapter.

9.1 SLENDERNESS RATIO FOR COLUMNS

The following are some general considerations for columns with an emphasis on simple sawn solid columns.

Slenderness Ratio

In wood construction the slenderness ratio of a freestanding simple solid column with a rectangular cross section is the ratio of its unbraced (laterally unsupported) length to the dimension of its least side, expressed as L/d. (See Figure 9.1*a*.) When members are braced so that the laterally unsupported length with respect to one face is less than that with respect to the other, L is the distance between the supports that prevent lateral movement in the direction along which the least dimension is measured. This is illustrated in Figure 9.1*b*. If the section is not square or round, it may be necessary to investigate two L/d conditions for such a column to determine which is the limiting one. The slenderness ratio for simple solid columns is limited to $L/d = 50$, although, during construction, before all bracing materials are installed, a temporary slenderness of 75 is permitted.

9.2 COMPRESSION CAPACITY OF SIMPLE SOLID COLUMNS

Figure 9.2 illustrates the typical form of the relationship between axial compression capacity and slenderness ratio for a linear compression member (column). The limiting conditions are those of the very short member and the very long member. The short member—such as a block of wood—fails in crushing (Zone 1 in Figure 9.2), which is limited by the mass of material and the stress limit for compression.

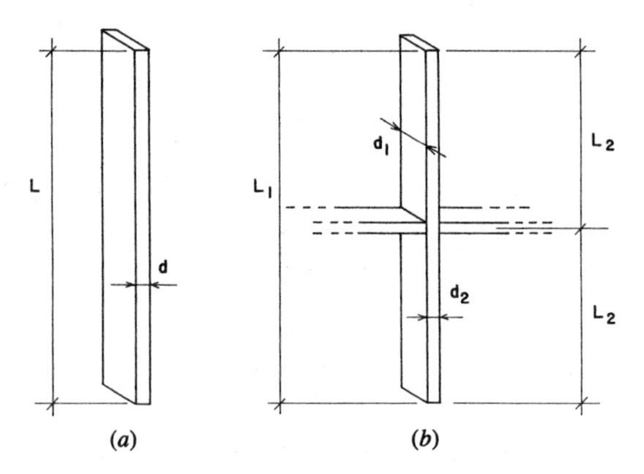

Figure 9.1 Determination of unbraced height for a column, as related to the critical column thickness dimension.

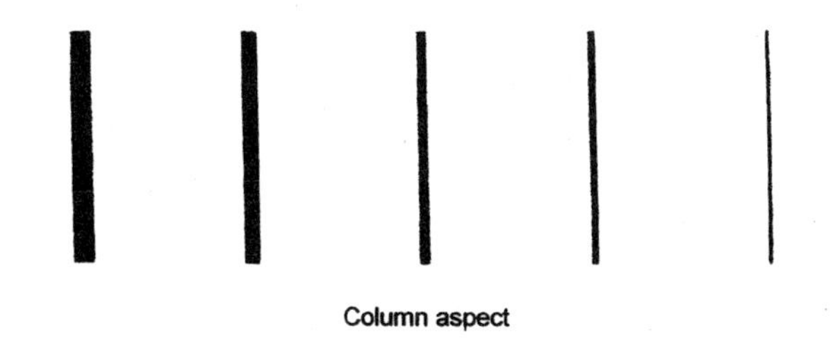

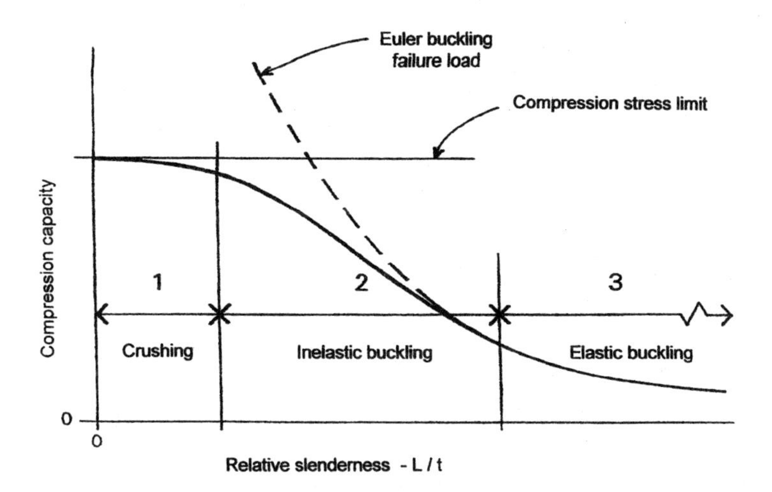

Figure 9.2 Effect of column slenderness on axial compression capacity.

The very long member—such as a yardstick—fails in elastic buckling (Zone 3 in Figure 9.2), which is determined by the stiffness of the member in bending resistance and the material stiffness property (modulus of elasticity). Between these two extremes (Zone 2 in Figure 9.2), which is where most wood columns fall, the behavior is determined by a transition between the two distinctly different responses.

Over the years, several methods have been employed to deal with this situation in the design of wood columns, or of any column for that matter. Earlier editions of the NDS used three separate formulas to cover the entire range of slenderness, reflecting the three distinct regions of the graph in Figure 9.2. In recent editions, however, a single formula is employed, effectively covering the whole range of the graph. The formula and its various factors are complex, and its use involves considerable computation; nevertheless, the basic process is essentially simplified through the use of a single defined relationship.

In design practice, use is commonly made of either tabulated data or computer-aided processes. The NDS formulas are basically analytical and inverting them to produce design formulas for direct use is not practical. Direct use of the formulas for design involves a trial and error process, with many runs through the complex formula before a good fit is found. Once the basic relationships are understood, design aids are very useful.

Column Load Capacity—ASD

The following discussion presents materials from the *NDS* (Ref. 1) for design of axially loaded columns. The basic formula for determination of the capacity of a wood column, based on the working stress method, is

$$P = (F_c^*)(C_p)(A)$$

where A = area of the column cross section
F_c^* = the allowable design value for compression parallel to the grain, as modified by applicable factors, except C_p
C_p = the column stability factor
P = the allowable column axial compression load

The column stability factor is determined as follows:

$$C_p = \frac{1 + (F_{cE}/F_c^*)}{2c} - \sqrt{\left[\frac{1 + F_{cE}/F_c^*)}{2c}\right]^2 - \frac{F_{cE}/F_c^*}{c}}$$

where F_{cE} = the Euler buckling stress, as determined by the formula below
c = 0.8 for sawn lumber, 0.85 for round poles, 0.9 for glued-laminated timbers

For the buckling stress:

$$F_{cE} = \frac{0.822E'_{\min}}{(L_e/d)^2}$$

where E'_{min} = modulus of elasticity for stability for the wood species and grade, adjusted by any applicable modifications

L_e = the effective length (unbraced height as modified by any factors for support conditions) of the column
d = the column cross section dimension (column width) measured in the direction that buckling occurs

The values to be used for the effective column length and the corresponding column width should be considered as discussed for the conditions displayed in Figure 9.1. For a basic reference, the buckling phenomenon typically uses a member that is pinned at both ends and prevented from lateral movement only at the ends, for which no modification for support conditions is made; this is a common condition for wood columns. The NDS presents methods for modified buckling lengths that are essentially similar to those used for steel design.

For solid-sawn columns, the formula for C_p is simply a function of the value of F_{cE}/F_c^* with the value of c being a constant of 0.8. It is therefore possible to plot a graph of the value for C_p as a function of the value of F_{cE}/F_c^*, as is done in Figure 9.3. Accuracy of values obtained from Figure 9.3 is low, but is usually acceptable for column design work. Of course, greater accuracy can always be obtained with the use of the formula.

The following examples illustrate the use of the NDS formulas for columns.

Example 1. A wood column consists of a 6 × 6 of Douglas fir-larch, No. 1 grade. Using the ASD method, find the safe axial compression load for unbraced lengths of: (1) 2 ft, (2) 8 ft, (3) 16 ft.

Solution: From Table 4.1 find values of $F_c = 1000$ psi and $E_{min} = 580{,}000$ psi. With no basis for adjustment given, the F_c value is used directly as the F_c^* value in the column formulas.

For (1): $L/d = 2(12)/5.5 = 4.36$. Then

$$F_{cE} = \frac{0.822\, E_{\min}}{(L_c/d)^2} = \frac{0.822 \times 580{,}000}{(4.36)^2} = 25{,}080 \text{ psi}$$

$$\frac{F_{cE}}{F_c^*} = \frac{25{,}080}{1000} = 25.08$$

$$C_p = \frac{1 + 25.08}{1.6} - \sqrt{\left[\frac{1 + 25.08}{1.6}\right]^2 - \frac{25.08}{0.8}} = 0.992$$

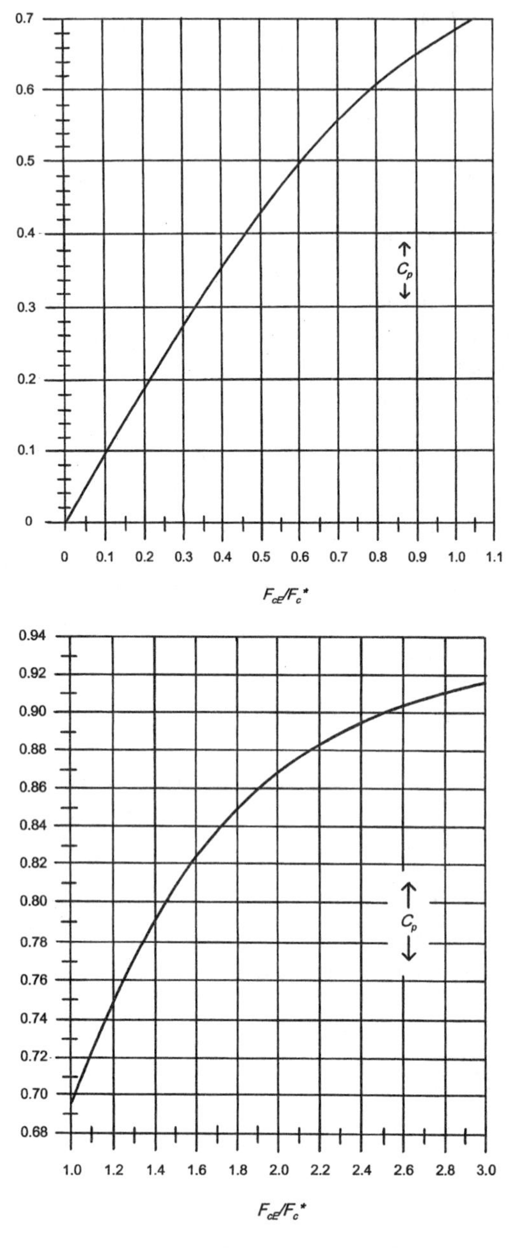

Figure 9.3 Column stability factor C_p as a function of F_{cE}/F_c^*.

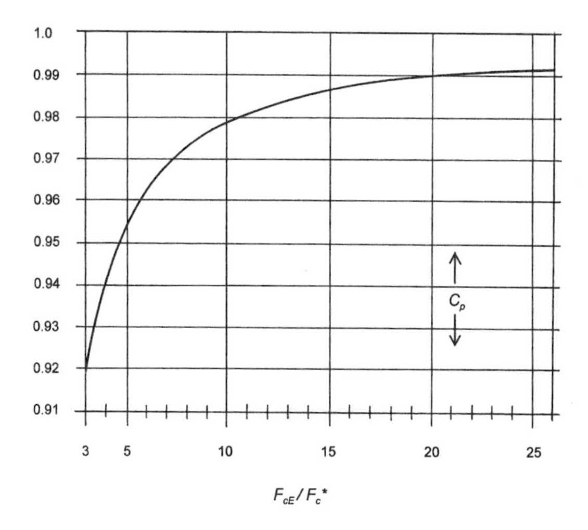

Figure 9.3 (*continued*)

And the allowable compression load is

$$P = (F_c^*)(C_p)(A) = (1000)(0.992)(5.5)^2 = 30{,}000\,\text{lb}$$

As mentioned previously, Figure 9.3 may be used to simplify the computation.

For (2): $L/d = 8(12)/5.5 = 17.45$. For which $F_{cE} = 1566$ psi, $F_{cE}/F_c^* = 1.566$, from Figure 9.3, $C_p = 0.82$, and thus

$$P = (1000)(0.82)(5.5)^2 = 24{,}800\,\text{lb}$$

For (3): $L/d = 16(12)/5.5 = 34.9$. For which $F_{cE} = 391$ psi, $F_{cE}/F_c^* = 0.391$, from Figure 9.3, $C_p = 0.35$, and thus

$$P = (1000)(0.35)(5.5)^2 = 10{,}590\,\text{lb}$$

Example 2. Wood 2 × 4 elements are to be used as vertical compression members to form a wall (ordinary stud construction). If the wood is Douglas fir-larch, stud grade, and the wall is 8.5 ft high, what is the column load capacity of a single stud?

Solution: It is assumed that the wall has a covering attached to the studs or blocking between the studs to brace them on their weak (1.5-in. dimension) axis. Otherwise, the practical limit for the height of the wall is 50 × 1.5 = 75 in. Therefore, using the larger dimension

$$\frac{L}{d} = \frac{8.5 \times 12}{3.5} = 29.14$$

From Table 4.1 F_c = 850 psi, E_{min} = 510,000 psi. From Table 4.2, the value for F_c is adjusted by a size factor to 1.05(850) = 892.5 psi. Then

$$F_{cE} = \frac{0.822(510,000)}{(29.14)^2} = 494\text{ psi}$$

$$\frac{F_{cE}}{F_c^*} = \frac{494}{892.5} = 0.554$$

From Figure 9.3, C_p 0.47, and the column capacity is

$$P = (F_c^*)(C_p)A = (892.5)(0.47)(1.5 \times 3.5) = 2202\text{ lb}$$

Problem 9.2.A-D. Using the ASD method, find the allowable axial compression load for the following wood columns. Use Douglas fir-larch, No. 2 grade.

	Nominal Size (in.)	Unbraced Length (ft)	(m)
A	4 × 4	8	2.44
B	6 × 6	10	3.05
C	8 × 8	18	5.49
D	10 × 10	14	4.27

Design of Wood Columns, ASD

The design of columns is complicated by the relationships in the column formulas. The allowable stress for the column is dependent upon the actual column dimensions, which are not known at the beginning of the design process. This does not allow for simply inverting the column formulas to derive required properties for the column. A trial and error process is therefore indicated. For this reason, designers typically use various design aids: graphs, tables, or computer-aided processes.

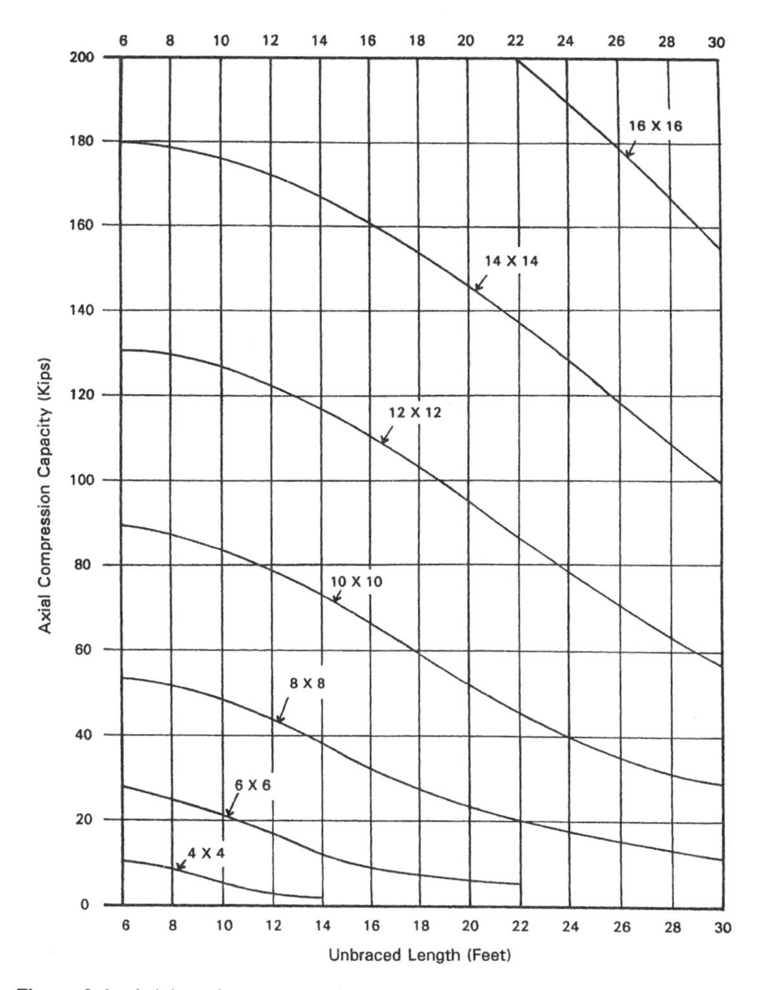

Figure 9.4 Axial service compression load capacity for wood members of square cross section. Derived from NDS requirements for Douglas fir-larch, No. 1 grade.

Because of the large number of wood species, resulting in many different values for allowable stress and modulus of elasticity, precisely tabulated capacities become impractical. Nevertheless, aids using average values are available and simple to use for design. Figure 9.4 is a graph on which the axial compression load capacity of some square column sections of a single species and grade are plotted. Table 9.1 yields the capacity for a range of columns. Note that the smaller size column sections fall into the classification in Table 4.1 for

TABLE 9.1 Safe Service Loads for Wood Columns — ASD[a]

Column Section		Unbraced Length (ft)										
Nominal Size	Area (in.^{2})	6	8	10	12	14	16	18	20	22	24	26
4 × 4	12.25	11.1	7.28	4.94	3.50	2.63						
4 × 6	19.25	17.4	11.4	7.76	5.51	4.14						
4 × 8	25.375	22.9	15.1	10.2	7.26	6.46						
6 × 6	30.25	27.6	24.8	20.9	16.9	13.4	10.7	8.71	7.17	6.53		
6 × 8	41.25	37.6	33.9	28.5	23.1	18.3	14.6	11.9	9.78	8.91		
6 × 10	52.25	47.6	43.0	36.1	29.2	23.1	18.5	15.0	13.4	11.3		
8 × 8	56.25	54.0	51.5	48.1	43.5	38.0	32.3	27.4	23.1	19.7	16.9	14.6
8 × 10	71.25	68.4	65.3	61.0	55.1	48.1	41.0	34.7	29.3	24.9	21.4	18.4
8 × 12	86.25	82.8	79.0	73.8	66.7	58.2	49.6	42.0	35.4	30.2	26.0	22.3
10 × 10	90.25	88.4	85.9	83.0	79.0	73.6	67.0	60.0	52.9	46.4	40.4	35.5
10 × 12	109.25	107	104	100	95.6	89.1	81.2	72.6	64.0	56.1	48.9	42.9
10 × 14	128.25	126	122	118	112	105	95.3	85.3	75.1	65.9	57.5	50.4
12 × 12	132.25	130	128	125	122	117	111	104	95.6	86.9	78.3	70.2
14 × 14	182.25	180	178	176	172	168	163	156	148	139	129	119
16 × 16	240.25	238	236	234	230	226	222	216	208	200	190	179

[a] Load capacity in kips for solid-sawn sections of No. 1 grade Douglas fir-larch with no adjustment for moisture or load duration conditions.

"Dimension Lumber" rather than for "Timbers." This makes for one more complication in the column design process.

Problem 9.2.E-H. Select square column sections of Douglas fir-larch, No. 1 grade, for the following data.

	Required Axial Load		Unbraced Length	
	(kips)	(kN)	(ft)	(m)
E	20	89	8	2.44
F	50	222	12	3.66
G	50	222	20	6.10
H	100	445	16	4.88

9.3 COLUMN LOAD CAPACITY, LRFD

For the LRFD process, the steps are essentially the same as for the ASD method described in the preceding sections. The principal differences consist of adjustments of values as achieved by various factors. The adjustments are as follows:

For loads: Load factors from Section 14.8 for various combinations.

Time effect factor, λ: see table 4.8.

Resistance factor: Table 4.7, $\phi_c = 0.90$ for compression, $\phi_s = 0.85$ for stability (E_{min}).

Reference values: for stress, $\dfrac{2.16}{1000\phi_c}$; for stability, $\dfrac{1.5}{1000\phi_s}$

Computations for investigation:

$$E'_{\min} = \phi_s \frac{1.5}{1000\phi_s} E_{\min}$$

$$F_{cE} = \frac{0.822 E'_{\min}}{(L/d)^2}$$

$$F_c^* = \lambda\phi_c \left(\frac{2.16}{1000\phi_c}\right) F_c$$

$$C_p = \frac{1 + F_{cE}/F_c^*}{1.6} - \sqrt{\left(\frac{1 + F_{cE}/F_c^*}{1.6}\right)^2 - \frac{F_{cE}/F_c^*}{0.8}}$$

For factored usable compression capacity:

$$P' = \lambda\phi_c \frac{2.16}{1000\phi_c} C_p F_c A$$

The following example illustrates this process. It uses the same data as in the previous example, Example 1, Part 2, that treats the ASD method.

Example 3. A wood column consists of a 6 × 6 of Douglas fir-larch, No. 1 grade. Using the LRFD method, find the factored usable compression capacity (factored resistance) for an unbraced length of 8 ft.

Solution: From Table 4.1 find values of F_c = 1000 psi and E_{min} = 580,000 psi. With no other information about conditions for modification, these values are subject only to the necessary adjustments for the LRFD method. Assume that the load is a typical combination of dead and live load, which yields a value for λ of 0.8 (Table 4.8).

$$E'_{\text{min}} = \phi_s \frac{1.5}{1000\phi_s} E_{\text{min}} = 0.85\left(\frac{1.5}{1000(0.85)}\right)(580{,}000) = 870\,\text{ksi}$$

$$\frac{L}{d} = \frac{8 \times 12}{5.5} = 17.45$$

$$F_{cE} = 0.822 \frac{E'_{\text{min}}}{(L/d)^2} = 0.822 \frac{870}{(17.45)^2} = 2.3486\,\text{ksi}$$

$$F_c^* = \lambda\phi_c \frac{2.16}{1000\phi_c} F_c = 0.8(0.9)\frac{2.16}{1000(0.9)} 1000 = 1.728\,\text{ksi}$$

$$\frac{F_{cE}}{F_c^*} = \frac{2.3486}{1.728} = 1.359$$

Using Figure 9.3, C_p = 0.78, and the capacity is

$$\lambda\phi_c P' = \lambda\phi_c \frac{2.16}{1000\phi_c} C_p F_c A = 0.8(0.9)\frac{2.16}{1000(0.9)}$$
$$\times (0.78)(1000)(5.5)^2 = 40.8\,\text{kips}$$

As with the ASD method, the column design process is quite laborious, unless some design aid or a computer-assisted procedure is used. These aids are indeed available, although they are not described in this book.

Problem 9.3.A-D. Using the LRFD method, find the factored usable compression capacity (factored resistance) for the following wood columns. Use Douglas fir-larch, No. 2 grade.

	Nominal Size (in.)	Unbraced Length (ft)	(m)
A	4 × 4	8	2.44
B	6 × 6	10	3.05
C	8 × 8	18	5.49
D	10 × 10	14	4.27

9.4 ROUND COLUMNS

Solid wood columns of circular cross section are not used extensively in building construction. As for load-bearing capacity, round and square wood columns of the same cross-sectional area will support the same axial loads and have the same degree of stiffness.

When designing a wood column of circular cross section, a simple procedure is to design a square column first and then select a round column with an equivalent cross-sectional area. To find the diameter of the equivalent round column, the side dimension of the square column is multiplied by 1.128.

Poles

Poles are round timbers consisting of the peeled logs of coniferous trees. In short lengths they may be relatively constant in diameter, but when long they are tapered in form, which is the natural form of the tree trunk. As columns, poles are designed with the same basic criteria used for rectangular sawn sections. For slenderness considerations the d used is taken as that of a square section of equal area. Thus, calling the diameter of the pole D,

$$\text{Area of the square} = \text{Area of the circle}$$

$$d^2 = \frac{\pi D^2}{4}$$

$$d = \sqrt{0.7854 D^2} = 0.886 D$$

For a tapered column, a conservative assumption for design is that the critical column diameter is the least diameter at the small end. If the column is very short, this is reasonable. However, for a slender column, with buckling occurring near the midheight of the column, this is very conservative, and the code provides for adjustment. Nevertheless, because of a typical lack of straightness and presence of many flaws, many designers prefer to use the unadjusted small-end diameter for design computations.

General use of poles is discussed in Section 12.9.

9.5 STUD WALL CONSTRUCTION

Studs are the vertical elements used for wall framing in light wood construction. Studs serve utilitarian purposes of providing for attachment of wall surfacing, but also serve as columns when the wall provides support for roof or floor systems. The most common stud is a 2 × 4 spaced at intervals of 12, 16, or 24 in.: the spacing deriving from the common 4 ft × 8 ft panels of wall coverings.

Studs of nominal 2 in. thickness must be braced on the weak axis when used for storyhigh walls, a simple requirement deriving from the limiting ratio of L/d of 50 for columns. If the wall is surfaced on both sides, the studs are usually considered to be adequately braced by the surfacing. If the wall is not surfaced, or is surfaced on only one side, horizontal blocking between studs must be provided, as shown in Figure 9.5. The number of rows of blocking and the spacing of the blocking will depend on the wall height and the need for column action by the studs.

Studs may also serve other functions, as in the case of an exterior wall subjected to wind forces. For this situation the studs must be designed for the combined actions of bending plus compression, as discussed in Section 9.8.

In colder climates it is now common to use studs with width greater than the nominal 4 in. in order to create a larger void space within the wall to accommodate insulation. This often results in studs with redundant strength for the ordinary tasks of one- and two-story buildings. Of course wider studs may be required for very tall walls as well.

If vertical loads are high or bending is great, it may be necessary to strengthen a stud wall. This can be done in a number of ways, such as:

1. Decreasing stud spacing from the usual 16 in. to 12. in.
2. Increasing the stud thickness from 2 in. nominal to 3 in. nominal.

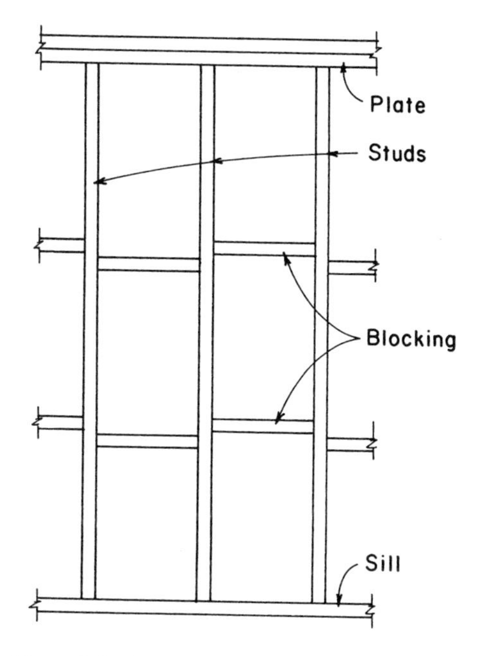

Figure 9.5 Stud wall construction with blocking.

3. Increasing the stud width from 4 in. nominal to 6 in. nominal or greater.
4. Using doubled studs or large timber sections as posts at locations of concentrated loads.

It is also sometimes necessary to use thicker studs or to restrict stud spacing for walls that function as shear walls, as discussed in Chapter 13.

In general, studs are columns and must comply with the various requirements for design of solid-sawn sections. Any appropriate grade of wood may be used, although special stud grades are commonly used for ordinary 2 × 4 studs.

Table 9.2, which is adapted from a table in the *International Building Code* (Ref. 4), provides data for the selection of studs for both bearing and nonbearing walls. The code sipulates that this data must be used in lieu of any engineering design for the studs, which means that other possibilities may be considered if computations can support a case for them.

TABLE 9.2 Requirements for Stud Wall Construction

Stud Size (inches)	Bearing Walls: Laterally Unsupported Stud Height[a] (feet)	Bearing Walls: Spacing (inches), Supporting Roof and Ceiling Only	Bearing Walls: Spacing (inches), Supporting One Floor, Roof and Ceiling	Bearing Walls: Spacing (inches), Supporting Two Floors, Roof and Ceiling	Nonbearing Walls: Laterally Unsupported Stud Height[a] (feet)	Nonbearing Walls: Spacing (inches)
2 × 3	—	—	—	—	10	16
2 × 4	10	24	16	—	14	24
3 × 4	10	24	24	16	14	24
2 × 5	10	24	24	—	16	24
2 × 6	10	24	24	16	20	24

[a] Listed heights are distances between points of lateral support placed perpendicular to the plane of the wall. Increases in unsupported height are permitted where justified by analysis.
[b] Shall not be used in exterior walls.
Source: Compiled from data in the *International Building Code* (Ref. 4), with permission of the publishers, International Code Council.

Stud wall construction is often used as part of a general light construction system described as *light wood frame construction*. This system has been highly refined over many years of usage in the United States, with its present most common form as shown in Figure 9.6. The joist and rafter construction discussed in Chapter 7, together with the stud wall construction discussed here, are the primary structural elements of this system. In most applications the system is almost entirely composed of 2 in. nominal dimension limber. Timber elements are sometimes used for freestanding columns and for heavily loaded or long span beams.

9.6 SPACED COLUMNS

A type of structural element sometimes used in wood structures is the *spaced column*. This is an element in which two or more wood members are fastened together to share load as a single compression unit. The design of such elements is quite complex, owing to the numerous code requirements. The following example shows the general procedure for analysis of a spaced column by the ASD method, but the reader should

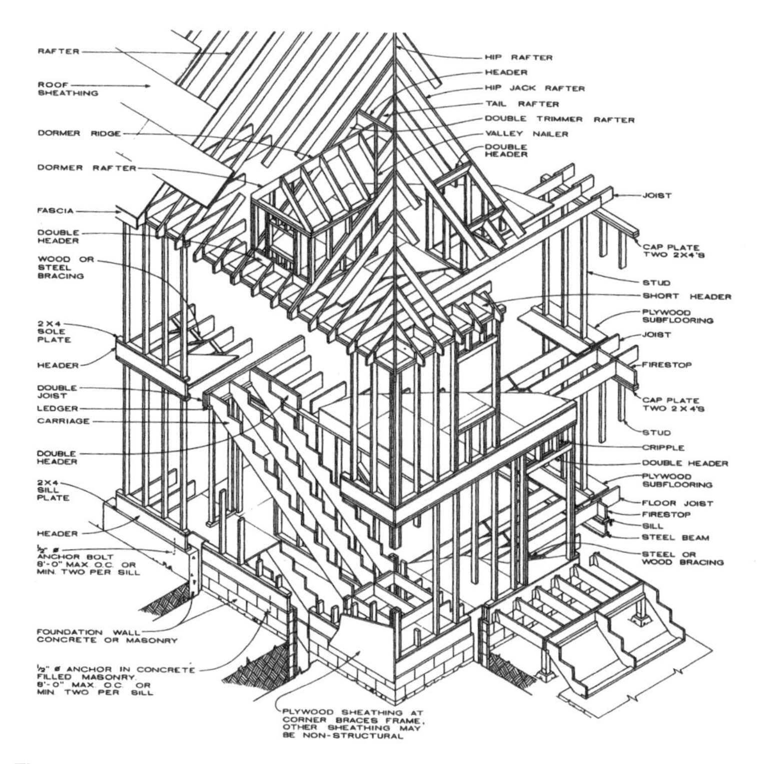

Figure 9.6 Typical light wood frame construction, western or platform type. (Reproduced from *Architectural Graphic Standards*, with permission of the publishers, John Wiley & Sons, New Jersey).

refer to the applicable code for the various requirements for any design work.

Example 4. A spaced column of the form shown in Figure 9.7 consists of three 3 × 12 pieces of Douglas fir-larch, No. 1 grade. Dimension L_l is 11 ft-8 in. and x is 6 in. Find the axial compression capacity.

Solution: From Table 4.1, $F_c = 1500$ psi, $E_{min} = 620{,}000$ psi. There are two separate conditions to be investigated for the spaced column. These relate to the effects of slenderness in the two directions, as designated by the x- and y-axes in Figure 9.6. In the y direction the column behaves simply as a set of solid-sawn columns. Thus the stress is limited by the dimensions d_2 and L_2 and their ratio. For this condition

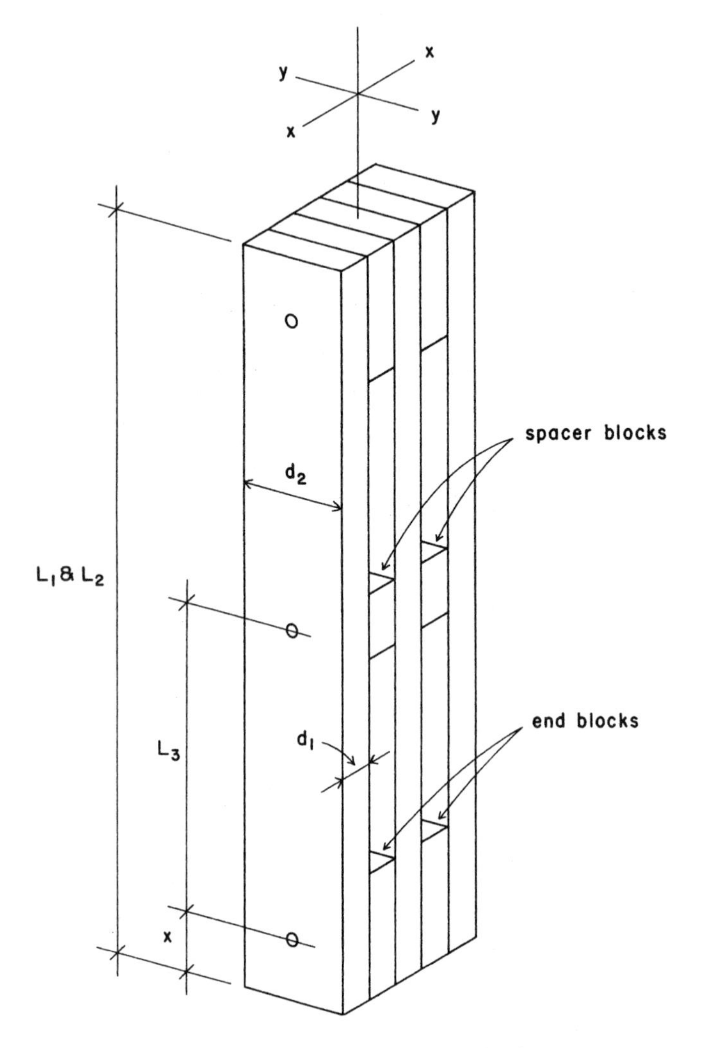

Figure 9.7 General form of a spaced column.

slenderness ratio is

$$\frac{L_2}{d_2} = \frac{(11.67)(12)}{11.25} = 12.5$$

Using this value for the slenderness ratio, the compression capacity for a single 3 × 12 is found using the procedure described in

Section 9.2; the total for this column is then three times that value. Application of the example data will yield a value of 0.88 for C_p. The capacity for this condition is then determined by multiplying this factor times the product of the reference value for compression stress and the total column cross-sectional area. In this example there is a second condition to consider, so the second value for C_p will be determined to see which is critical.

For the condition of buckling in the x direction, there are limits for the slenderness ratio, as follows:

$$\text{Maximum value for } L_3/d_1 = 40$$

$$\text{Maximum value for } L_1/d_1 = 80$$

Then, using the data for the example,

$$\frac{L_3}{d_1} = \frac{64}{2.5} = 25.6 \quad \text{(less than 40)}$$

and

$$\frac{L_1}{d_1} = \frac{140}{2.5} = 56 \quad \text{(less than 80)}$$

So the limits are not exceeded.

The capacity for this condition depends on the value of L_1/d_1 and is determined in a manner similar to that for a solid section, except that a modified form is used for the buckling value F_{cE} as follows:

$$F_{cE} = \frac{0.822 K_x E_{\text{min}}}{(L/d)^2}$$

The value for K_x is based on the situation at the end blocks. In the illustration in Figure 9.7 the distance x is the distance from the end of the column to the centroid of the connectors that are used to fasten the end blocks into the column. Two values for K_x are given, based on the relation of the x distance to the overall length of the column (L_1 in Figure 9.7). Thus

1. $K_x = 2.5$ when x is equal to or less than $L_1/20$.
2. $K_x = 3.0$ when x is between $L_1/20$ and $L_1/10$.

For the example, $x = 6$ in. and $L_1/20 = 140/20 = 7$. Thus $K_x = 2.5$, and the value for F_{cE} is

$$F_{cE} = \frac{0.822 K_x E_{\min}}{(L/d)^2} = \frac{(0.822)(2.5)(620{,}000)}{(56)^2} = 406 \text{ psi}$$

$$\frac{F_{cE}}{F_c^*} = \frac{406}{1500} = 0.271$$

From Figure 9.3, $C_p = 0.25$

As this yields a value for C_p that is less than that for the condition of y direction buckling, the capacity is limited by this condition. Thus the capacity is determined to be (as illustrated in Section 9.2):

$$P = (F_c^*)(C_p)(A) = (1500)(0.25)(3 \times 28.125) = 31{,}600 \text{ lb}$$

Design of spaced columns by the LRFD method essentially involves the same steps as in the ASD method. Adjustments are made in the general manner illustrated for solid-sawn columns in Section 9.3.

Problem 9.6.A. A spaced column of the form shown in Figure 9.7 consists of two pieces of 2 × 8 lumber of Douglas fir-larch, select structural grade. Overall height is 10 ft and the centroid of the end block connectors is 5 in. from the end of the column. Using the ASD method find the axial compression capacity for the column.

Problem 9.6.B. Same as Problem 9.6.A, except column consists of three 2 × 10s and overall height is 9 ft.

9.7 BUILT-UP COLUMNS

In various situations single columns may consist of multiple elements of solid-sawn sections. Although the description includes glued-laminated and spaced columns, the term *built-up column* is generally used for multiple-element columns such as those shown in Figure 9.8. Glued-laminated columns are essentially designed as solid sections with some special qualifying conditions.

Built-up columns usually have the elements attached to each other by mechanical devices such as nails, spikes, lag screws, or machine bolts. The *NDS* (Ref. 1) has data and procedures for evaluation of built-up columns using fasteners of nails and bolts.

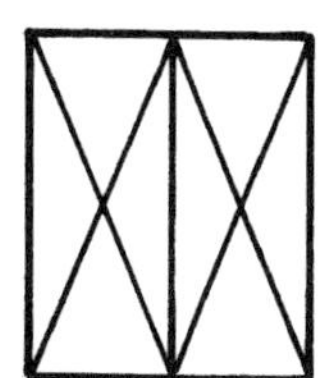

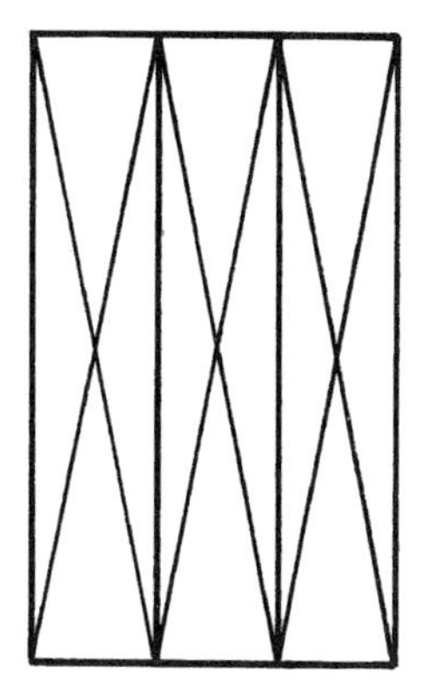

Figure 9.8 Cross sections of built-up columns with multiple lumber member, nailed or glued-laminated.

9.8 COLUMNS WITH BENDING

There are a number of situations in which structural members are subjected to combined effects of axial compression and bending. Stresses developed by these two actions are both of the direct type (tension and compression) and can be combined for consideration of a net stress condition. However, the basic actions of a column and a bending member are essentially different in character, and it is therefore customary to consider this combined activity by what is called *interaction*.

The classic form of interaction is represented by the graph in Figure 9.9. Referring to the notation on the graph:

The maximum axial load capacity of the column (without bending) is P_0.

The maximum bending capacity of the member (without compression) is M_0.

At some applied compression load below P_0 the column has some capacity for bending in combination with the load. This combination is indicated as P_n and M_n.

The classic form of the interaction relationship is expressed in the formula

$$\frac{P_n}{P_0} + \frac{M_n}{M_0} = 1$$

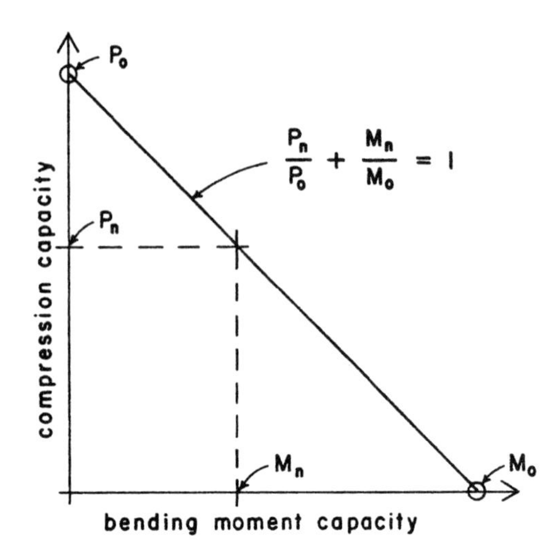

Figure 9.9 Idealized condition for column interaction, axial compression plus bending.

The plot of this equation is the straight line connecting P_0 and M_0 as shown on Figure 9.9.

A graph similar to that in Figure 9.9 can be produced using stresses rather than loads and moments, since the stresses are directly proportional to the loads and moments. This is the procedure generally used in wood and steel design, with the graph taking a form expressed as

$$\frac{f_a}{F_a} + \frac{f_b}{F_b} \leq 1$$

where f_a = computed stress due to load
F_a = allowable column action stress
f_b = computed stress due to bending
F_b = allowable bending stress

Various effects cause deviation from the pure, straight line form of interaction, including inelastic behavior, effects of lateral stability or torsion, and general effects of member cross section form. One major effect is the so-called *P-delta effect*. Figure 9.10*a* shows a common situation that occurs in buildings when an exterior wall functions as a bearing wall or contains a column. The combination of gravity load and lateral load due to wind or seismic action can produce the loading

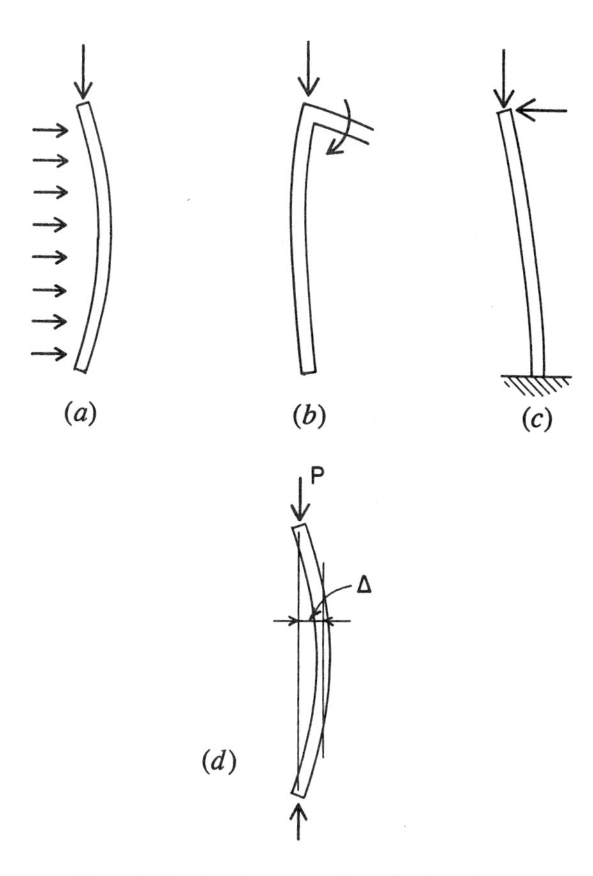

Figure 9.10 Development of bending in columns.

condition shown. If the member is quite flexible and the deflection due to bending is significant in magnitude, an additional bending moment is developed as the axis of the member deviates from the action line of the vertical compression load. The resulting additional moment is the product of the load (P) and the deflection (Δ), yielding the term used to describe the phenomenon. (See Figure 9.10*d*).

Various other situations can result in the P-delta effect. Figure 9.10 *b* shows an end column in a rigid frame structure, where moment is induced at the top of a column by the moment-resistive connection to the beam. Although the form of the deflection is different in this case, the P-delta effect is similar. The vertically cantilevered column in Figure 9.10*c* presents potentially an extreme case of this effect.

The P-delta effect may or may not be a critical concern; a major factor is the relative slenderness and flexibility of the column. Relatively stiff columns will both tolerate the eccentric load effect and sustain little deflection due to the bending, making the P-delta effect quite insignificant. In a worst-case scenario, however, the P-delta effect can be an accelerating one in which the added moment due to the P-delta effect causes additional deflection, which in turn results in additional P-delta effect, which then causes more deflection, and so on. Potentially critical situations are those involving very slender compression members, for which the phenomenon should be carefully studied—a condition seldom occurring in wood structures.

In wood structures columns with bending occur most frequently as shown in Figure 9.11. Studs in exterior walls represent the situation shown in Figure 9.11*a*, with a loading consisting of vertical gravity plus horizontal wind loads. Due to use of common construction details, columns carrying only vertical loads may sometimes be loaded eccentrically, as shown in Figure 9.11*b*.

Investigation of Columns with Bending, ASD Method

Present design of wood columns uses the straight line interaction relationship and then adds considerations for buckling due to bending, P-delta effects, and so on. For solid-sawn columns the NDS provides

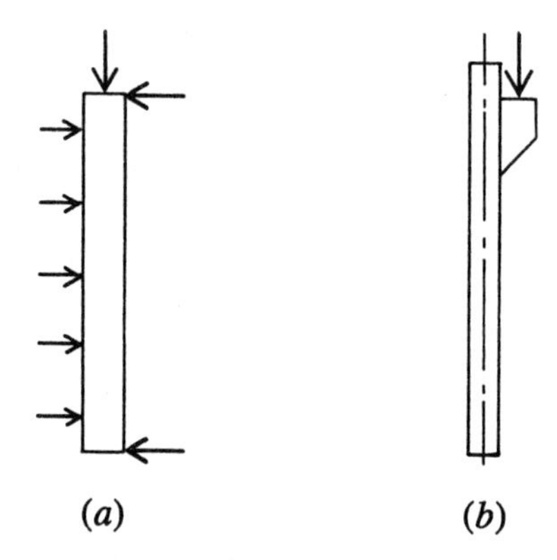

Figure 9.11 Common cases involving combined axial compression and bending in columns: (a) exterior stud or truss chord, (b) column with bracketed support for spanning member.

the following formula for investigation

$$\left(\frac{f_c}{F'_c}\right)^2 + \frac{f_b}{F_b\left(1 - \dfrac{f_c}{F_{cE}}\right)} \leq 1$$

in which f_c = computed compressive stress due to column load
F'_c = tabulated reference design value for compressive stress, adjusted by all modification factors
f_b = computed bending stress due to bending moment
F_b = tabulated reference design stress for bending
F_{cE} = the value determined for buckling of solid-sawn columns, as discussed in Section 9.2

This equation assumes that bending is in one direction only. The NDS provides additional equations to be used for bending on two axes (biaxial bending), although we will not address the problem here.

The following examples demonstrate some applications for the procedure.

Example 5. An exterior wall stud of Douglas fir-larch, stud grade, is loaded as shown in Figure 9.12*a*. Investigate the stud for the combined loading. (*Note:* This is the wall stud from the building example in Chapter 15.)

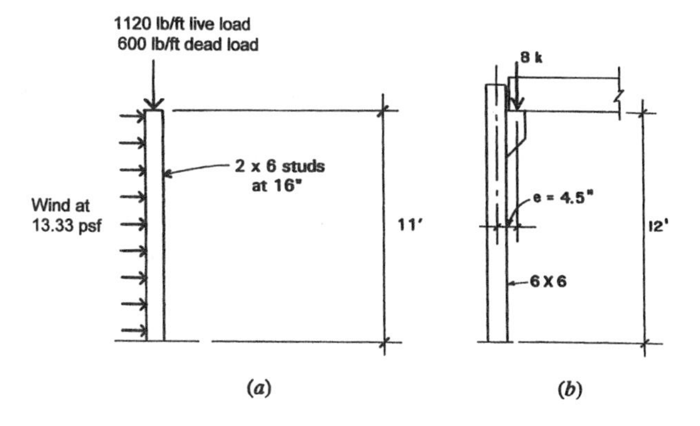

Figure 9.12 Reference for Examples 5 and 6.

Solution: From Table 4.1, $F_b = 700$ psi, $F_c = 850$ psi, and $E_{min} =$ 510,000 psi. Note that the allowable stresses are not changed by Table 4.2, as the table factors are 1.0. With inclusion of the wind loading, the stress values (but not E) may be increased by a factor of 1.6 (see Table 4.4).

Assume that wall surfacing braces the 2 × 6 studs adequately on their weak axis ($d = 1.5$ in.), so the critical value for d is 5.5 in. Thus

$$\frac{L}{d} = \frac{11 \times 12}{5.5} = 24$$

$$F_{cE} = \frac{0.822 E_{\min}}{\left(\dfrac{L}{d}\right)^2} = \frac{0.822 \times 510,000}{(24)^2} = 728 \text{ psi}$$

The first investigation involves the gravity load without the wind, for which the stress increase factor of 1.6 is omitted. Thus

$$F_c^* = 850 \text{ psi}$$

$$\frac{F_{cE}}{F_c^*} = \frac{728}{850} = 0.856$$

From Figure 9.3, $C_p = 0.63$, and the stud compression capacity is

$$P = (F_c^*)(C_p)(A) = (850)(0.63)(8.25) = 4418 \text{ lb}$$

This is compared to the given load for the 16-in. stud spacing, which is

$$P = (16/12)(1720) = 2293 \text{ lb}$$

which demonstrates that the gravity-only load is not a critical concern.

Proceeding with consideration for the combined loading, we determine that

$$F_c^* = 1.6\, F_c = 1.6(850) = 1360 \text{ psi}$$

$$\frac{F_{cE}}{F_c^*} = \frac{728}{1360} = 0.535$$

From Figure 9.3,

$$C_p = 0.45$$

For the load combination with wind, the adjusted vertical load is

$$P = \frac{16}{12}[\text{Dead load} + 0.75(\text{Live load})] = \frac{16}{12}(600 + 840) = 1920\,\text{lb}$$

$$F_c' = C_p F_c^* = 0.45 \times 1360 = 612\,\text{psi}$$

$$f_c = \frac{P}{A} = \frac{1920}{8.25} = 233\,\text{psi}$$

For the wind load use $w = 0.75(13.33) = 10$ psf. Then

$$M = \frac{16}{12}\frac{wL^2}{8} = \frac{16}{12}\frac{10(11)^2}{8} = 202\,\text{lb-ft}$$

$$f_b = \frac{M}{S} = \frac{202 \times 12}{7.563} = 320\,\text{psi}$$

$$\frac{f_c}{F_{cE}} = \frac{233}{728} = 0.320$$

Then, using the code formula for the interaction,

$$\left(\frac{f_c}{F_c'}\right)^2 + \frac{f_b}{F_b\left(1 - \dfrac{f_c}{F_{cE}}\right)} \le 1$$

$$\left(\frac{233}{612}\right)^2 + \frac{320}{1.6 \times 700(1 - 0.320)} = 0.139 + 0.420 = 0.559$$

As the result is less than 1, the stud is adequate.

Example 6. The column shown in Figure 9.12*b* is of Douglas fir-larch, dense No. 1 grade. Investigate the column for combined column action and bending.

Solution: From Table 4.1, $F_b = 1400$ psi, $F_c = 1200$ psi, $E_{min} = 620{,}000$ psi. From Table A.3, $A = 30.25\,\text{in.}^2$ and $S = 27.7\,\text{in.}^3$. Then

$$\frac{L}{d} = \frac{12 \times 12}{5.5} = 26.18$$

$$F_{cE} = \frac{0.822 \times 620{,}000}{(26.18)^2} = 744\,\text{psi}$$

$$\frac{F_{cE}}{F_c} = \frac{744}{1200} = 0.62$$

From Figure 9.3, $C_p = 0.51$

$$f_c = \frac{8000}{30.25} = 264\,\text{psi}$$

$$F'_c = C_p F_c = (0.51)(1200) = 612\,\text{psi}$$

$$\frac{f_c}{F_{cE}} = \frac{264}{744} = 0.355$$

$$f_b = \frac{M}{S} = \frac{8000 \times 4.5}{27.7} = 1300\,\text{psi}$$

and for the column interaction

$$\left(\frac{f_c}{F'_c}\right)^2 + \frac{f_b}{F_b\left(1 - \dfrac{f_c}{F_{cE}}\right)} \leq 1$$

$$\left(\frac{264}{612}\right)^2 + \frac{1300}{1400(1 - 0.355)} = 0.186 + 1.440 = 1.626$$

As this exceeds one, the column is inadequate. Since bending is the main problem, a second try might be for a 6 × 8 or a 6 × 10, or for an 8 × 8 if a square section is required.

Problem 9.8.A. Nine feet high 2 × 4 studs of Douglas fir-larch, No. 1 grade, are used in an exterior wall. Wind load is 17 psf on the wall surface; studs are 24 in. on center; the gravity load on the wall is 400 lb/ft of wall length. Investigate the studs for combined action of compression plus bending, using the ASD method.

Problem 9.8.B. Ten feet high 2 × 4 studs of Douglas fir-larch, No. 1 grade, are used in an exterior wall. Wind load is 25 psf on the wall surface; studs are 16 in. on center; the gravity load on the wall is 500 lb/ft of wall length. Investigate the studs for combined action of compression plus bending, using the ASD method.

Problem 9.8.C. A 10 × 10 column of Douglas fir-larch, No. 1 grade, is 9 ft high and carries a compression load of 20 kips that is 7.5 in. eccentric from the column axis. Investigate the column for combined compression and bending, using the ASD method.

Problem 9.8.D. A 12 × 12 column of Douglas fir-larch, No. 1 grade, is 12 ft high and carries a compression load of 24 kips that is 9.5 in. eccentric from the column axis. Investigate the column for combined compression plus bending, using the ASD method.

Investigation of Columns with Bending, LRFD

The process for investigation of columns with bending in the LRFD method uses essentially the same steps as in the ASD method. The usual adjustments are made with load factors, resistance factors, and conversions of reference values. The following example illustrates the process, using the same data as in the ASD work for Example 6.

Example 7. The column shown in Figure 9.12*b* is of Douglas fir-larch, dense No. 1 grade. Investigate the column for combined column action and bending, using the LRFD method. The applied compression load is one half live load and one half dead load.

Solution: From Table 4.1, $F_b = 1400$ psi, $F_c = 1200$ psi, $E_{min} = 620{,}000$ psi. From Table A.3, $A = 30.25$ in.^{2}, $S = 27.7$ in.^{3}.

$$P_u = 1.2(DL) + 1.6(DL) = 1.2(4000) + 1.6(4000) = 11,200 \text{ lb, or } 11.2 \text{ kips}$$

$$M_u = 11.2 \times 4.5 = 50.4 \text{ kip-in.}$$

$$\frac{L}{d} = \frac{12 \times 12}{5.5} = 26.18$$

$$E'_{\text{min}} = \frac{1.5}{\phi_s}\phi_s E_{\text{min}} = \frac{1.5}{0.85}0.85(620{,}000) = 930{,}000 \text{ psi}$$

$$F_{cE} = \frac{0.822 E_{\text{min}}}{(L/d)^2} = \frac{0.822(930{,}000)}{(26.18)^2} = 1115 \text{ psi, or } 1.115 \text{ ksi}$$

$$F_c^* = \lambda\phi_c\left(\frac{2.16}{1000\phi_c}\right)F_c = 0.8(0.90)\left(\frac{2.16}{1000(0.90)}\right) \times 1200 = 2.074 \text{ ksi}$$

$$\frac{F_{cE}}{F_c^*} = \frac{1.115}{2.074} = 0.5376$$

From Figure 9.3, $C_p = 0.46$

$$F_c' = \lambda\phi_c\left(\frac{2.16}{1000\phi_c}\right)C_pF_c = 0.8(0.90)\left(\frac{2.16}{1000(0.90)}\right) \times (0.46)(1200) = 0.954\,\text{ksi}$$

For a consideration of the value to be used for bending stress, an investigation should ordinarily be done of the effects of lateral and torsional buckling, as in the case of a beam. This is usually not critical unless the depth to width ratio of the section is greater than 3. In this case the square section has a ratio of 1.0, and the issue is not a concern. What remains to be done to establish the limit for bending stress is simply to make the appropriate adjustments, thus

$$F_b' = \lambda\phi_b\frac{2.16}{1000\phi_b}F_b = 0.8(0.85)\left(\frac{2.16}{1000(0.85)}\right)1400 = 2.419\text{ ksi}$$

$$f_c = \frac{P}{A} = \frac{11.2}{(5.5)^2} = 0.370\text{ ksi}$$

$$f_b = \frac{M}{S} = \frac{50.4}{27.7} = 1.819\text{ ksi}$$

And, for the interaction analysis:

$$\left(\frac{f_c}{F_c'}\right)^2 + \left(\frac{f_b}{F_b'\left[1-\dfrac{f_c}{F_{cE}}\right]}\right) = \left(\frac{0.370}{0.954}\right)^2 + \left(\frac{1.819}{2.419\left[1-\dfrac{0.370}{1.115}\right]}\right) = 0.150 + 1.126 = 1.276$$

As this exceeds 1.0, the column is not adequate. A second try might use a 6 × 8 which has a significantly larger section modulus to reduce the bending stress.

Problem 9.8.E. A 10 × 10 column of Douglas fir-larch, No. 1 grade, is 9 ft high and carries a compression load of 10 kips dead load plus 10 kips live load that is 7.5 in. eccentric from the column axis. Investigate the column for combined compression and bending, using the LRFD method.

Problem 9.8.F. A 12 × 12 column of Douglas fir-larch, No. 1 grade, is 12 ft high and carries a compression load of 12 kips dead load plus 12 kips live load that is 9.5 in. eccentric from the column axis. Investigate the column for combined compression plus bending, using the LRFD method.

10

CONNECTIONS FOR WOOD STRUCTURES

Structures of wood typically consist of large numbers of separate pieces that must be joined together. Fastening of pieces is rarely achieved directly, as in the fitted and glued joints of furniture, except for the production of glued products, such as plywood. For assemblage of building construction, fastening is most often achieved by using some steel device, common ones being nails, screws, bolts, and specially formed steel fasteners. A major portion of the *NDS* (Ref. 1) is devoted to concerns for structural fastenings for wood. This chapter presents a highly condensed treatment of the simple cases for some very common fasteners.

10.1 BOLTED JOINTS

When steel bolts are used to connect wood members, there are several design concerns. Some of the principal concerns are the following:

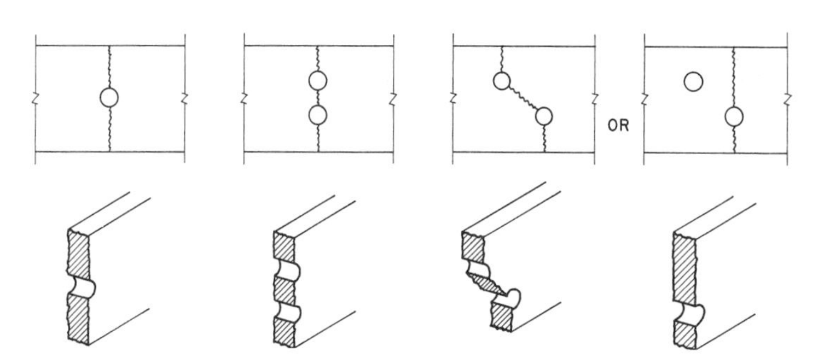

Figure 10.1 Effect of bolt holes on reduction of cross section for tension members.

1. *Net Cross Section in Member.* Holes made for the placing of bolts reduce the wood member cross section. For this investigation the hole diameter is assumed to be 1/16 in. larger than that of the bolt. Common situations are shown in Figure 10.1. When bolts in multiple rows are staggered, it may be necessary to make two investigations, as shown in the illustration.
2. *Bearing of the Bolt on the Wood.* This compressive stress limit varies with the angle of the wood grain to the load direction.
3. *Bending of the Bolt.* Long thin bolts in thick wood members will bend considerably, causing a concentration of bearing at the edge of the hole.
4. *Number of Members Bolted at a Single Joint.* The worst case, as shown in Figure 10.2, is that of the two-member joint. In this case the lack of symmetry in the joint produces considerable twisting. This situation is referred to as *single shear*, since the bolt is subjected to shear on a single cross section of the bolt. With more members in the joint, twisting may be eliminated and the bolt is sheared at multiple cross sections.
5. *Ripping Out the Bolt When Too Close to an Edge.* This problem, together with that of the minimum spacing of bolts, is dealt with

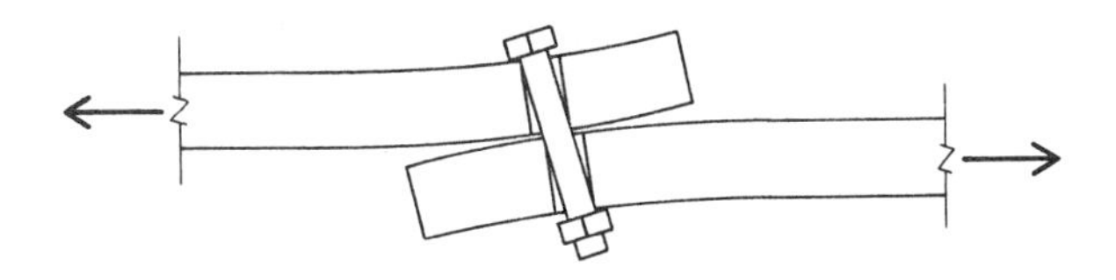

Figure 10.2 Twisting in the two-member bolted joint.

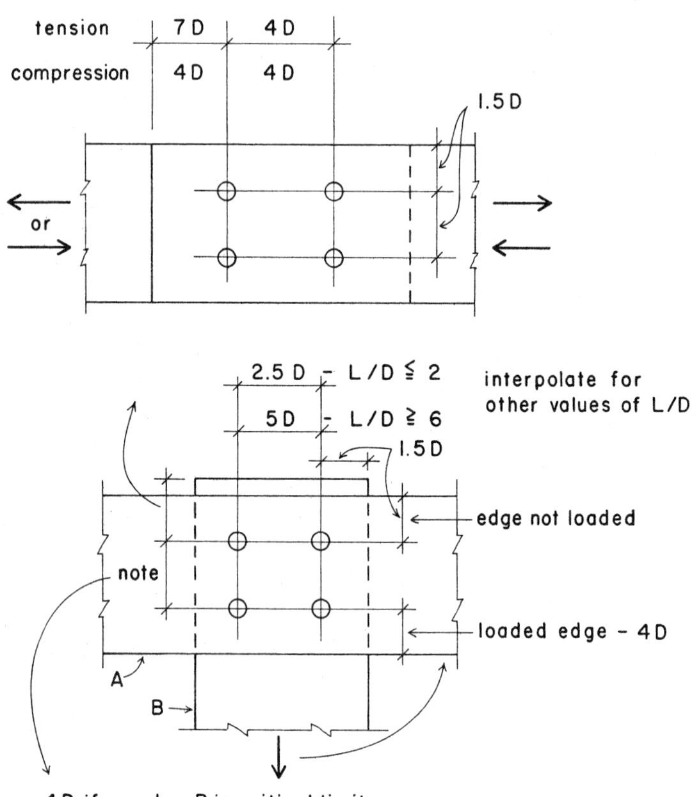

Figure 10.3 Edge, end, and center-to-center spacing distances for bolts in wood construction.

by using the criteria given in Figure 10.3. Note that the limiting dimensions involve the consideration of the bolt diameter *D*, the bolt length *L*, the type of force (tension or compression), and the angle of load to the grain of the wood.

The NDS presents materials for design of bolted joints in layered stages. The first stage involves general concerns for the basic wood construction and the particular species and grade of the wood.

The second layer of concern has to do with some considerations for structural fastenings in general. Some of the general concerns are for the form of connections, arrangements of multiple fasteners, and the eccentricity of forces in connections.

The third layer of concern involves considerations for the particular type of fasteners. The NDS provides reference values for bolts and for adjustments based on a number of considerations. For a single bolted joint, the capacity of a bolt may be expressed as

$$Z' = Z(C_D)(C_M)(C_t)(C_g)(C_\Delta)$$

where Z' = the adjusted bolt design value
Z = the reference bolt design value, based on one of many possible modes of failure
C_D = the load duration factor from Table 4.3
C_M = the factor for any special moisture condition
C_t = the temperature factor for extreme climate conditions
C_g = the group action factor for joints with more than one bolt in a row in the direction of the load
C_Δ = a factor related to the geometry (general form) of the joint (single shear, double shear, eccentrically loaded, etc.)

For all of this complexity, additional adjustments may be necessary. Two additional concerns have to do with the angle of the load to the grain direction in the wood and the adequacy of dimensions in the bolt layout.

For the direction of loads with respect to the grain, there are two major limiting positions: that with the load parallel to the grain (0°) and that with the load perpendicular to the grain (90°). Between these limits is a so-called angle-to-grain loading, for which an adjustment is made using the Hankinson formula (see discussion in Section 4.4). Figure 10.4 illustrates the application of the Hankinson formula in the form of a graph. Considering the P direction as zero and the Q direction as 90°, angles between zero and 90 are expressed as the specific angle in degrees, designated by the Greek letter theta in the illustration. The designations of P and Q are used for split-ring connectors, but the designations of $Z_{\|}$ and $Z_{\perp}$ are used for bolts. An example of the use of the graph is shown in Figure 10.4.

Bolt layout becomes critical when a number of bolts must be used in members of relatively narrow dimensions. Major dimensional concerns

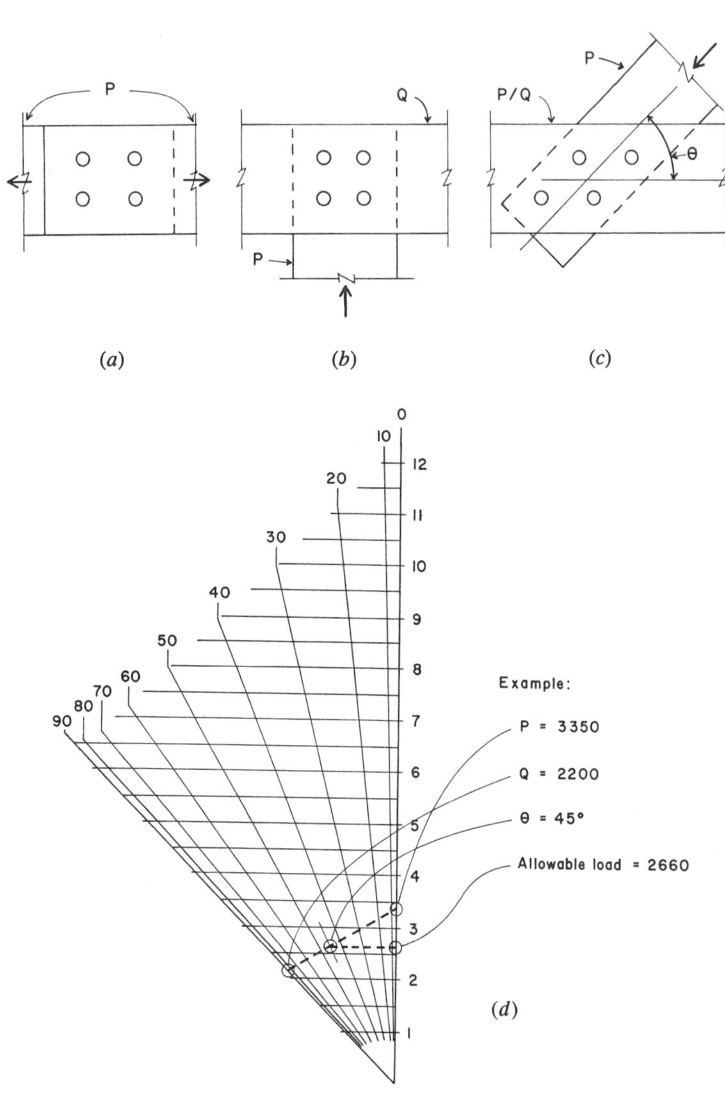

Figure 10.4 Relation of load to grain direction in bolted joints: (a) parallel members (angle of 0°), (b) members at right angles (angle of 90°), (c) members with angle between 0 and 90°, (d) Hankinson graph for adjusted design values for angle-to-grain loadings.

are illustrated in Figure 10.3. The NDS establishes two limiting dimensions. The first limit is the minimum required for design use of the full value of the bolt capacity. The second limit is an absolute minimum, for which some reduction of capacity is specified. In general, any dimensions falling between these two limits in real situations may be used to establish bolt capacity values by direct interpolation between the limiting capacity values.

Using all of the NDS requirements in a direct way in practical design work is a mess. Consequently, the NDS, or others, provide some shortcuts. One such aid is represented by a series of tables that allow the determination of a bolt capacity value by direct selection from the table. Care must be taken in using these tables to be aware of the data that is incorporated in the table. However, even more care must be taken in paying attention to what is not incorporated into the tables: notably, the various adjustments for moisture, load duration, etc.

Table 10.1 presents a sampling from much larger tables in the NDS, data being given here only for Douglas fir-larch, whereas data for several species of wood is included in the reference source. Table 10.1 is compiled from two separate tables in the NDS, where one table provides data only for single shear joints and another table provides data only for double shear joints. These tables provide values for joints achieved with all wood members; other NDS tables also provide data for joints achieved with combinations of wood members and steel plates.

Values are listed in three columns in the table for the following designations:

$Z_{||}$ = value for load parallel to the grain of the member

$Z_{s\perp}$ = value for load perpendicular to the grain of the side member

$Z_{m\perp}$ = value for load perpendicular to the grain of the middle member, or to the grain of the main member in a single shear joint

The following examples illustrate the use of the materials presented here from the NDS.

Example 1. A three-member (double shear) joint is made with members of Douglas fir-larch, select structural grade lumber. (See Figure 10.5.) The joint is loaded as shown, with the load parallel to the grain direction in the members. The tension force is 9 kips. The

TABLE 10.1 Bolt Reference Lateral Design Values for Wood Joints with Douglas Fir-Larch (lb/bolt)

Member Thickness (in.)			Loading Condition					
			Single Shear (lb)			Double Shear (lb)		
Main Member t_m	Side Member t_s	Bolt Diameter D(in.)	$Z_{\|\|}$	$Z_{s\perp}$	$Z_{m\perp}$	$Z_{\|\|}$	$Z_{s\perp}$	$Z_{m\perp}$
1.5	1.5	½	480	300	300	1050	730	470
		5/8	600	360	360	1310	1040	530
		¾	720	420	420	1580	1170	590
		7/8	850	470	470	1840	1260	630
2.5	1.5	5/8	850	520	430	1760	1040	880
		¾	1020	590	500	2400	1170	980
		7/8	1190	630	550	3060	1260	1050
3.5	1.5	5/8	880	520	540	1760	1040	1190
		¾	1200	590	610	2400	1170	1370
		7/8	1590	630	680	3180	1260	1470
	3.5	5/8	1120	700	700	2240	1410	1230
		¾	1610	870	870	3220	1750	1370
		7/8	1970	1060	1060	4290	2130	1470
5.5	1.5	¾	1200	590	790	2400	1170	1580
		7/8	1590	630	980	3180	1260	2030
		1	2050	680	1060	4090	1350	2480
	3.5	¾	1610	870	1030	3220	1750	2050
		7/8	2190	1060	1260	4390	2130	2310
		1	2660	1290	1390	5330	2580	2480
7.5	1.5	¾	1200	590	790	2400	1170	1580
		7/8	1590	630	1010	3180	1260	2030
		1	2050	680	1270	4090	1350	2530
	3.5	¾	1610	870	1030	3220	1750	2050
		7/8	2190	1060	1360	4390	2130	2720
		1	2660	1290	1630	5330	2580	3380

Source: Developed from data in the *National Design Specification® for Wood Construction* (Ref. 1), with permission of the publishers, American Forest & Paper Association.

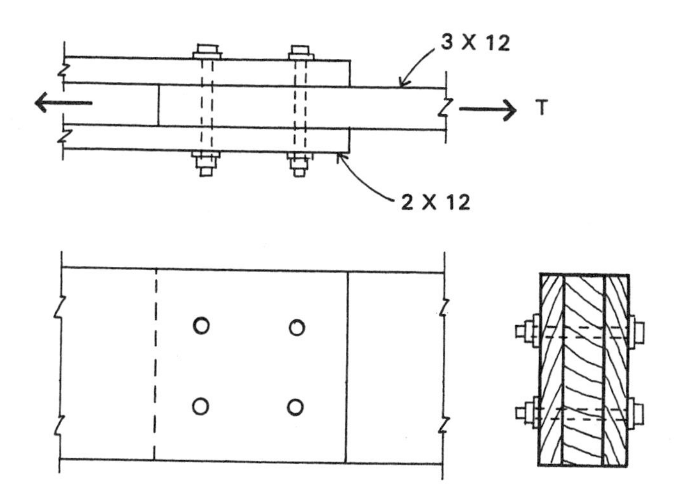

Figure 10.5 Reference for Example 1.

middle member (designated main member in Table 10.1) is a 3 × 12 and the outer members (side members in Table 10.1) are each 2 × 12. Is the joint adequate for the required load if it is made with four ¾-in. bolts as shown? No adjustment is required for load duration or moisture, and it is assumed that the layout dimensions are adequate for use of the full design values for the bolts.

Solution: From Table 10.1, the allowable load per bolt is 2400 lb ($Z_{||}$ in table). With four bolts, the total capacity of the bolts is thus

$$T = 4(2400) = 9600\,\text{lb}$$

based only on the bolt actions in bearing and shear in the joint.

For tension stress in the wood, the critical section is that for the main member, since its thickness (2.5 in.) is less than the total thickness of the side members (2 × 1.5 = 3.0 in.). With the holes typically being considered as 1/16-in. larger than the bolts, the net section through the two bolt holes across the member is thus

$$A = (2.5)[11.25 - 2(13/16)] = 24.06\,\text{in.}^2$$

From Table 4.1, the allowable tension stress is 1000 psi with no adjustment required from Table 4.2. The maximum tension capacity of the main member is thus

$$T = (\text{allowable stress})(\text{net area})$$
$$= (1000)(24.06) = 24,060 \text{ lb}$$

And the joint is adequate for the load.

It should be noted that the NDS provides for a reduction of capacity in joints with multiple connectors. However, the factor becomes negligible for joints with as few as two bolts per row and other details as given for the preceding example.

Example 2. A bolted two-member joint consists of two 2 × 10 members of Douglas fir-larch, select structural grade lumber, attached at right angles to each other, as shown in Figure 10.6. What is the maximum capacity of the joint if it is made with two 7/8-in. bolts?

Solution: This is a single shear joint with both members of 1.5-in.-thick side pieces. From Table 10.1, the value for load perpendicular to the side member is 470 lb per bolt, and the total joint capacity is thus

$$C = (2)(470) = 940 \text{ lb}$$

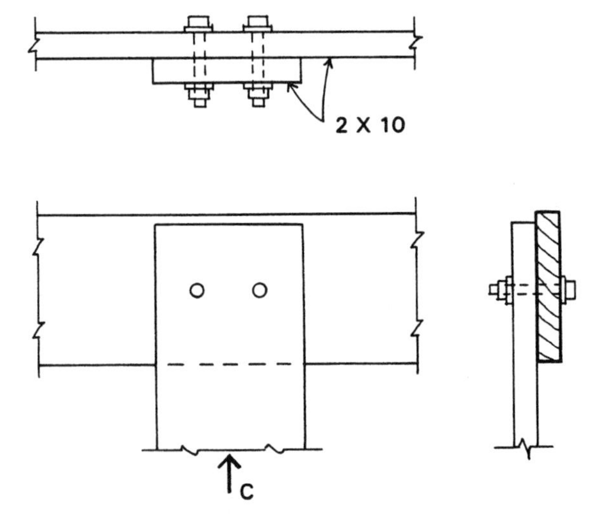

Figure 10.6 Reference for Example 2.

Example 3. A three-member joint consists of two outer members, each 2 × 10, bolted to a 4 × 12 middle member with two $^3/_4$-in. bolts. The outer members are arranged at an angle to the middle member, as shown in Figure 10.7. Wood of all members is Douglas fir-larch, select structural grade. Find the maximum compression force that can be transmitted through the joint by the outer members.

Solution: In this case, investigations must be made for both the outer and middle members. The load is parallel to the grain in the outer members and at an angle of 45° to the grain in the middle member.

For the outer members, Table 10.1 yields a value of 2400 lb per bolt. (Main member 3.5-in., side members 1.5-in., double shear, load parallel to the grain.)

For the middle member, the table yields values of 2400 lb and 1370 lb for the directions parallel and perpendicular to the grain, respectively. Entering these values on the graph in Figure 10.4, an approximate value of 1600 lb/bolt is found for the angle of 45°. Since this is less than the limit for the outer members, the capacity of the joint is thus

$$C = (2)(1600) = 3200\ \text{lb}$$

For the following problems, use Douglas fir-larch, No. 1 grade, assuming no adjustments for moisture or load duration.

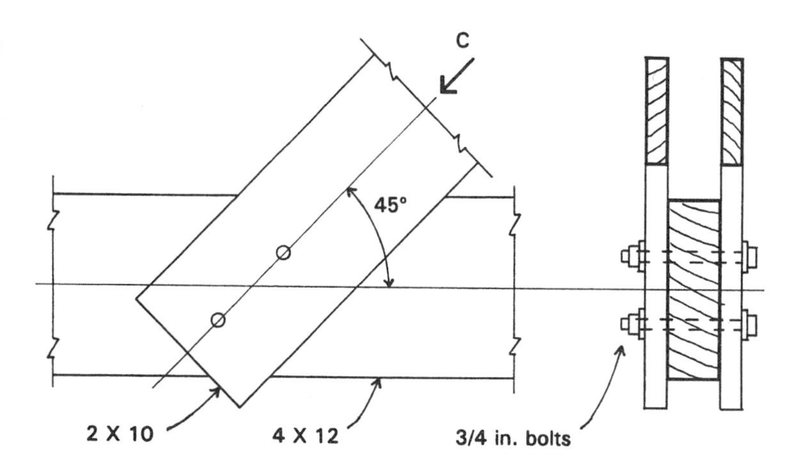

Figure 10.7 Reference for Example 3.

Problem 10.1.A. A three-member tension joint has 2 × 12 outer members and a 4 × 12 middle member (form as shown in Figure 10.5). The joint is made with six 3/4-in. bolts in two rows. Find the total load capacity of the joint.

Problem 10.1.B. A two-member tension joint consists of two parallel 2 × 6 members bolted with two 3/4-in. bolts. What is the limit for the tension force?

Problem 10.1.C. Two outer members, each 2 × 8, are bolted with 3/4-in. bolts to a middle member consisting of a 3 × 12 (form as shown in Figure 10.7). The members form an angle of 45°. What is the maximum compressive force that can be transmitted through the joint by the outer members?

10.2 NAILED JOINTS

Nails are used in great variety in building construction. For structural fastening, the nail most commonly used is called—appropriately—the *common wire nail*. As shown in Figure 10.8, the critical concerns for such nails are the following:

1. *Nail Size.* Critical dimensions are the diameter and length (see Figure 10.8*a*). Sizes are specified in pennyweight units, designated as 4d, 6d, and so on, and referred to as four penny, six penny, and so on.
2. *Load Direction.* Pullout loading in the direction of the nail shaft is called *withdrawal*; shear loading perpendicular to the nail shaft is called *lateral load.*
3. *Penetration.* Nailing is typically done through one element and into another, and the load capacity is essentially limited by the amount of the length of embedment of the nail in the second member (see Figure 10.8*b*). The length of this embedment is called the penetration.
4. *Species and Grade of Wood.* The heavier the wood (indicating generally harder, tougher material), the greater is the load resistance capability.

Design of good nailed joints requires a little engineering and a lot of good carpentry. Some obvious situations to avoid are those shown in

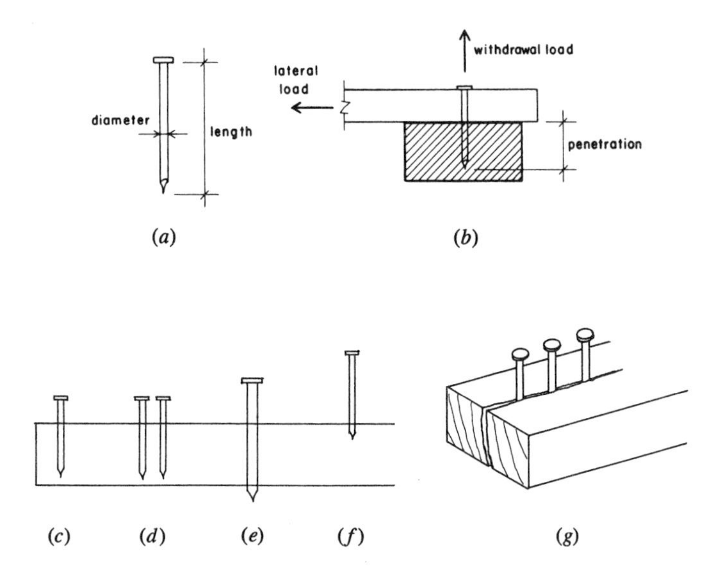

Figure 10.8 Use of common wire nails. (a) Critical dimensions. (b) Loading considerations. (c) through (g) Poor nailing practices: (c) too close to edge, (d) nails too close together, (e) nail too large for wood piece, (f) too little penetration of nail into holding wood piece, (g) too many closely-spaced nails in a single row parallel to wood grain and/or nails too close to member end.

Figures 10.8*c* through *g*. A little actual carpentry experience is highly desirable for anyone who designs nailed joints.

Withdrawal load capacities of nails are given in units of force per inch of nail penetration length. This unit load is multiplied by the actual penetration length to obtain the total force capacity of the nail. For structural connections, withdrawal resistance is relied on only when the nails are perpendicular to the wood grain direction.

Lateral load capacities for common wire nails are given in Table 10.2 for joints with both plywood and lumber side pieces. The NDS contains very extensive tables for many wood types as well as metal side pieces. The following example illustrates the design of a nailed joint using the ASD method with data from Table 10.2.

Example 4. A structural joint is formed as shown in Figure 10.9, with the wood members connected by 16d common wire nails. Wood is Douglas fir-larch. What is the maximum value for the compression force in the two side members?

TABLE 10.2 Reference Lateral Load Values for Common Wire Nails (lb/in.)

Side Member Thickness t_s (in.)	Nail Length L (in.)	Nail Diameter D (in.)	Nail Pennyweight	Load per Nail Z (lb)
Part 1—with Wood Structural Panel Side Members[a] ($G = 0.42$)				
	2	0.113	6d	48
3/8	2½	0.131	8d	63
	3	0.148	10d	76
	2	0.113	6d	50
15/32	2½	0.131	8d	65
	3	0.148	10d	78
	3½	0.162	16d	92
	2	0.113	6d	58
23/32	2½	0.131	8d	73
	3	0.148	10d	86
	3½	0.162	16d	100
Part 2—with Sawn Lumber Side Members[b] ($G = 0.50$)				
	2½	0.131	8d	90
3/4	3	0.148	10d	105
	3½	0.162	16d	121
	4	0.192	20d	138
	3	0.148	10d	118
	3½	0.162	16d	141
1½	4	0.192	20d	170
	4½	0.207	30d	186
	5	0.225	40d	205
	5½	0.244	50d	211

[a] Values for single shear joints with wood structural panel side members with $G = 0.42$ and nails anchored in sawn lumber of Douglas fir-larch with $G = 0.50$.
[b] Values for single shear joints with both members of sawn lumber of Douglas fir-larch with $G = 0.50$.
Source: Adapted from the *National Design Specification® for Wood Construction*, 2005 edition (Ref. 1), with permission of the publishers, American Forest & Paper Association.

Solution: From Table 10.2, we read a value of 141 lb per nail. (Side member thickness of 1.5-in., 16d nails). As shown in the illustration, there are five nails on each side or a total of 10 nails in the joint. The total joint load capacity is thus

$$C = (10)(141) = 1410\,\text{lb}$$

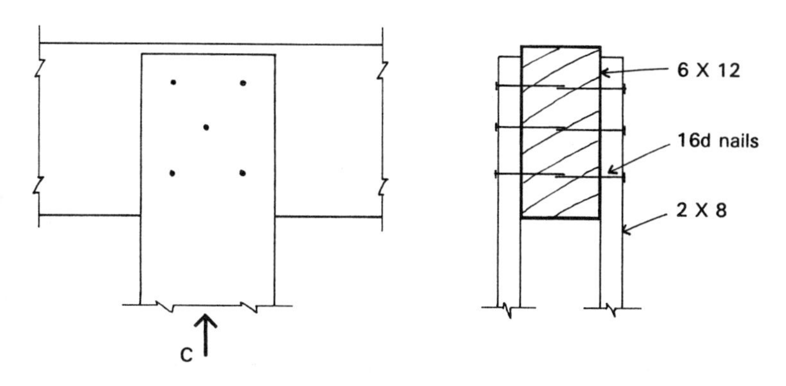

Figure 10.9 Reference for Example 4.

No adjustment is made for direction of load to the grain. However, the basic form of nailing assumed here is so-called side grain nailing, in which the nail is inserted at 90 degrees to the grain direction and the load is perpendicular (lateral) to the nails.

Minimum adequate penetration of the nails into the supporting member is a necessity, but use of the combinations given in Table 10.2 assures adequate penetration if the nails are fully buried in the members.

Example 6 in Section 10.9 treats a joint for a light wood truss using lumber members for the truss and connecting panels (called gusset plates) of structural plywood.

Problem 10.2.A. A joint similar to that in Figure 10.9 is formed with outer members of one inch nominal thickness (3/4-in. actual thickness) and 10d common wire nails. Find the compression force that can be transferred to the two side members.

Problem 10.2.B. Same as Problem 10.2.A, except outer members are 2 × 10, middle member is 4 × 10, nails are 20d.

10.3 SCREWS

When loosening or popping of nails is a problem, a more positive grabbing of a nail by the surrounding wood can be achieved by having some formed surface on the nail shaft. Many special nails with formed surfaces are used, but another means for achieving this effect is to use threaded screws in place of nails. Screws can be tightened to squeeze connected members together in a manner that is not usually possible with nails.

Whereas nails may slip within the wood, with resulting "popping" of the nail heads and loosening of connected parts, slipping of the screw within the wood seldom occurs. For dynamic loading, such as shaking by an earthquake, the tight, positively-anchored connection achieved with screws is a distinct advantage.

Screws are produced in great variety, although three of the most common types used for wood structures are those shown in Figure 10.10. The flat head screw is designed to be driven so that the head sinks entirely into the wood, resulting in a surface with no protrusions. The round head screw is usually used with a washer, or may be used to

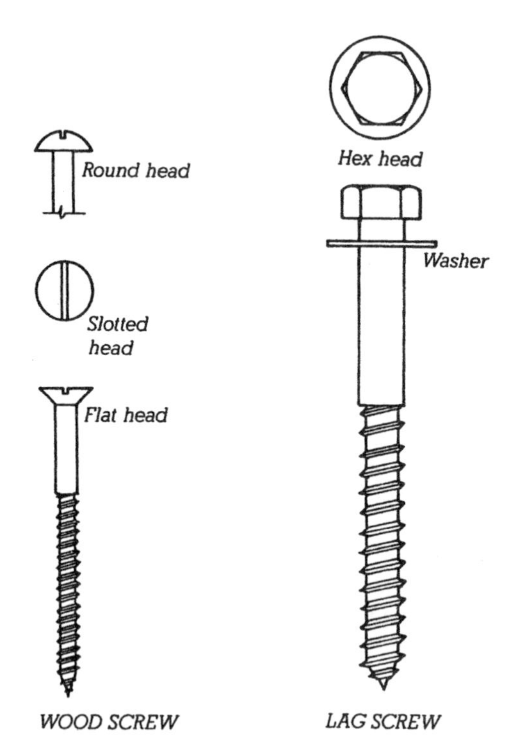

Figure 10.10 Types of wood screws. (Reproduced from *Fundamentals of Building Construction* by Edward Allen, with permission of the publishers, John Wiley & Sons, New Jersey.)

attach metal objects to wood surfaces. The hex head screw is called a *lag screw* or *lag bolt* and is designed to be tightened by a wrench rather than a screwdriver. Lag screws are made in a considerable size range and can be used for some major structural connections, replacing bolts in some situations.

Screws work essentially the same as nails, being used to resist either withdrawal or lateral shear-type loading. Screws must usually be installed by drilling a guide hole, called a *pilot hole*, with a diameter slightly smaller than that of the screw shaft. Specifications for structural connections with screws give requirements for the limit of pilot hole diameters and for various other details for proper installation. Capacities for withdrawal and lateral loading are given as a function of the screw size and the type of wood holding the screw.

As with nailed joints, the use of screws involves much judgment that is more craft than science. Choice of the screw type, size, spacing, length, and other details of a good joint may be controlled by some specifications, but it is also a matter of experience.

Although it is generally not recommended that nails be relied on for computed loading in withdrawal, screws are not so limited and are often chosen where details of the connection require such loading.

10.4 MECHANICALLY DRIVEN FASTENERS

Although the hand hammer, screwdriver, and wrench are still in every carpenter's toolbox, many structural fastenings are now routinely achieved with powered devices. In some cases this has produced fasteners that do not fit the old classifications, and some codes provide criteria for the most common methods of mechanical fastening methods. The simple hand staple gun has been extrapolated into a range of devices that can install some significantly strong structural fasteners. Field attachment of structural panel materials is now mostly accomplished with mechanical driving equipment. Code-approved load capacities for such attachments are rated individually for proprietary equipment and fastening materials.

10.5 SHEAR DEVELOPERS

When flat wood members are lapped at a joint and bolted, it is often difficult to prevent some joint movement in the form of slipping between the lapped members. If force reversals, such as those that occur with wind and earthquake effects, cause back-and-forth stress on the joint, this lack of tightness in the connection may be especially objectionable. Various types of devices, called *shear developers*, are sometimes inserted between the lapped members so that when the bolts are tightened, there is some form of resistance to slipping besides the simple friction between the lapped members.

Toothed or ridged devices of metal are sometimes used for this development of enhanced shear resistance in the lapped joint. They are simply placed between the members, and the tightening of the bolts causes them to bite into both members. Hardware products of various forms and sizes are available and are commonly used, most notably for heavy, rough timber construction.

A slightly more sophisticated shear developer for lapped joints is the *split-ring connector* consisting of a steel ring that is installed by cutting matching circular grooves in the faces of the lapped members. When the ring is inserted into the grooves and the bolt is tightened, the ring is squeezed tightly into the grooves, and the resulting connection has a shear resistance considerably greater than that with the bolt alone. Design of split-ring connectors in discussed in the next section.

10.6 SPLIT-RING CONNECTORS

The ordinary form of the split-ring connector and the method of its installation are shown in Figure 10.11. Design considerations for this device include the following:

1. *Size of the rings.* Rings are available in the two sizes shown in the figure with nominal diameters of 2.5 and 4 in.
2. *Stress on the net section of the wood members.* As shown in Figure 10.11, the cross section of the wood piece is reduced by the ring profile (*A* in the figure) and the bolt hole. If rings are placed on both sides of a wood piece, there will be two reductions for the ring profile.
3. *Thickness of the wood piece.* If the wood piece is too thin, the cut for the ring will bite excessively into the cross section. Rated load capacities reflect concern for this condition.
4. *Number of faces of the wood member having rings.* As shown in Figure 10.12, the outside members in a joint will have rings in only one face, whereas the inside members will have rings on both faces. Thickness considerations therefore are more critical for the inside members.

D	=	2.5"	4"
bolt size	=	1/2"	3/4"
d	=	9/16"	13/16"
b	=	3/8"	1/2"
A	=	1.10 in^2	2.24 in^2

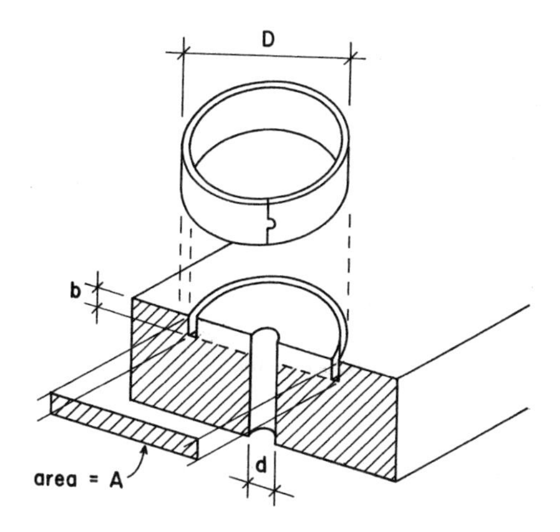

Figure 10.11 Split-ring connectors for bolted joints in wood construction.

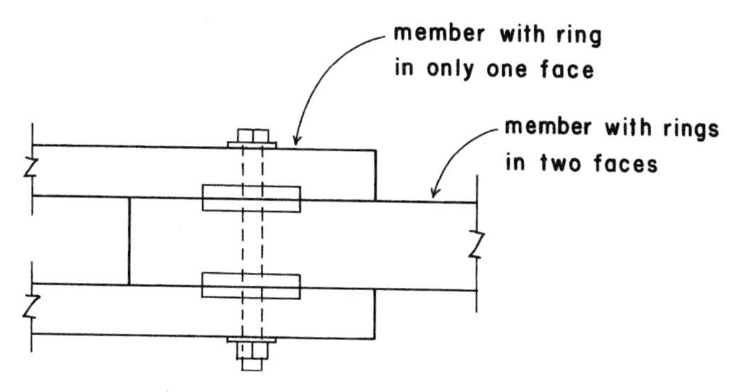

Figure 10.12 Determination of the number of faces of a member with split-ring connectors.

5. *Edge and end distances.* These must be sufficient to permit the placing of the rings and to prevent splitting out from the sides of the wood piece when the joint is loaded. Concern is greatest for an edge in the direction of loading—called the loaded edge. (See Figure 10.13.)
6. *Spacing of rings.* Spacing must be sufficient to permit the placing of the rings and the full development of the ring capacity in the wood piece.

Figure 10.14 shows the four placement dimensions that must be considered. The limits for these dimensions are given in Table 10.3. In some cases, two limits are given. One limit is that required for the full development of the ring capacity (100% of the reference design values). The other limit is the minimum dimension permitted for the ring for which some reduction factor is given for the ring capacity. Load capacities for dimensions between these limits can be directly proportioned.

Table 10.4 gives reference design values for split-ring connectors for both regular and dense grades of Douglas fir-larch. As with bolts, values are given for load directions both parallel and perpendicular to the wood grain. Values for loadings at some angle to the grain can be determined with the use of the Hankinson formula (Section 4.4) or the graph in Figure 10.4.

The following example illustrates the ASD procedure for the investigation of a joint using split-ring connectors.

Example 5. The joint shown in Figure 10.15, using $2\frac{1}{2}$ in. split-rings and wood of Douglas fir-larch, No. 1 grade, sustains the loading indicated. Find the limiting value for the load.

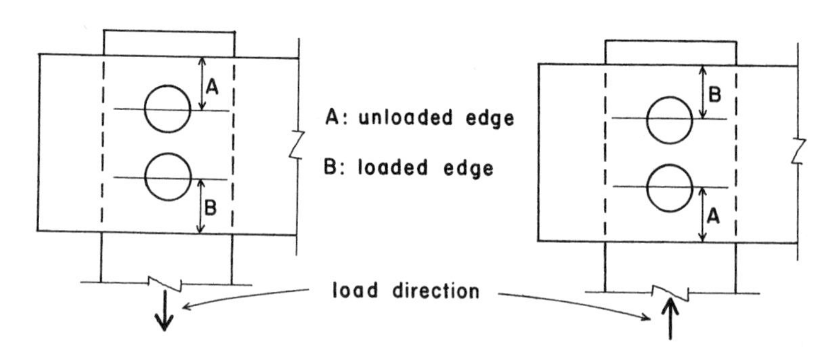

Figure 10.13 Determination of the loaded edge condition.

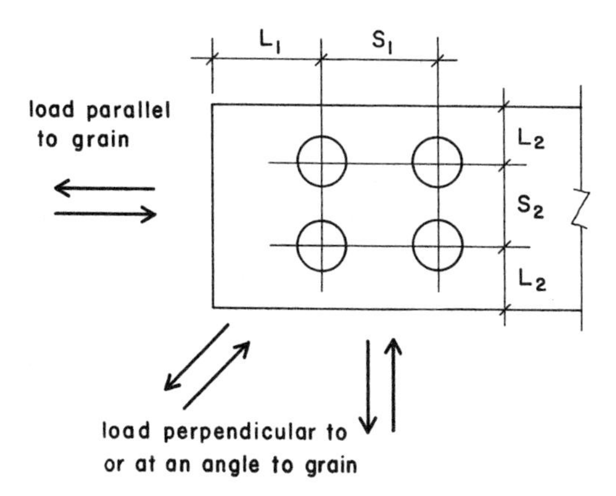

Figure 10.14 Reference figure for the end, edge, and spacing requirements for split-ring connectors. See Table 10.3.

TABLE 10.3 Spacing, Edge Distances, and End Distances for Split-Ring Connectors

		Distances (in inches), and Corresponding Percentages of Reference Design Values			
		Load Parallel to Grain		Load Perpendicular to Grain	
Dimension[a]		2.5 in. Ring	4 in. Ring	2.5 in. Ring	4 in. Ring
L_1	Tension	5.5, 100% 2.75 min., 62.5%	7, 100% 3.5 min., 62.5%	5.5, 100% 2.75 min., 62.5%	7, 100% 3.5 min., 62.5%
	Compression	4, 100% 2.5 min., 62.5%	5.5, 100% 3.25 min., 62.5%	5.5, 100% 2.75 min., 62.5%	7, 100% 3.5 min., 62.5%
L_2	Unloaded	1.75 min., 100%	2.75 min., 100%	1.75 min., 100%	2.75 min., 100%
	Loaded	1.75 min., 100%	2.75 min., 100%	2.75 100% 1.75 min., 83%	3.75, 100% 2.75 min., 83%
S_1		6.75, 100% 3.5 min., 50%	9, 100% 5 min., 50%	3.5 min., 100%	5 min., 100%
S_2		3.5 min.	5 min.	4.25, 100% 3.5 min., 50%	6, 100% 5 min., 50%

[a] See Figure 10.14.

Source: Adapted from data in *National Design Specification® for Wood Construction* (Ref. 1), with permission of the publishers, American Forest & Paper Association.

TABLE 10.4 Reference Design Values for Split-Ring Connectors with Douglas Fir-Larch Wood (lb/ring)

				Load Parallel to Grain P		Load Perpendicular to Grain Q	
Ring Size (in.) Grades	Bolt Diameter (in.)	Faces with Connectors[a]	Actual Thickness of Pieces (in.)	Dense Grades	Non-Dense Grades	Dense Grades	Non-Dense
2.5	½	1	1 min.	2630	2270	1900	1620
			1.5 or more	3160	2730	2280	1940
		2	1.5 min.	2430	2100	1750	1500
			2 or more	3160	2730	2280	1940
4	3/4	1	1 min.	4090	3510	2840	2440
			1.5	6020	5160	4180	3590
		2	1.5 min.	4110	3520	2980	2450
			2	4950	4250	3440	2960
			2.5	5830	5000	4050	3480
			3 or more	6140	5260	4270	3660

[a] See Figure 10.12.

Source: Adapted from data in *National Design Specification® for Wood Construction* (Ref. 1), with permission of the publishers, American Forest & Paper Association.

Solution: Separate investigations must be made for the members in this joint. For the 2 × 6,

Load is parallel to the grain.

Rings are in two faces.

Critical dimensions are member thickness of 1.5 in. and end distance of 4 in.

From Table 10.3, the end distance required for use of the full ring capacity is 5.5 in. and if the minimum distance of 2.75 in. is used, the capacity must be reduced to 62.5% of the full value. The value to

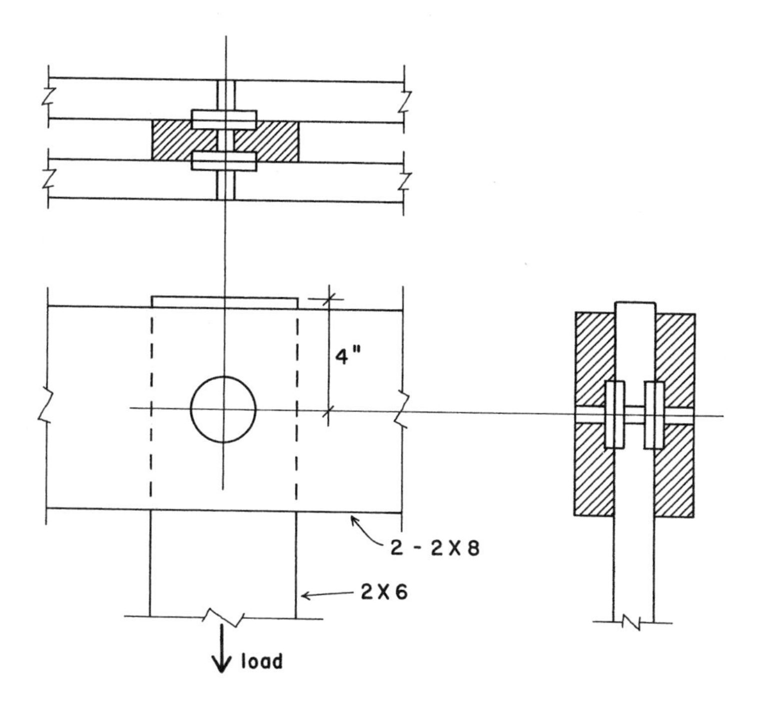

Figure 10.15 Reference for Example 5.

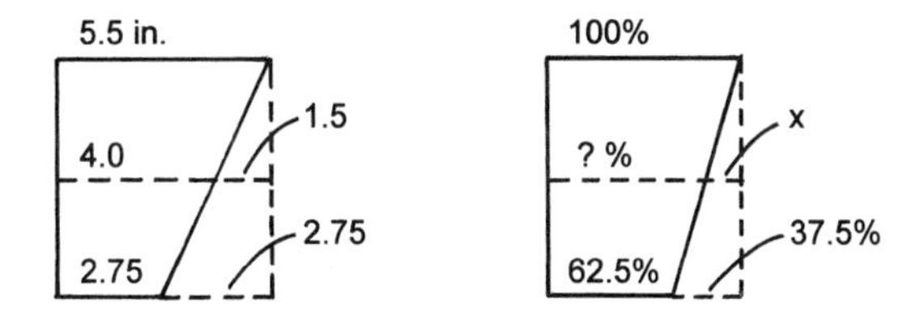

Figure 10.16 Adjusted capacity for the rings.

be used for the 4 in. end distance must be interpolated between these limits, as shown in Figure 10.16. Thus

$$\frac{1.5}{2.75} = \frac{x}{37.5}, \qquad x = \frac{1.5}{2.75}(37.5) = 20.45\%$$

$$y = 100 - 20.45 = 79.55\%, \text{ or approximately } 80\%$$

From Table 10.4 the full capacity of a ring is 2100 lb. Therefore the usable capacity is

$$(0.80)(2100) = 1680\,\text{lb/ring}$$

For the 2 × 8,

Load is perpendicular to the grain.
Rings are in only one face.
Loaded edge distance is one half of 7.25 in., or 3.625 in.

For this situation the load value from Table 10.4 is 1940 lb. From Table 10.3, the required edge distance for use of 100% of the reference value is 2.75 in. As the distance to the edge is 7.25/2 = 3.625 in, no reduction is required. Therefore the joint is limited by the conditions for the 2 × 6, and the capacity of the joint with the two rings is

$$T = (2)(1680) = 3360\,\text{lb}$$

It should be verified that the 2 × 6 is capable of sustaining this load in tension stress on the net section at the joint. As shown in Figure 10.17, the net area is

$$A = 8.25 - (2)(1.10) - \left(\frac{9}{16}\right)(0.75) = 5.63\text{ in.}^2$$

From Table 4.1 the reference value for tension is 675 psi. From Table 4.2, this value is increased by a factor of 1.3. Therefore the capacity of the member is

$$T = 1.3(675)(5.63) = 4940\,\text{lb}$$

And the member is not critical in tension stress.

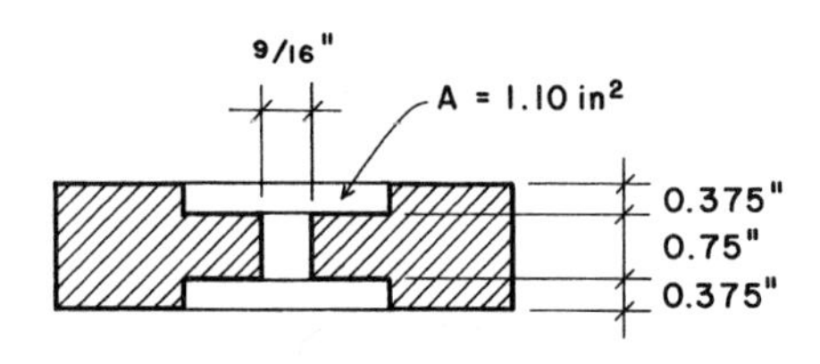

Figure 10.17 Determination of the net cross-sectional area for Example 5.

Problem 10.6.A. A joint similar to that in Figure 10.15 is made with $2^1/_2$ in. split-rings and wood members with Douglas fir-larch, No. 1 grade. End distance for the middle member is 5.5 in. Find the limit for the tension load if the outer members are 2 × 12 and the middle member is a 2 × 8.

Problem 10.6.B. A joint similar to that in Figure 10.15 is made with 4 in. split-rings and wood members of Douglas fir-larch, No. 1 grade. End distance for the middle members is 5 in. Find the limit for the tension load if the outer members are 3 × 10 and the middle member is a 4 × 10.

10.7 FORMED STEEL FRAMING ELEMENTS

Formed metal framing devices have been used for many centuries for the assembly of structures of heavy timber. In ancient times elements were formed of bronze or cast iron or wrought iron. Later they were formed of forged or bent and welded steel elements. (See Figure 10.18.) Some of the devices commonly used today are essentially the same in function and detail as those used long ago.

For large timber members, connecting elements are now mostly formed of steel plate that is bent and welded to produce the desired shape. (See Figure 10.19.) The ordinary tasks of attaching beams to columns and columns to foundations continue to be required, and the simple means of achieving the tasks evolved from practical concerns.

For resistance to gravity loads, connections such as those shown in Figure 10.19 sometimes have no direct structural functions. In theory, it is possible to simply rest a beam on top of a column, as is done in some rustic construction. However, for resistance to lateral loads from wind or earthquakes, the tying and anchoring functions of these connecting devices are often quite essential. They also serve a practical function of simply holding the parts together during the construction process.

A development of more recent times is the extension of the use of metal devices for the assembly of light wood frame construction. Devices of thin sheet metal, such as those shown in Figure 10.20, are now commonly used for stud and joist construction employing predominantly wood members of 2 in. nominal dimension thickness. As with the devices used for heavy timber construction, these lighter connectors often serve useful functions of tying and anchoring the structural members. Load transfers between basic elements of a building's lateral

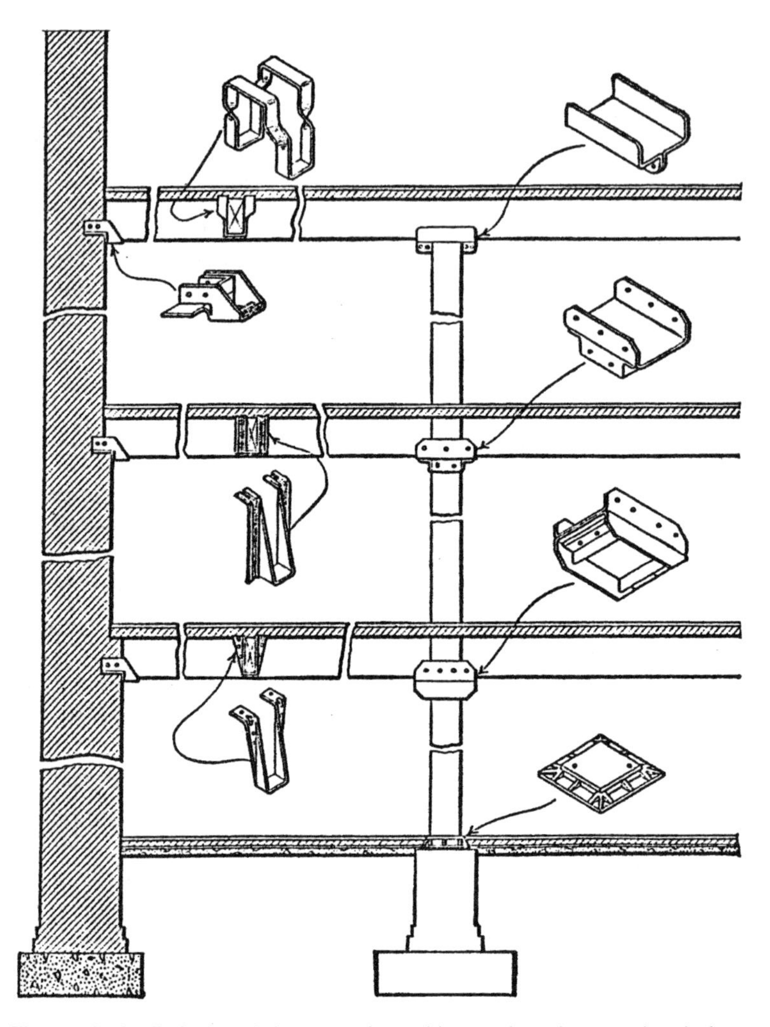

Figure 10.18 Early twentieth century formed iron and steel connecting devices in timber construction. (Reproduced from *Architects and Builders Handbook*, with permission of the publishers, John Wiley & Sons, New Jersey.)

bracing system are often achieved with these elements. See discussion in Chapter 13.

Commonly used connection devices of both the light sheet steel type and the heavier steel plate type are readily available from building material suppliers. Many of these devices are approved by building

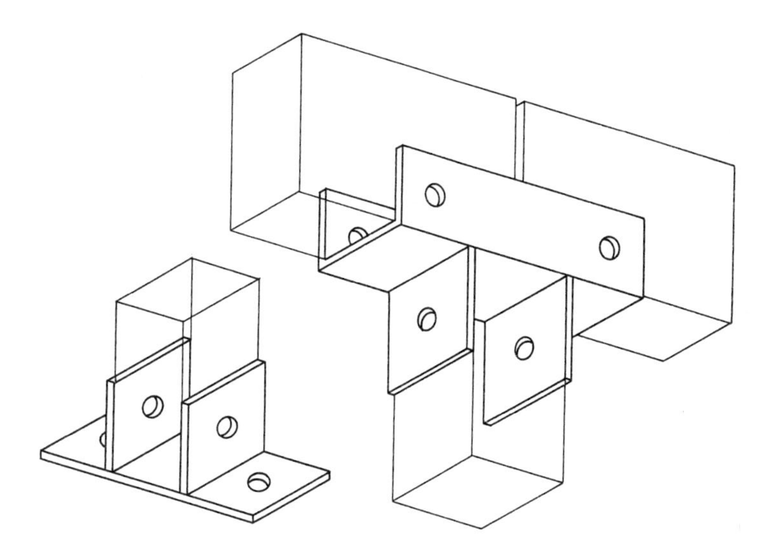

Figure 10.19 Simple connecting devices formed from bent and welded steel plates.

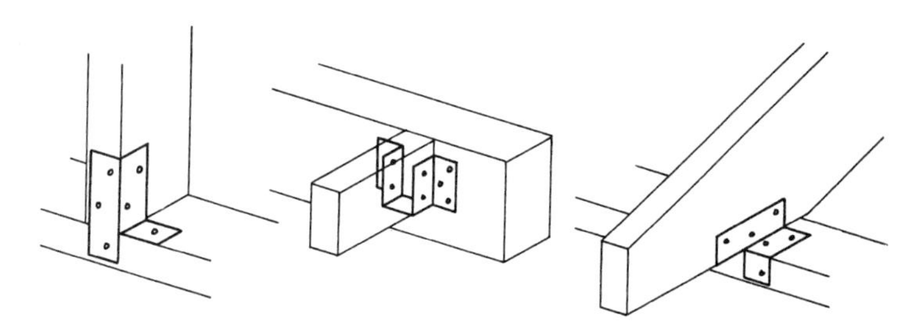

Figure 10.20 Connecting devices used for light wood frame construction, formed from bent sheet steel.

codes for rated structural capacity functions. Information should be obtained from local suppliers for these items.

For special situations it may be necessary to design a custom-formed framing device. Such devices can often be made by local metal fabricating shops. However, the catalogs of manufacturers of these devices are filled with a considerable variety of products for all kinds of situations, and it is wise to first determine that the required device is not available as a standard hardware item. Custom-designed devices are likely to be expensive and may present a problem for code approval.

10.8 CONCRETE AND MASONRY ANCHORS

Wood members supported by concrete or masonry structures must usually be anchored through some intermediate device. The most common attachment is with steel bolts cast into the concrete or masonry. However, there is also a wide variety of devices that may be directly cast into the supports or attached with drilled-in, dynamically-anchored, or other elements.

Two common situations are those shown in Figure 10.21. The sill member for a wood stud is typically attached directly with steel anchor bolts that are cast into the supports. These bolts serve to hold the wall securely in position during the construction process. However, they may also serve to anchor the wall against lateral or uplift forces.

Figure 10.21*b* shows a common situation in which a wood-framed roof or floor is attached to a masonry wall through a member bolted to the face of the wall, called a *ledger*. For vertical load transfer the shear effect on the bolt is essentially as described in Section 10.1. For lateral force a problem is the pullout or tension effect on the bolt, although another problem may be the cross-grain bending in the ledger. In zones of high seismic risk, it is usually required to have a separate *horizontal anchor*, such as the strap shown in Figure 10.21*b*.

10.9 PLYWOOD GUSSETS

Cut pieces of plywood are sometimes used as connecting devices, although the availability of manufactured metal devices is widespread. Light trusses consisting of a single plane of wood members of 2 in. nominal thickness are sometimes assembled with gussets of plywood.

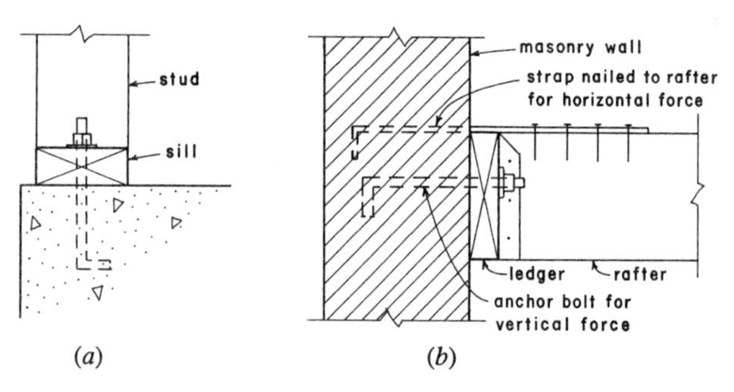

Figure 10.21 Devices for anchoring wood structures to concrete and masonry supports.

Although such connections may have considerable load resistance, it is best to be conservative in using them for computed structural forces, especially with regard to tension stress in the plywood. The following example treats a joint for a light truss using lumber members for the truss and connecting panels of structural grade plywood.

Example 6. The truss heel joint shown in Figure 10.22 is made with 2 in. nominal thickness lumber and gusset plates of $^1/_2$-in.-thick plywood. Nails are 6d common wire with the nail layout shown occurring

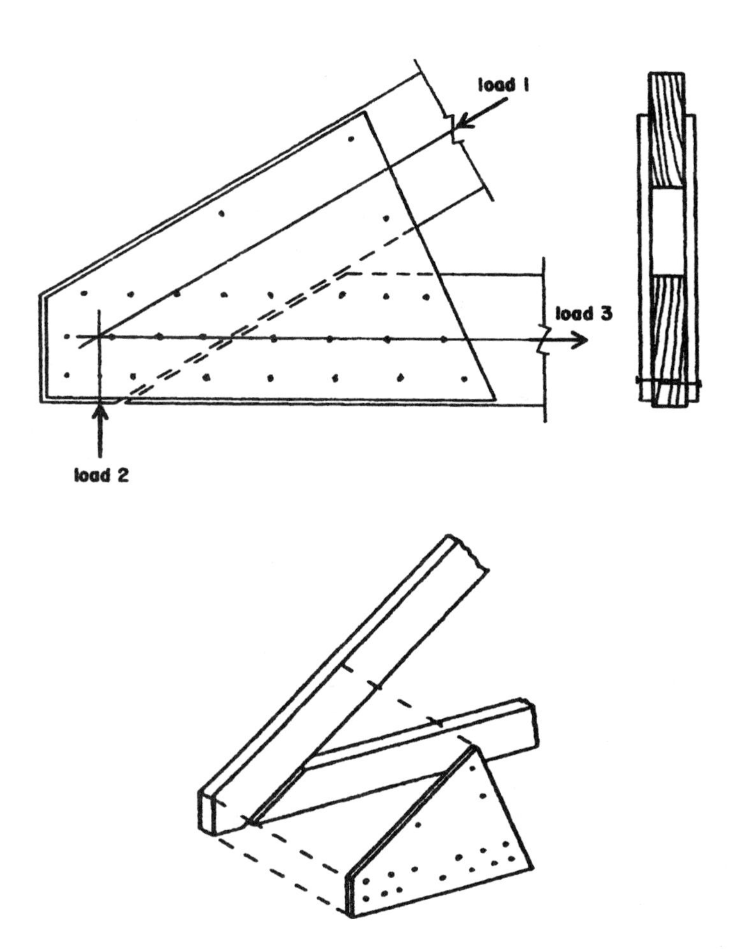

Figure 10.22 Reference for Example 6.

on both sides of the joint. Find the tension load capacity for the bottom chord member (load 3 in the figure).

Solution: From Table 10.2 the capacity of one nail is 50 lb. With 12 nails on each side of the joint, the total capacity of the joint is thus

$$T = (24)(50) = 1200\,\text{lb}$$

Problem 10.9.A. A truss heel joint similar to that in Figure 10.22 is made with gusset plates of $^1/_2$-in. plywood and 8d nails. Find the tension force limit for the bottom chord.

Problem 10.9.B. A truss heel joint similar to that in Figure 10.22 is made with $^3/_4$-in. plywood and 10d nails. Find the tension force limit for the bottom chord.

10.10 INVESTIGATION OF CONNECTIONS, LRFD

Use of the LRFD method for connections involves the same basic procedures as for the ASD method. Reference values are adjusted by the format conversion factor, $K_F = 2.16/1000\phi_z$, for which $\phi_z = 0.65$. Other adjustment factors for loads and resistance are used appropriate to the load combinations, moisture conditions, etc.

The following example illustrates the process for a bolted joint. Data for this example is the same as that used for the ASD solution in Example 1.

Example 7. A three-member (double shear) joint is made with members of Douglas fir-larch, select structural grade lumber. (See Figure 10.5.) The joint is loaded as shown, with the load parallel to the grain direction in the members. The tension force is 4 kips dead load and 5 kips live load. The middle member (designated main member in Table 10.1) is a 3 × 12 and the outer members (side members in Table 10.1) are each 2 × 12. Is the joint adequate for the required load if it is made with four $^3/_4$ in. bolts as shown? No adjustment is required for load duration or moisture, and it is assumed that the layout dimensions are adequate for use of the full design values for the bolts.

Solution: From the ASD work in Example 1, $Z = 2400$ lb/bolt, $F_t = 1000$ psi, and the net cross section of the middle member is 24.06 in.^{2}.

For the LRFD work, $\lambda = 0.8$, $\phi_z = 0.65$, and the stress value adjustment factor is $2.16/1000\phi_z$. For the bolts:

$$\lambda\phi_Z Z' = \lambda\phi_Z\left(\frac{2.16}{1000\phi_Z}\right)Z = 0.8(0.65)\left(\frac{2.16}{1000(0.65)}\right)(2400)$$
$$= 4.147 \text{ kips/bolt}$$

The total usable load based on the four bolts is thus

$$T = 4(4.147) = 16.588 \text{ kips}$$

On the net section of the middle member the usable tension is

$$\lambda\phi_t T' = \lambda\phi_t\left(\frac{2.16}{1000\phi_t}\right)F_t A_{net} = 0.8(0.80)\left(\frac{2.16}{1000(0.80)}\right)(1000)(24.06)$$
$$= 41.58 \text{ kips}$$

These capacities are compared to the required ultimate tension load, determined as

$$T_u = 1.2(4) + 1.6(5) = 12.8 \text{ kips}$$

And, as with the ASD solution, the joint is adequate.

11

TRUSSES

Wood trusses are used for a wide range of applications. They are used mostly for roof structures owing to the lighter loads, more frequent use of longer spans, generally reduced requirements for fire resistance, and the ability of trusses to accommodate many roof profile forms. This chapter presents a general discussion of issues relating to the use of trusses that utilize wood elements. Design of trusses for various situations is presented in the example in Chapter 15. Trussing is also used for lateral bracing for wind and earthquakes, which is discussed in Chapter 13.

11.1 GENERAL CONSIDERATIONS

A historically common use of the truss is to achieve the simple, double-slope, gabled roof form. This is typically done by use of sloping top members and a horizontal bottom member, as shown in Figure 11.1. Depending on the size of the span, the interior of the simple triangle formed by these three members may be filled by various arrangements

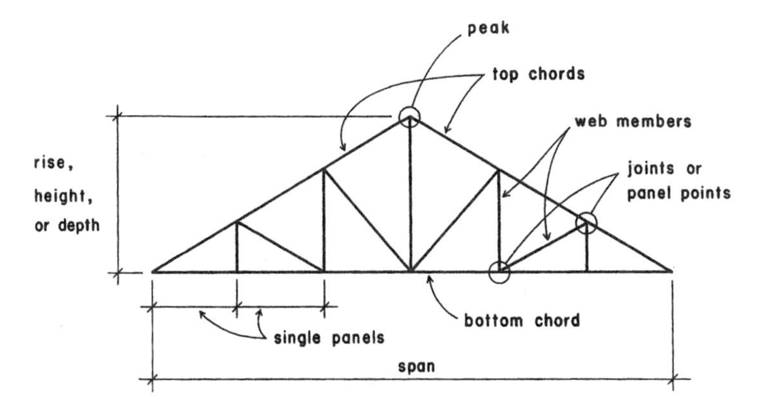

Figure 11.1 Elements of planar trusses.

of triangulated members. Some of the terminology used for the components of such a truss, as indicated in Figure 11.1, are as follows:

Chord Members. These are the top and bottom boundary members of the truss, analogous to the top and bottom flanges of a steel beam. For trusses of modest size these members are often made of a single element that is continuous through several joints, with a total length limited only by the maximum piece ordinarily obtainable from suppliers for the element selected.

Web Members. The interior members of the truss are called web members. Unless there are interior joints, these members are of a single piece between chord joints.

Panels. Most trusses have a pattern that consists of some repetitive modular unit. This unit ordinarily is referred to as the panel of the truss; joints are sometimes referred to as panel points.

A critical dimension of a truss is its overall height, which is sometimes referred to as its rise or its depth. For the truss illustrated, this dimension relates to the establishment of the roof pitch and also determines the length of the web members. A critical concern with regard to the efficiency of the truss as a spanning structure is the ratio of the span of the truss to its height. Although beams and joists may be functional with span/height ratios as high as 20 to 30, trusses generally require much lower ratios.

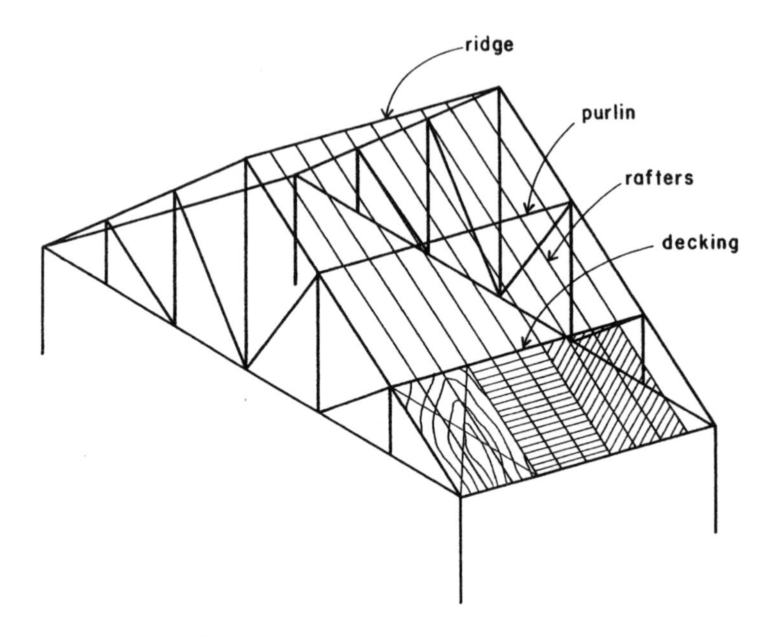

Figure 11.2 A roof structure with trusses.

Trusses may be used in a number of ways as part of the total structural system for a building. Figure 11.2 shows a series of single-span, planar trusses of the form shown in Figure 11.1 with the other elements of the building structure that develop the roof system and provide support for the trusses. In this example the trusses are spaced a considerable distance apart. In this situation it is common to use purlins to span between the trusses, usually supported at the top chord joints of the trusses to avoid bending in the chords. The purlins, in turn, support a series of closely spaced rafters that are parallel to the trusses. The roof deck is then attached to the rafters so that the roof surface actually floats above the level of the top of the trusses.

Figure 11.3 shows a similar structural system of trusses with parallel chords. This system may be used for a floor or a flat roof.

When trusses are closer together, it may be more practical to eliminate the purlins and to increase the size of the top chords to accommodate the additional bending due to the rafters. As an extension of this idea, if the trusses are really close, it may be possible to eliminate the rafters as well and to place the deck directly on the top chords of the trusses.

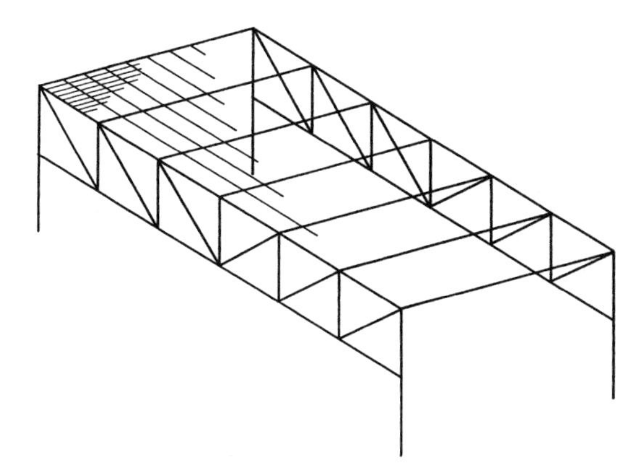

Figure 11.3 Flat-spanning, parallel-chorded trusses.

For various situations additional elements may be required for the complete structural system. If a ceiling is required, another framing system is used at the level of the bottom chords or suspended some distance below it. If the supported roof or floor deck and framing and the ceiling framing do not provide it adequately, it may be necessary to use some bracing system perpendicular to the trusses in order to brace the laterally unstable trusses.

11.2 TYPES OF TRUSSES

Figure 11.4 shows some of the common forms for roof trusses. While the bottom chord of a truss is most often horizontal, the top chords most often need to accommodate water drainage, and the common truss profiles reflect this need. The height or *rise* of a truss divided by the span is called the *pitch*; the rise of a symmetrical gable-form truss (of triangular profile) divided by half the span is the *slope*, which is the tangent of the angle of slope of the top chord. Unfortunately, these two terms are often used interchangeably to define the slope. Another way of expressing the slope is to give the amount of rise per foot of the span. A roof that rises 6 in. in a horizontal distance of 12 in. has a slope of "6 in 12." Reference to Table 11.1 should clarify this terminology.

11.3 BRACING FOR TRUSSES

Single planar trusses are very thin structures that require some form of lateral bracing. The compression chord of the truss must be designed

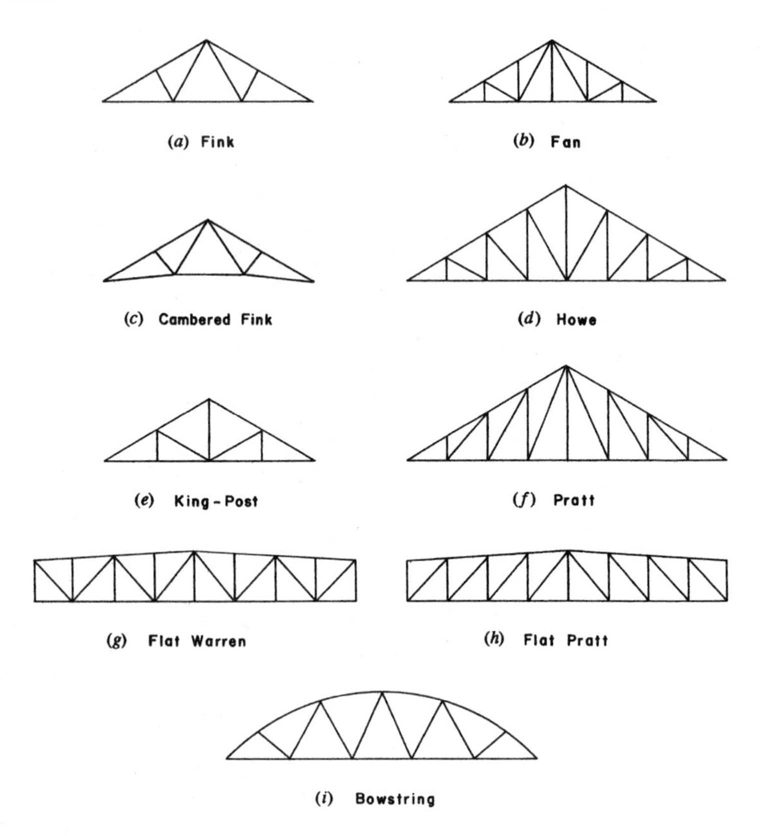

Figure 11.4 Common forms for wood trusses.

TABLE 11.1 Roof Pitches and Slopes

Pitch Rise/Span	1/8	1/6	1/5	1/4	1/3.46	1/3	1/2
Slope Degrees	14° 3′	18° 26′	21° 48′	26° 34′	30° 0′	33° 0′	45° 0′
Slope Ratio	3 in 12	4 in 12	4.8 in 12	6 in 12	6.92 in 12	8 in 12	12 in 12

for its laterally unbraced length. In the plane of the truss, the chord is braced by other truss members at each joint. However, if there is no lateral bracing, the unbraced length of the chord in a direction perpendicular to the plane of the truss becomes the full length of the truss. Obviously, it is not feasible to design a slender compression member for this unbraced length.

In most buildings other elements of the construction ordinarily provide some or all of the necessary bracing for the trusses. In the structural system shown in Figure 11.5*a*, the top chord of the truss is braced at each truss joint by the purlins. If the roof deck is a reasonably rigid planar structural element and is adequately attached to the purlins, this constitutes a very adequate bracing of the compression chord—which is the main problem for the truss. However, it is also necessary to brace the truss generally for out-of-plane movement throughout its height. In Figure 11.5*a* this is done by providing a vertical plane of X-bracing at every other panel point of the truss. The purlin does an additional service by serving as part of this vertical plane of trussed bracing. One panel of this bracing is actually capable of bracing a pair of trusses, so that it would be possible to place it only in alternate bays between the trusses. However, the bracing may be part of the general bracing system for the building, as well as providing for the bracing of the individual trusses. In the latter case, it would probably be continuous.

Light trusses that directly support a deck, as shown in Figure 11.5*b*, are usually adequately braced at the top chord level by the deck. This constitutes continuous bracing, so that the unbraced length of the chord in this case may be virtually zero (depending on the form of attachment of the deck). Additional bracing in this situation often is limited to a series of continuous steel rods or single small angles that are attached to the bottom chords as shown in the illustration.

Another form of bracing that is used is that shown in Figure 11.5*c*. In this case a horizontal plane of X-bracing is placed between two trusses at the level of the bottom chords. This single braced bay may be used to brace several other bays of trusses by connecting them to the X-braced trusses with horizontal struts. As in the previous example, with vertical planes of bracing, the top chord is braced by the roof construction. It is likely that bracing of this form is also part of the general lateral bracing system for the building so that its use, location, and details are not developed strictly for the bracing of the trusses.

11.4 LOADS ON TRUSSES

The first step in the design of a roof truss consists of computing the loads the truss will be required to support. These are dead and live loads. The former includes the weight of all construction materials supported by the truss; the latter includes loads resulting from snow and wind, and, on flat roofs, occupancy loads and an allowance for the possible ponding of water due to impaired drainage.

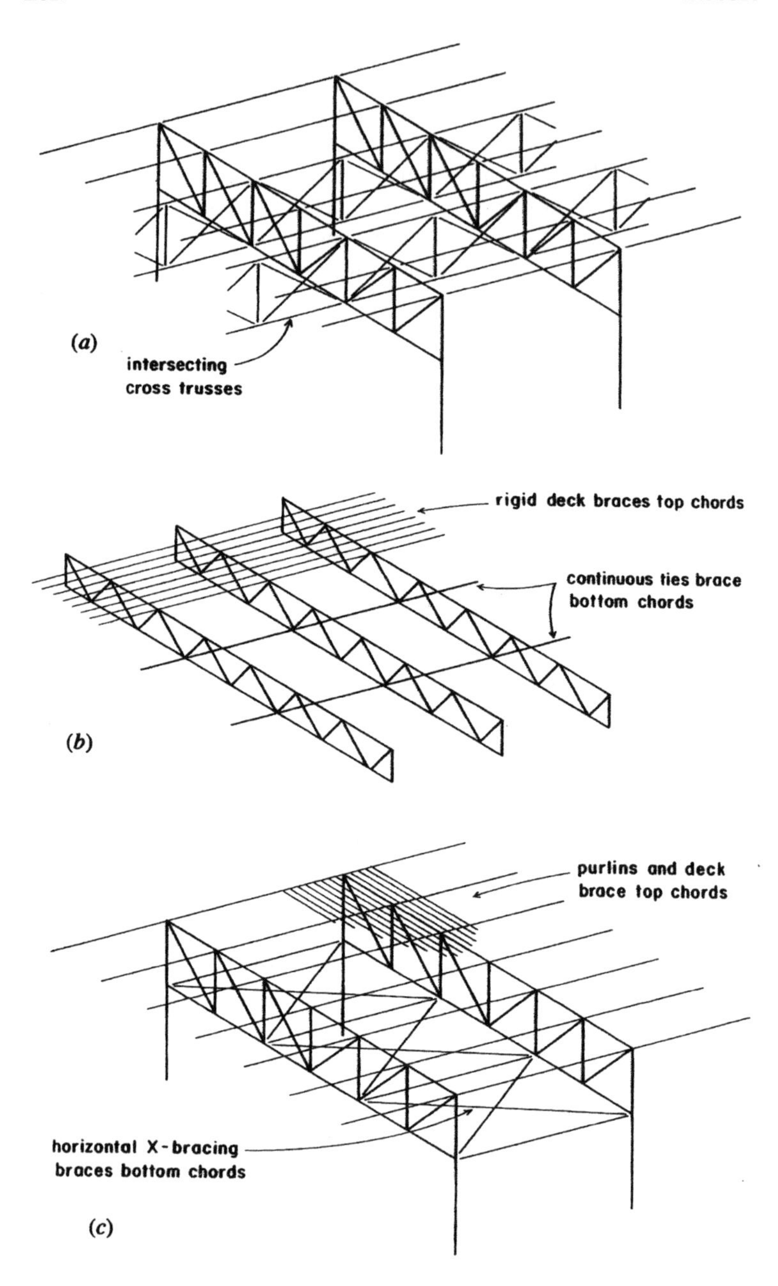

Figure 11.5 Forms of lateral bracing for trusses.

TABLE 11.2 Approximate Weight of Wood Trusses in Pounds per Square Foot of Supported Roof Surface

Span		Slope of Roof			
ft	m	45°	30°	20°	Flat
Up to 40	Up to 12	5	6	7	8
40–50	12–15	6	7	7	8
50–65	15–20	7	8	9	10
65–80	20–25	9	9	10	11

Table 11.2 provides estimated weights of wood trusses for various spans and pitches. With respect to the latter, one procedure is to establish an estimate in pounds per square foot of roof surface and consider this load as acting at the panel points of the upper chord. After the truss has been designed, its actual weight may be computed and compared with the estimated weight.

Required general design live loads for roofs are specified by local building codes. When snow is a potential problem, the load is usually based on anticipated snow accumulation. Otherwise the specified live load is intended essentially to provide some capacity for sustaining loads experienced during construction and maintenance of the roof. The basic required load can usually be modified when the roof slope is of some significant angle and on the basis of the total roof surface area supported by the structure.

For a general explanation of the analysis and design of the effects of wind and earthquake forces on buildings, the reader is referred to *Simplified Building Design for Wind and Earthquake Forces* (Ref.11).

11.5 INVESTIGATION FOR INTERNAL FORCES IN PLANAR TRUSSES

Planar trusses, composed of linear elements assembled in triangulated frameworks, have been used for spanning structures in buildings for many centuries. Investigation for internal forces in trusses may be accomplished by using either algebraic or graphical methods of solution.

Graphical Analysis of Planar Trusses

When the so-called *method of joints* is used, finding the internal forces in the members of a planar truss consists of solving a series of

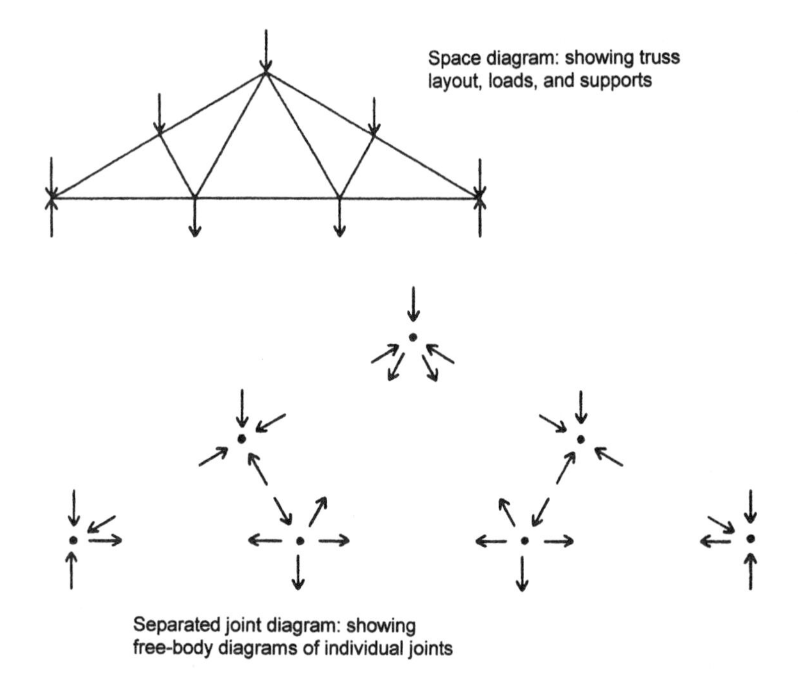

Figure 11.6 Examples of diagrams used to represent trusses and their actions.

concurrent force systems. Figure 11.6 shows a truss with the truss form, the loads, and the reactions displayed in a *space diagram*. Below the space diagram is a figure consisting of the free-body diagrams of the individual joints of the truss. These are arranged in the same manner as they are in the truss in order to show their interrelationships. However, each joint constitutes a complete concurrent planar force system that must have its independent equilibrium. Solving the problem consists of determining the equilibrium conditions for all of the joints. The procedures used for this solution will now be illustrated.

Figure 11.7 shows a single-span, planar truss subjected to gravity loads. This example will be used to illustrate the procedures for determining the internal forces in the truss: that is, the tension and compression forces in the members of the truss. The space diagram in the figure shows the truss form and dimensions, the support conditions, and the loads. The letters on the space diagram identify forces at the truss joints. The sequence of placement of the letters is arbitrary, the only consideration being to place a letter in each space between

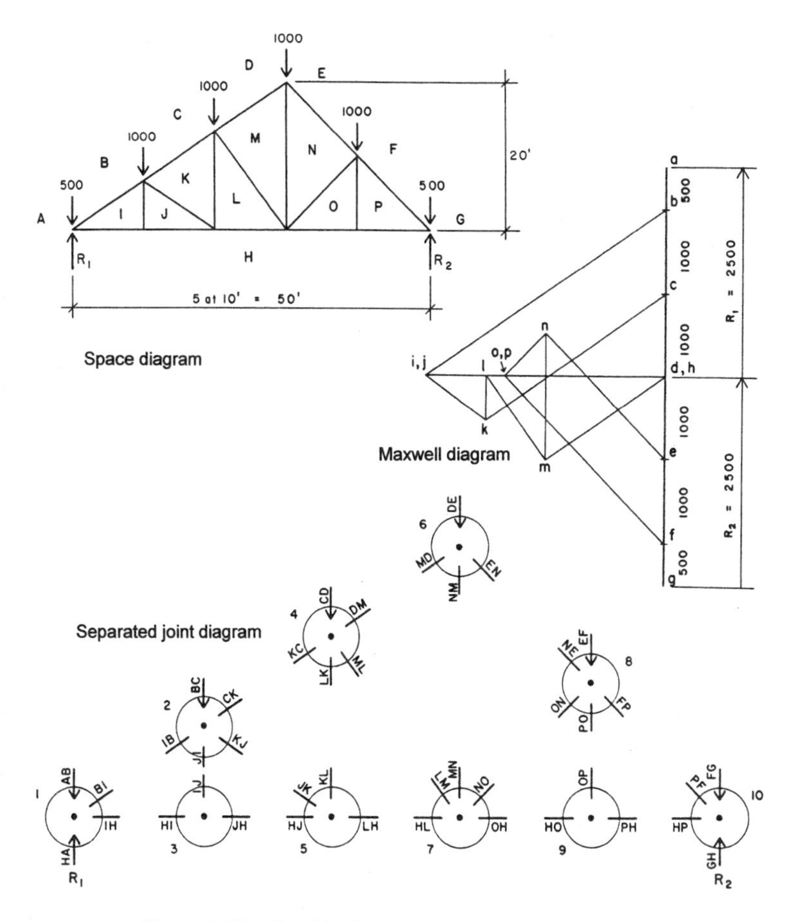

Figure 11.7 Graphic diagrams for the sample problem.

the loads and the individual truss members so that each force at a joint can be identified by a two-letter symbol.

The separated joint diagram in the figure provides a useful means for visualization of the complete force system at each joint as well as the interrelation of the joints through the truss members. The individual forces at each joint are designated by two-letter symbols that are obtained by simply reading around the joint in the space diagram in a clockwise direction. Note that the two-letter symbols are reversed at the opposite ends of each of the truss members. Thus the top chord member at the left end of the truss is designated as *BI* when shown in the joint at the left support (joint 1) and is designated as *IB* when shown in the first interior upper chord joint (joint 2). The purpose of this procedure will be demonstrated in the following explanation of the graphical analysis.

The third diagram in Figure 11.7 is a composite force polygon for the external and internal forces in the truss. It is called a Maxwell diagram after one of its early promoters, James Maxwell, a British engineer. The construction of this diagram constitutes a complete solution for the magnitudes and senses of the internal forces in the truss. The procedure for this construction is as follows.

Construct the force polygon for the external forces. Before this can be done, the values for the reactions must be found. Next a convenient scale is selected for construction of the force polygons—such as one inch equals 1000 lb. There are graphic techniques for finding the reactions, but it is usually much simpler and faster to find them with an algebraic solution. In this example, although the truss is not symmetrical, the loading is, and it may simply be observed that the reactions are each equal to one half of the total load on the truss, or $5000/2 = 2500$ lb. Since the external forces in this case are all in a single direction, the force polygon for the external forces is actually a straight line. Using the two-letter symbols for the forces and starting with the letter *A* at the left end, we read the force sequence by moving in a clockwise direction around the outside of the truss. The loads are thus read as *AB, BC, CD, DE, EF*, and *FG*, and the two reactions are read as *GH* and *HA*. Beginning at *A* on the Maxwell diagram, the force vector sequence for the external forces is read from *A* to *B, B* to *C, C* to *D*, and so on, ending back at *A*, which shows that the force polygon closes and the external forces are in the necessary state of static equilibrium. Note that we have pulled the vectors for the reactions off to the side in the diagram to indicate them more clearly. Note also that we have used lowercase letters for the vector ends in the Maxwell diagram, whereas uppercase letters are used on the space diagram. The alphabetic correlation is thus retained (*A* to *a*), while any possible confusion between the two diagrams is prevented. The letters on the space diagram designate open spaces, while the letters on the Maxwell diagram designate points of intersection of lines.

Construct the force polygons for the individual joints. The graphic procedure for this consists of locating the points on the Maxwell diagram that correspond to the remaining letters, *I* through *P*, on the space diagram in Figure 11.7. When all the lettered points on the diagram are located, the complete force polygon for each joint may be read on the diagram. In order to locate these points, we use two relationships. The first is that the truss members can resist only forces that are parallel to the members' positioned directions. Thus we know the directions of all the internal forces. The second relationship is a simple one from plane geometry: a point may be located at the intersection of two lines. Consider the forces at joint 1, as shown in the separated joint diagram in Figure 11.7. Note that there are four forces and that two of them are known (the load and the reaction) and two are unknown (the internal forces in the truss members). The force polygon for this joint, as shown on the Maxwell diagram, is read as *ABIHA*. *AB* represents the load, *BI* the force in the upper chord member, *IH* the force in the lower chord member, and *HA* the reaction. Thus the location of point *i* on the Maxwell diagram is determined by noting that i must be in a horizontal direction from *h* (corresponding to the horizontal position of the lower chord) and in a direction from *b* that is parallel to the position of the upper chord.

The remaining points on the Maxwell diagram are found by the same process, using two known points on the diagram to project lines of known direction whose intersection will determine the location of an unknown point. Once all the points are located, the diagram is complete and can be used to find the magnitude and sense of each internal force. The process for construction of the Maxwell diagram typically consists of moving from joint to joint along the truss. Once one of the letters for an internal space is determined on the Maxwell diagram, it may be used as a known point for finding the letter for an adjacent space on the space diagram. The only limitation of the process is that it is not possible to find more than one unknown point on the Maxwell diagram for any single joint. Consider joint 7 on the separated joint diagram in Figure 11.7. To solve this joint first, knowing only the locations of letters *a* through *h* on the Maxwell diagram, it is necessary to locate four unknown points: *l, m, n*, and *o*. This is three more unknowns than can be determined in a single step, so three of the unknowns must be found by using other joints.

Solving for a single unknown point on the Maxwell diagram corresponds to finding two unknown forces at a joint, since each letter on the space diagram is used twice in the force identification for the internal forces. Thus for joint 1 in the previous example, the letter *I* is part of the identity of forces *BI* and *IH*, as shown on the separated joint diagram. The graphic determination of single points on the Maxwell diagram, therefore, is analogous to finding two unknown quantities in an algebraic solution. Two unknowns are the maximum that can be solved for the equilibrium of a coplanar, concurrent force system, which is the condition of the individual joints in the truss.

When the Maxwell diagram is completed, the internal forces can be read from the diagram as follows:

The magnitude is determined by measuring the length of the line in the diagram, using the scale that was used to plot the vectors for the external forces.

The sense of individual forces is determined by reading the forces in clockwise sequence around a single joint in the space diagram and tracing the same letter sequences on the Maxwell diagram.

Figure 11.8*a* shows the force system at joint 1 and the force polygon for these forces as taken from the Maxwell diagram. The forces known initially are shown as solid lines on the force polygon, and the unknown forces are shown as dashed lines. Starting with letter *A* on the force system, we read the forces in a clockwise sequence as *AB, BI, IH*, and *HA*. Note that on the Maxwell diagram moving from *a* to *b* is moving in the order of the sense of the force, that is from tail to head of the force vector arrow that represents the external load on the joint. Using this sequence on the Maxwell diagram, this force sense flow will be a continuous one. Thus reading from *b* to *i* on the Maxwell diagram is reading from tail to head of the force vector, which indicates that force *BI* has its head at the left end. Transferring this sense indication from the Maxwell diagram to the joint diagram indicates that force *BI* is in compression; that is, it is pushing, rather than pulling, on the joint. Reading from *i* to *h* on the Maxwell diagram shows that the arrowhead for this vector is on the right, which translates to a tension effect on the joint diagram.

Having solved for the forces at joint 1 as described, the fact that the forces in truss members *BI* and *IH* are known can be used to consider the adjacent joints, 2 and 3. However, it should be noted that the sense reverses at the opposite ends of the members in the joint diagrams. Referring to the separated joint diagram in Figure 11.7, if the upper chord member shown as force *BI* in joint 1 is in compression, its arrowhead is at the lower left end in the diagram for joint 1, as shown in

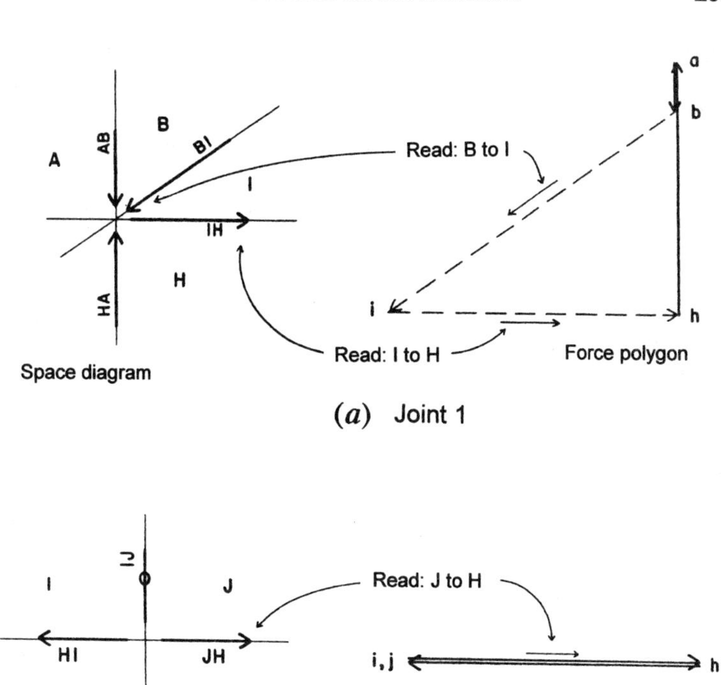

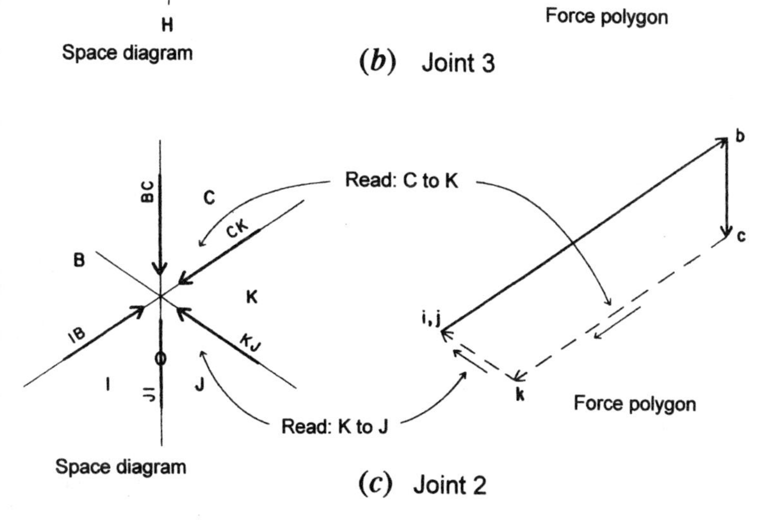

Figure 11.8 Graphic solutions for joints 1, 2, and 3.

Figure 11.8*a*. However, when the same force is shown as *IB* at joint 2, its pushing effect on the joint will be indicated by having the arrowhead at the upper right end in the diagram for joint 2. Similarly, the tension effect of the lower chord is shown in joint 1 by placing the arrowhead on the right end of the force *IH*, but the same tension force will be indicated in joint 3 by placing the arrowhead on the left end of the vector for force *HI*.

If the solution sequence of solving joint 1 and then joint 2 is chosen, it is now possible to transfer the known force in the upper chord to joint 2. Thus the solution for the five forces at joint 2 is reduced to finding three unknowns, since the load *BC* and the chord force *IB* are now known. However, it is still not possible to solve joint 2, since there are two unknown points on the Maxwell diagram (*k* and *j*) corresponding to the three unknown forces. An option, therefore, is to proceed from joint 1 to joint 3, at which there are presently only two unknown forces. On the Maxwell diagram the single unknown point *j* can be found by projecting vector *IJ* vertically from *i* and projecting vector *JH* horizontally from point *h*. Since point *i* is also located horizontally from point *h*, this shows that the vector *IJ* has zero magnitude, since both *i* and *j* must be on a horizontal line from *h* in the Maxwell diagram. This indicates that there is actually no stress in this truss member for this loading condition and that points *i* and *j* are coincident on the Maxwell diagram. The joint force diagram and the force polygon for joint 3 are as shown in Figure 11.8*b*. In the joint force diagram place a zero, rather than an arrowhead, on the vector line for *IJ* to indicate the zero stress condition. In the force polygon in Figure 11.8*b*, the two force vectors are slightly separated for clarity, although they are actually coincident on the same line.

Having solved for the forces at joint 3, proceed to joint 2, since there remain only two unknown forces at this joint. The forces at the joint and the force polygon for joint 2 are shown in Figure 11.8*c*. As for joint 1, read the force polygon in a sequence determined by reading clockwise around the joint: *BCKJIB*. Following the continuous direction of the force arrows on the force polygon in this sequence, it is possible to establish the sense for the two forces *CK* and *KJ*.

It is possible to proceed from one end and to work continuously across the truss from joint to joint to construct the Maxwell diagram in this example. The sequence in terms of locating points on the Maxwell diagram would be *i-j-k-l-m-n-o-p*, which would be accomplished by solving the joints in the following sequence: 1,3,2,5,4,6,7,9,8. However, it is advisable to minimize the error in graphic construction by working from both ends of the truss. Thus a better procedure would be to find points *i-j-k-l-m*, working from the left end of the truss, and then to find points *p-o-n-m*, working from the right end. This would result in finding two locations for *m*, whose separation constitutes the error in drafting accuracy.

Problems 11.5.A,B. Using a Maxwell diagram, find the internal forces in the truss in Figure 11.9.

Algebraic Analysis of Planar Trusses

Graphical solution for the internal forces in a truss using the Maxwell diagram corresponds essentially to an algebraic solution by the *method of joints*. This method consists of solving the concentric force systems at the individual joints using simple force equilibrium equations. The process will be illustrated using the previous example.

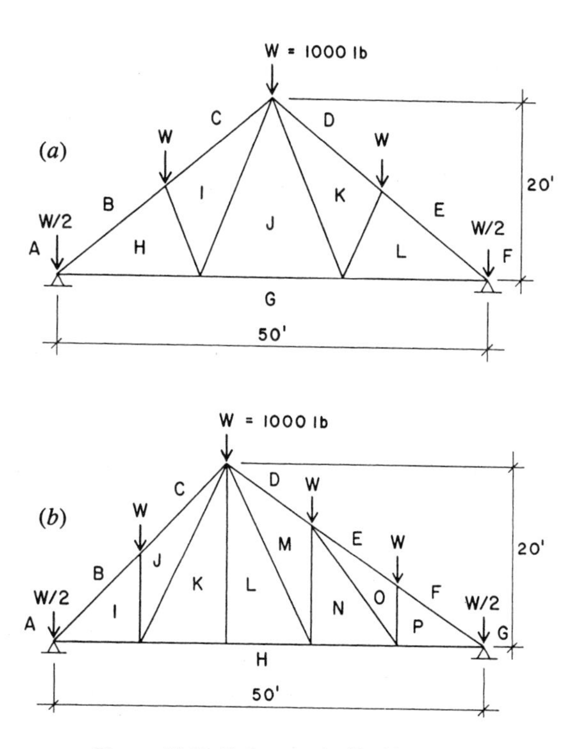

Figure 11.9 Reference for Problem 11.5.

As with the graphic solution, first determine the external forces, consisting of the loads and the reactions. Then proceed to consider the equilibrium of the individual joints, following a sequence as in the graphic solution. The limitation of this sequence, corresponding to the limit of finding only one unknown point in the Maxwell diagram, is that only two unknown forces at any single joint can be found in a single step. (Two conditions of equilibrium produce two equations.) Referring to Figure 11.10, the solution for joint 1 is as follows.

The force system for the joint is drawn with the sense and magnitude of the known forces shown, but with the unknown internal forces represented by lines without arrowheads, since their senses and magnitudes initially are unknown. For forces that are not vertical or horizontal, replace the forces with their horizontal and vertical components. Then consider the two conditions necessary for the equilibrium of the system: the sum of the vertical forces is zero, and the sum of the horizontal forces is zero.

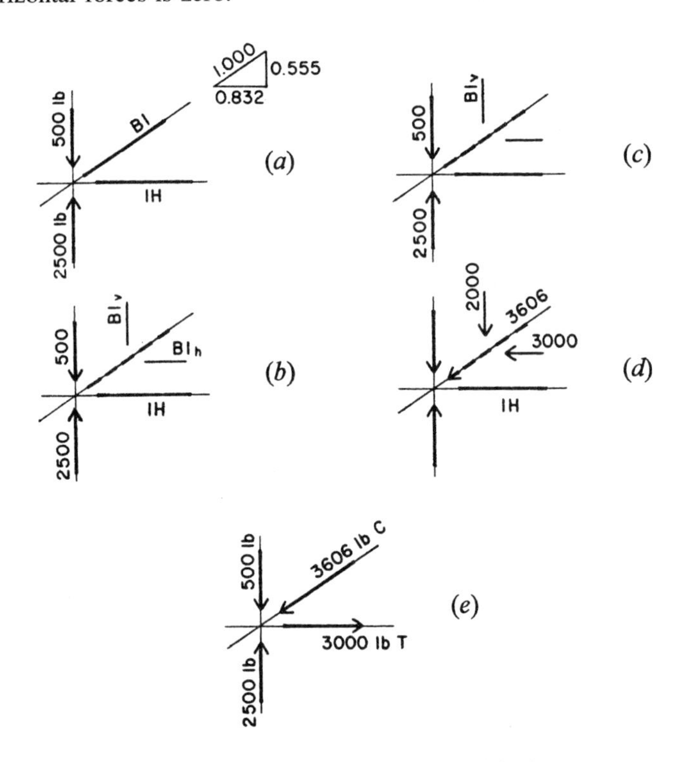

Figure 11.10 Algebraic solution for joint 1.

If the algebraic solution is performed carefully, the sense of the forces will be determined automatically. However, it is recommended that whenever possible the sense be predetermined by simple observations of the joint conditions, as will be illustrated in the solutions.

The problem to be solved at joint 1 is as shown in Figure 11.10*a*. In Figure 11.10*b* the system is shown with all forces expressed as vertical and horizontal components. Note that although this now increases the number of unknowns to three (*IH*, BI_v, and BI_h), there is a numeric relationship between the two components of *BI*. When this condition is added to the two algebraic conditions for equilibrium, the number of usable relationships totals three, so that the necessary conditions to solve for the three unknowns are present.

The condition for vertical equilibrium is shown at (*c*) in Figure 11.10. Since the horizontal forces do not affect the vertical equilibrium, the balance is between the load, the reaction, and the vertical component of the force in the upper chord. Simple observation of the forces and the known magnitudes makes it obvious that force BI_v must act downward, indicating that *BI* is a compression force. Thus the sense of *BI* is established by simple visual inspection of the joint, and the algebraic equation for vertical equilibrium (with upward force considered positive) is

$$\Sigma F_v = 0 = +2500 - 500 - BI_v$$

From this equation BI_v is determined to have a magnitude of 2000 lb. Using the known relationships between *BI*, BI_v, and BI_h, the values of these three quantities can be determined if any one of them is known. Thus

$$\frac{BI}{1.000} = \frac{BI_v}{0.555} = \frac{BI_h}{0.832}$$

from which

$$BI_h = \left(\frac{0.832}{0.555}\right) BI_v = \left(\frac{0.832}{0.555}\right)(2000) = 3000 \text{ lb}$$

and

$$BI = \left(\frac{1.000}{0.555}\right) BI_v = \left(\frac{1000}{0.555}\right)(2000) = 3606 \text{ lb}$$

The results of the analysis to this point are shown at (*d*) in Figure 11.10, from which it may be observed that the conditions for equilibrium of the horizontal forces can be expressed. Stated algebraically, (with force sense toward the right considered positive) the condition is

$$\Sigma F_h = 0 = IH - 3000$$

from which it is established that the force in *IH* is 3000 lb.

The final solution for the joint is then as shown at (*e*) in the figure. On this diagram the internal forces are identified as to sense by using *C* to indicate compression and *T* to indicate tension.

As with the graphic solution, proceed to consider the forces at joint 3. The initial condition at this joint is as shown at (*a*) in Figure 11.11, with the single known force in member *HI* and the two unknown forces in *IJ* and *JH*. Since the forces at this joint are all vertical and horizontal, there is no need to use components. Consideration of vertical equilibrium makes it obvious that it is not possible to have a force in member *IJ*. Stated algebraically, the condition for vertical equilibrium is

$$\Sigma F_v = 0 = IJ \text{ (since } IJ \text{ is the only force)}$$

It is equally obvious that the force in *JH* must be equal and opposite to that in *HI*, since they are the only two horizontal forces. That is, stated algebraically

$$\Sigma F_v = 0 = +JH - 3000$$

The final answer for the forces at joint 3 is as shown at (*b*) in Figure 11.11. Note the convention for indicating a truss member with no internal force.

Now proceed to consider joint 2; the initial condition is as shown at (*a*) in Figure 11.12. Of the five forces at the joint only two remain

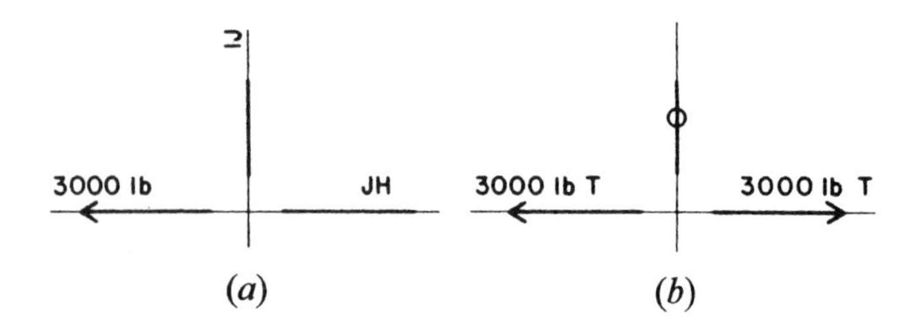

Figure 11.11 Algebraic solution for joint 3.

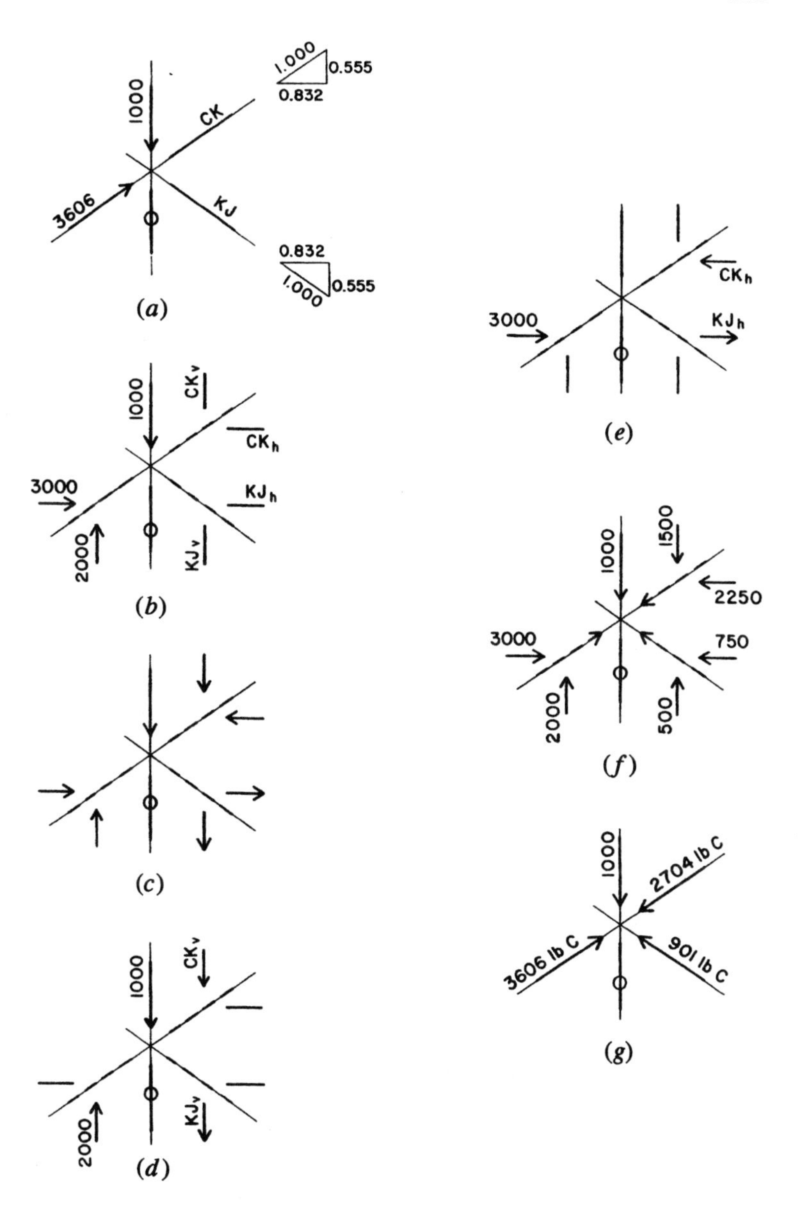

Figure 11.12 Algebraic solution for joint 2.

unknown. Following the procedure for joint 1, first resolve the forces into their vertical and horizontal components, as shown at (*b*) in Figure 11.12.

Since the sense of forces *CK* and *KJ* is unknown, use the procedure of considering them to be positive until proven otherwise. That is, if they are entered into the algebraic equations with an assumed sense, and the solution produces a negative answer, then the assumption was wrong. However, be careful to be consistent with the sense of the force vectors, as the following solution will illustrate.

Arbitrarily assume that force *CK* is in compression and force *KJ* is in tension. If this is so, the forces and their components will be as shown at (*c*) in Figure 11.12. Then consider the conditions for vertical equilibrium; the forces involved will be those shown at (*d*) in Figure 11.12, and the equation for vertical equilibrium will be

$$\Sigma F_v = 0 = -1000 + 2000 - CK_v - KJ_v$$

or

$$0 = +1000 - 0.555CK - 0.555KJ \qquad (11.5.1)$$

Now consider the conditions for horizontal equilibrium; the forces will be as shown at (*e*) in Figure 11.12, and the equation will be

$$\Sigma F_h = 0 = +3000 - CK_h + KJ_h$$

or

$$0 = +3000 - 0.832CK + 0.832KJ \qquad (11.5.2)$$

Note the consistency of the algebraic signs and the sense of the force vectors, with positive forces considered as upward and toward the right. Now solve these two equations simultaneously for the two unknown forces as follows

1. Multiply equation 11.5.1 by $\dfrac{0.832}{0.555}$

$$0 = \left(\frac{0.832}{0.555}\right)(+1000) + \left(\frac{0.832}{0.555}\right)(-0.555CK) + \left(\frac{0.832}{0.555}\right)(-0.555KJ)$$

or

$$0 = +1500 - 0.832CK - 0.832KJ$$

2. Add this equation to equation 11.5.2 and solve for *CK*.

$$0 = +4500 - 1.664CK, \qquad CK = \frac{4500}{1.664} = 2704 \text{ lb}$$

Note that the assumed sense of compression in *CK* is correct, since the algebraic solution produces a positive answer. Substituting this value for *CK* in equation 11.5.1,

$$0 = +1000 - 0.555(2704) - 0.555(KJ)$$

and

$$KJ = \frac{-500}{0.555} = -901 \text{ lb}$$

Since the algebraic solution produces a negative quantity for *KJ*, the assumed sense for *KJ* is wrong and the member is actually in compression.

The final answers for the forces at joint 2 are as shown at (*g*) in Figure 11.12. In order to verify that equilibrium exists, however, the forces are shown in the form of their vertical and horizontal components at (*f*) in the illustration.

When all of the internal forces have been determined for the truss, the results may be recorded or displayed in a number of ways. The most direct way is to display them on a scaled diagram of the truss, as shown in Figure 11.13*a*. The force magnitudes are recorded next to each member with the sense shown as *T* for tension or *C* for compression. Zero stress members are indicated by the conventional symbol consisting of a zero placed directly on the member.

When solving by the algebraic method of joints, the results may be recorded on a separated joint diagram, as shown in Figure 11.13*b*. If the values for the vertical and horizontal components of force in sloping members are shown, it is a simple matter to verify the equilibrium of the individual joints.

Problem 11.5.C, D. Using the algebraic method of joints, find the internal forces in the truss in Figure 11.9.

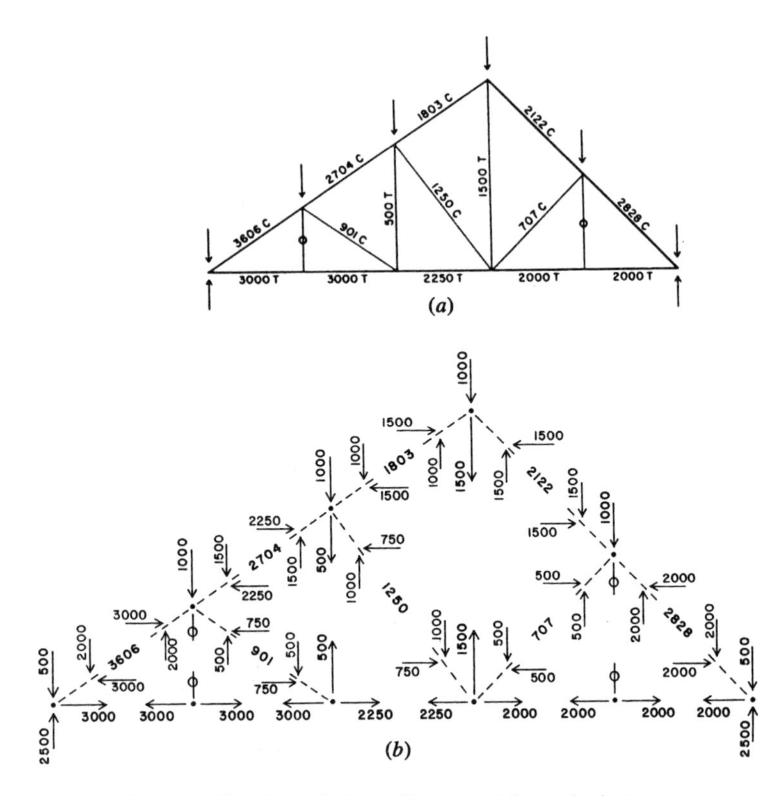

Figure 11.13 Presentation of the internal forces in the truss.

Internal Forces Found by Coefficients

Figure 11.14 shows a number of simple trusses of both parallel-chorded and gable form. Table 11.3 lists coefficients that may be used to find the values for the internal forces in these trusses. For the gable-form trusses, coefficients are given for three different slopes of the top chord: 4 in 12, 6 in 12, and 8 in 12. For the parallel-chorded trusses, coefficients are given for two different ratios of the truss depth to the truss panel length: 1 to 1 and 3 to 4. Loading results from gravity loads and is assumed to be applied symmetrically to the truss.

Values in Table 11.3 are based on a unit load of 1.0 for W, thus true forces may be simply proportioned for specific values of W once actual loading is determined. Since all the trusses are symmetrical, values are given for only half of each truss.

11.6 DESIGN FORCES FOR TRUSS MEMBERS

The primary concern in analysis of trusses is the determination of the critical forces for which each member of the truss must be designed. The first step in this process is the decision about which combinations of loading must be considered. In some cases the potential combinations may be quite numerous. When both wind and seismic actions are potentially critical and more than one type of live loading occurs (e.g., roof loads plus hanging loads), the theoretically possible combinations of loadings can be overwhelming. However, designers are usually able to exercise judgment in reducing the sensible combinations to a reasonable number; for example, it is statistically improbable that a violent windstorm will occur simultaneously with a major earthquake shock.

Once the required design loading conditions are established, the usual procedure is to perform separate analyses for each of the loadings. The values obtained can then be combined at will for each member to ascertain the particular combination that establishes the critical result for the member. This means that in some cases certain members will be designed for one combination, and others for different combinations.

What also must be considered are the factored values for the various load combinations, as described in Section 14.8. For the LRFD method, consideration must also be given to possible different values for the load combination factor λ and different values for resistance factors.

11.7 COMBINED ACTIONS IN TRUSS MEMBERS

When analyzing trusses the usual procedure is to assume that the loads will be applied to the truss joints. This results in the members themselves being loaded only through the joints and thus having only direct tension or compression forces. In some cases, however, truss members may be directly loaded; for example, when the top chord of a truss supports a roof deck without benefit of purlins or rafters. Thus the chord member is directly loaded with a linear uniform load and functions as a beam between its end joints.

The usual procedure in these situations is to accumulate the loads at the truss joints and analyze the truss as a whole for the typical joint-loading arrangement. The truss members that sustain the direct loading are then designed for the combined effects of the axial force caused by the truss action and the bending caused by the direct loading.

A typical situation for a roof truss is one in which the actual loading consists of the roof load distributed continuously along the top chords

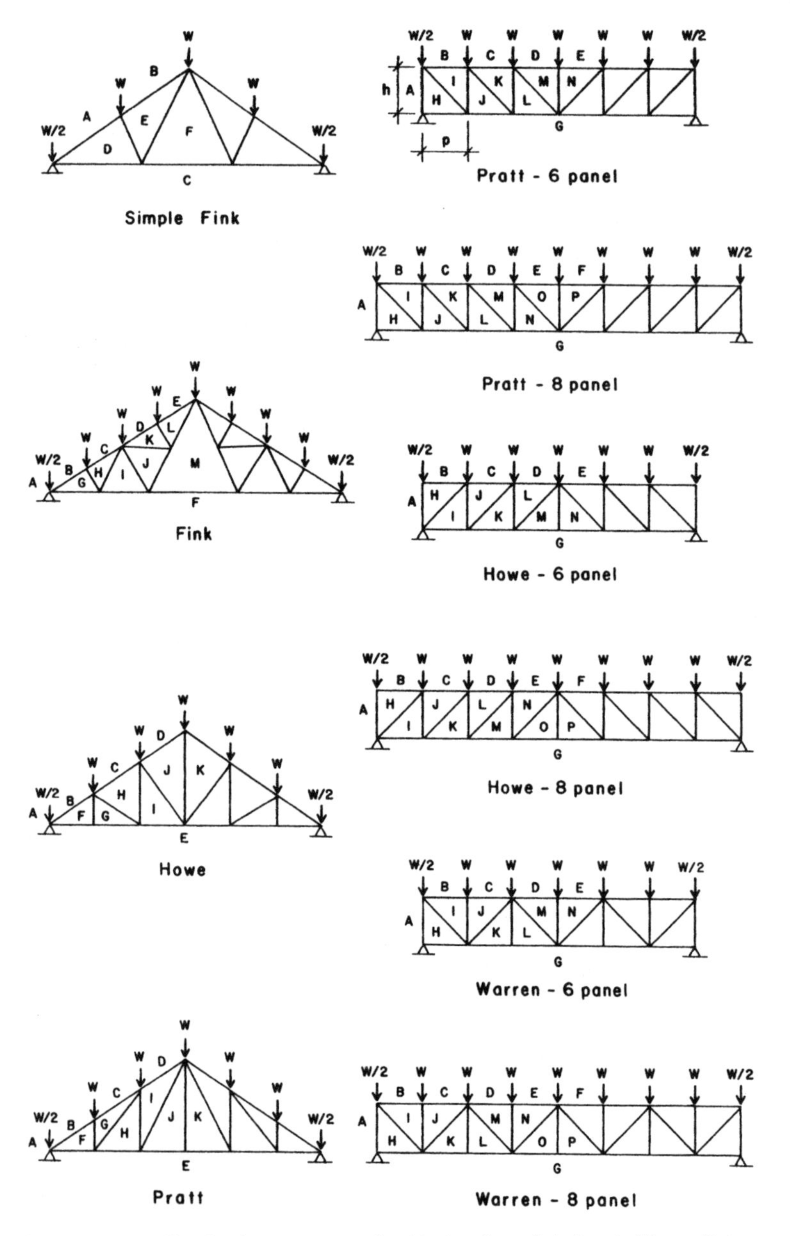

Figure 11.14 Simple planar trusses of gabled and parallel-chorded form. Reference for Table 11.3.

TABLE 11.3 Coefficients for Internal Forces in Simple Trusses[a]

Force in members = (table coefficient) X (panel load, W)

T indicates tension, C indicates compression

Gable Form Trusses

Truss Member	Type of Force	Roof Slope 4/12	6/12	8/12
Truss 1 - Simple Fink				
AD	C	4.74	3.35	2.70
BE	C	3.95	2.80	2.26
DC	T	4.50	3.00	2.25
FC	T	3.00	2.00	1.50
DE	C	1.06	0.90	0.84
EF	T	1.06	0.90	0.84
Truss 2 - Fink				
BG	C	11.08	7.83	6.31
CH	C	10.76	7.38	5.76
DK	C	10.44	6.93	5.20
EL	C	10.12	6.48	4.65
FG	T	10.50	7.00	5.25
FI	T	9.00	6.00	4.50
FM	T	6.00	4.00	3.00
GH	C	0.95	0.89	0.83
HI	T	1.50	1.00	0.75
IJ	C	1.90	1.79	1.66
JK	T	1.50	1.00	0.75
KL	C	0.95	0.89	0.83
JM	T	3.00	2.00	1.50
LM	T	4.50	3.00	2.25
Truss 3 - Howe				
BF	C	7.90	5.59	4.51
CH	C	6.32	4.50	3.61
DJ	C	4.75	3.35	2.70
EF	T	7.50	5.00	3.75
EI	T	6.00	4.00	3.00
GH	C	1.58	1.12	0.90
HI	T	0.50	0.50	0.50
IJ	C	1.81	1.41	1.25
JK	T	2.00	2.00	2.00
Truss 4 - Pratt				
BF	C	7.90	5.59	4.51
CG	C	7.90	5.59	4.51
DI	C	6.32	4.50	3.61
EF	T	7.50	5.00	3.75
EH	T	6.00	4.00	3.00
EJ	T	4.50	3.00	2.25
FG	C	1.00	1.00	1.00
GH	T	1.81	1.41	1.25
HI	C	1.50	1.50	1.50
IJ	T	2.12	1.80	1.68

Flat - Chorded Trusses

Truss Member	Type of Force	6 Panel Truss $\frac{h}{p}=1$	6 Panel Truss $\frac{h}{p}=\frac{3}{4}$	8 Panel Truss $\frac{h}{p}=1$	8 Panel Truss $\frac{h}{p}=\frac{3}{4}$
Truss 5 - Pratt					
BI	C	2.50	3.33	3.50	4.67
CK	C	4.00	5.33	6.00	8.00
DM	C	4.50	6.00	7.50	10.00
EO	C	–	–	8.00	10.67
GH	O	0	0	0	0
GJ	T	2.50	3.33	3.50	4.67
GL	T	4.00	5.33	6.00	8.00
GN	T	–	–	7.50	10.00
AH	C	3.00	3.00	4.00	4.00
IJ	C	2.50	2.50	3.50	3.50
KL	C	1.50	1.50	2.50	2.50
MN	C	1.00	1.00	1.50	1.50
OP	C	–	–	1.00	1.00
HI	T	3.53	4.17	4.95	5.83
JK	T	2.12	2.50	3.54	4.17
LM	T	0.71	0.83	2.12	2.50
NO	T	–	–	0.71	0.83
Truss 6 - Howe					
BH	O	0	0	0	0
CJ	C	2.50	3.33	3.50	4.67
DL	C	4.00	5.33	6.00	8.00
EN	C	–	–	7.50	10.00
GI	T	2.50	3.33	3.50	4.67
GK	T	4.00	5.33	6.00	8.00
GM	T	4.50	6.00	7.50	10.00
GO	T	–	–	8.00	10.67
AH	C	0.50	0.50	0.50	0.50
IJ	T	1.50	1.50	2.50	2.50
KL	T	0.50	0.50	1.50	1.50
MN	T	0	0	0.50	0.50
OP	O	–	–	0	0
HI	C	3.53	4.17	4.95	5.83
JK	C	2.12	2.50	3.54	4.17
LM	C	0.71	0.83	2.12	2.50
NO	C	–	–	0.71	0.83
Truss 7 - Warren					
BI	C	2.50	3.33	3.50	4.67
DM	C	4.50	6.00	7.50	10.00
GH	O	0	0	0	0
GK	T	4.00	5.33	6.00	8.00
GO	T	–	–	8.00	10.67
AH	C	3.00	3.00	4.00	4.00
IJ	C	1.00	1.00	1.00	1.00
KL	O	0	0	0	0
MN	C	1.00	1.00	1.00	1.00
OP	O	–	–	0	0
HI	T	3.53	4.17	4.95	5.83
JK	C	2.12	2.50	3.54	4.17
LM	T	0.71	0.83	2.12	2.50
NO	C	–	–	0.71	0.83

[a] See Figure 11.14 for truss forms and member identifications.

and a ceiling loading distributed continuously along the bottom chords for a combination of axial tension plus bending. This will of course result in somewhat larger members being required for both chords, and any estimate of the truss weight should account for this anticipated additional requirement.

11.8 TRUSS MEMBERS AND JOINTS

The three common forms of truss configuration are those shown in Figure 11.15. The single-member type, with all members in one plane as shown in Figure 11.15*a*, is that used most often to produce the simple W-form truss (Figure 11.4*a*), with members usually of 2 in. nominal thickness. Joints may use plywood gussets as shown in Figure 11.15*a*, but are more often made with metal connecting devices when trusses are produced as products by a manufacturer. In the latter case the joint performance is certified by load testing of prototypes. Design of a plywood gusset plate for a truss is illustrated in Example 6 in Section 10.9.

For larger trusses the form shown in Figure 11.15*b* may be used, with members consisting of multiples of standard lumber elements. If the member carries compression, it will usually be designed as a spaced column, as described in Section 9.6. Joints are usually made with bolts and some form of shear developers such as split rings.

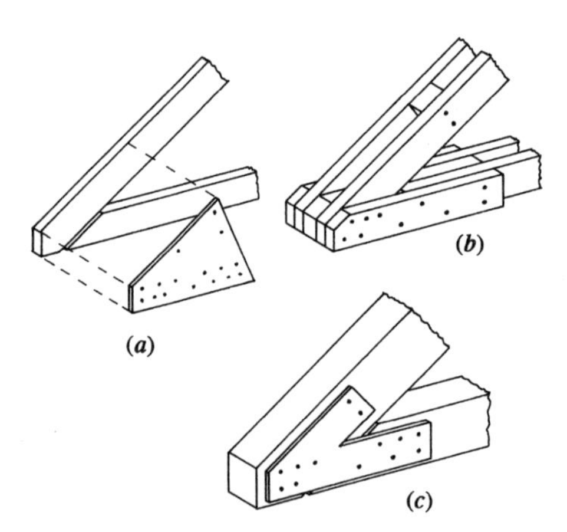

Figure 11.15 Common forms for construction of wood trusses: (a) single-piece members in light wood trusses, with plywood gussets, (b) multiple-element members with lapped and bolted joints, (c) heavy timber members with steel plate gussets and bolted joints.

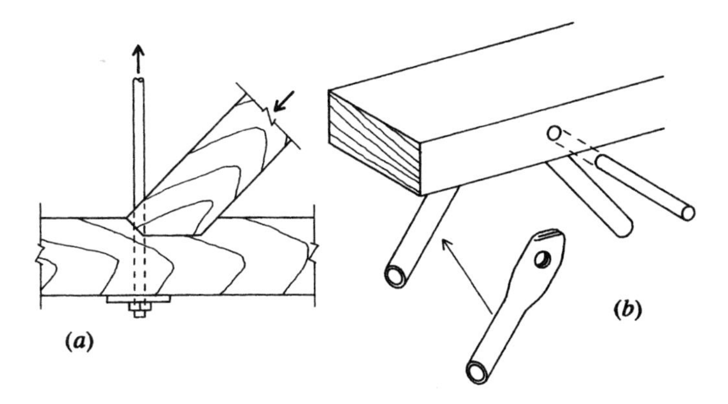

Figure 11.16 Forms for construction of wood and steel combination trusses.

In the so-called *heavy timber truss* the individual members are of large timber or glued-laminated elements, usually occurring in a single plane, as shown in Figure 11.15*c*. A common type of joint in this case is one using steel plates attached by lag screws or through-bolts. Depending on the truss pattern and loading, it may be possible to make some joints without gussets, as shown in Figure 11.16*a* for the diagonal compression member and the bottom chord connection. This was common in the past, but it requires carpentry work not easy to obtain today.

Although wood members have considerable capacity to resist tension, the achieving of tension connections is not so easy, especially in heavy timber trusses. Thus in some trusses the tension members are made of steel, as shown for the vertical member in Figure 11.16*a*. A common form today for manufactured trusses is that shown in Figure 11.16*b* where a flatchorded truss has chords of solid wood and all interior members of steel.

The selection of truss members and jointing methods depends on the size of the truss and the loading conditions. Unless trusses are exposed to view and appearance is a major concern, the specific choices of members and details of the fabrication are often left to the discretion of the fabricators of the trusses.

11.9 TIMBER TRUSSES

Single members of large-dimension timber were often used for large trusses many years ago. As iron and steel technology developed, cast iron and steel elements were employed for various tasks in this type of

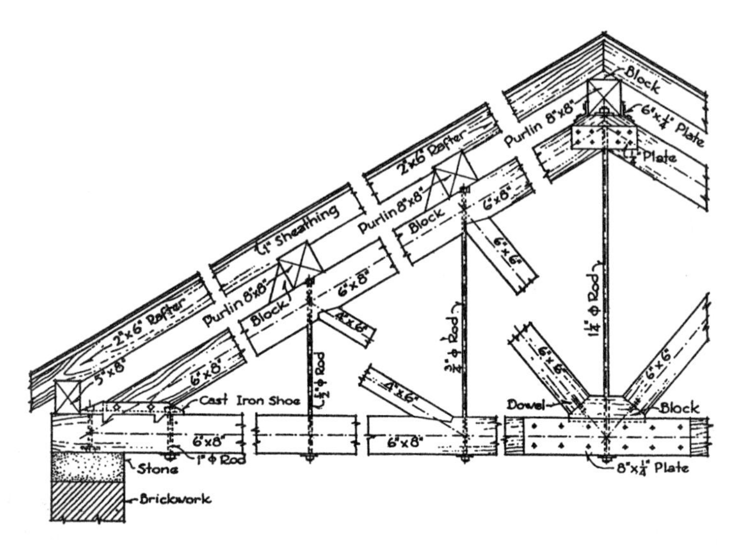

Figure 11.17 Early twentieth-century combination truss. (Reproduced from *Materials and Methods of Architectural Construction* by Gay and Parker, 1932, with permission of the publishers, John Wiley & Sons, New Jersey.)

construction. Figure 11.17 shows an example of this type of structure as it was developed in the early part of the nineteenth century. This type of construction—and especially the jointing methods—is seldom used today except in restoration or reproduction of historic buildings. The special parts, such as the cast iron support shoe and the highly crafted notched joints, are generally unattainable.

Trusses of the size and general form of that shown in Figure 11.17 are now mostly made with multiple-element members and bolted joints, as shown in Figure 11.15*b*. Although steel elements may be used for some tension members, this is done less frequently today because joints using shear developers are capable of sufficient load resistance to permit use of wood tension members. It was chiefly the connection problem that inspired the use of steel rods in the timber truss. The one vestige of this composite construction today is the use of steel members in the manufactured trusses of the form shown in Figure 11.16*b*.

11.10 MANUFACTURED TRUSSES

The majority of trusses used for building structures today are produced as manufactured products, and the detailed engineering design is done largely by the engineers in the employ of the manufacturers. Since shipping of trusses over great distances is not generally feasible, the use of these products is chiefly limited to a region within reasonable distance from a particular manufacturing facility. If use of such products is planned, it should first be determined which particular products are available in the region of the building site.

There are three principal types of manufactured truss. Manufacturers vary in size of operation; some produce only one type, and others have a range of products. The three common types are as follows:

1. One type is the simple gable-form, W-pattern truss (Figure 11.4*a*) with truss members of single-piece, 2 in. nominal lumber. These are quite simple to produce and can be turned out in small quantities by small local suppliers in many cases.
2. Second is the composite truss, usually consisting of a combination of wood and steel elements, as shown in Figure 11.16*b*. These are more sophisticated in fabrication detail than the simple W-trusses, and they are usually produced as proprietary products by larger companies. These are available in a wide range of sizes for uses ranging from simple floor joists to long-spanning roof trusses. They usually compete with open web steel joists in regions where they are readily available.
3. The third type is the large, long-span truss, usually using multiple-element members, as shown in Figure 11.15*b*. These may be produced with some standardization by a specific manufacturer, but are often customized to some degree for a particular building. One form is the bowstring truss, which in reality is essentially a tied arch (see Figure 11.4*i*).

Suppliers of manufactured trusses usually have some standardized models, but always have some degree of variability to accommodate the specific usage conditions for a particular building. The possible range of these variations can only be established by working with the suppliers.

12

MISCELLANEOUS WOOD PRODUCTS AND ELEMENTS

The preceding chapters of this book have dealt primarily with wood structures comprised of sawn lumber. While this basic form of wood product is still in wide use, there is an ever-increasing array of other structural products that use wood as a primary material. This chapter treats a number of these manufactured products as well as some special elements used for wood structures.

12.1 ENGINEERED WOOD PRODUCTS

The general term *engineered wood products* is now used to describe all manufactured structural wood products, other than sawn lumber. The name implies that something is done to the raw wood material as it comes from the tree. Indeed, sawn wood is also processed to some extent, possibly involving finishing of surfaces or impregnating with chemicals to enhance resistance to decay, insect damage, or fire. However, most engineered products begin with some reconstitution of the basic material: slicing it into thin veneers, shredding it into fibers, or forming it into chips or strands. These reconstituted materials are then subjected to some forming and bonding processes to produce laminated veneer lumber, plywood, oriented strand board, or composite panels. Combinations of sawn lumber and engineered products can also be assembled to produce glued-laminated lumber and I-joists. The following sections describe these products and their applications in building structures.

12.2 GLUED-LAMINATED STRUCTURAL MEMBERS

The gluing together of multiple laminations of standard 2 in. nominal thickness lumber has been used for many years to produce large beams and girders. This is really the only option for using sawn wood for large members that are beyond the feasible range of size for single sawn pieces. However, there are other reasons for using the laminated beam that include the following:

1. *Higher Strength.* Lumber used for laminating has a moisture content described as kiln-dried. This is the opposite end of the quality range from the green wood condition ordinarily assumed for solid-sawn members. This, plus the minimizing effect of flaws due to lamination, permits use of stresses for flexure and shear that are much higher than those allowed for single-piece members. The result is that much smaller sections can often be used, which helps to offset the usually higher cost of the laminated products.
2. *Better Dimensional Stability.* This refers to the tendency for wood to warp, split, shrink, and so on. Both the use of the kiln-dried materials and the laminating process itself tend to create a very stable product. This is often a major consideration where shape change can adversely affect the building construction.
3. *Shape Variability.* Lamination permits the production of curved, tapered, and other special profile forms for beams, as shown in Figure 12.1. Cambering as compensation for service load deflection, as sloping for roof drainage, and other useful custom profiling can be done with relative ease. This is otherwise possible only with a truss or a built-up section.

Laminated beams have seen wide use for many years, and industry-wide standards are well established. Cross-sectional sizes are

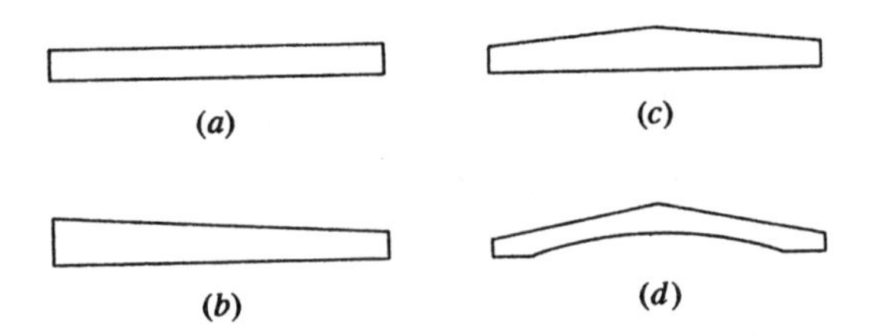

Figure 12.1 Profile variations of glued-laminated timber elements.

derived from the number of laminations and the size of the individual pieces used. Thus depths are multiples of 1.5 in. and widths are slightly less than the lumber size as a result of the finishing of the product. Minor misalignments and the unavoidable sloppiness of the gluing process result in an unattractive surface. Finishing may simply consist of smoothing off, although special finishes, such as rough-sawn ones, are also available.

Investigation and design of glued-laminated timber members is done primarily with the procedures explained for solid-sawn beams as described in Section 7.1. Criteria for design are provided in most building codes, in the *NDS* (Ref. 1), in texts such as Ref. 2, and in literature provided by manufacturers and suppliers of the products.

These elements are manufactured products and are mostly not able to be transported great distances, so information about them should be obtained from local suppliers.

Laminated Arches and Bents

Individual elements of glued-laminated timber can be custom-profiled to produce a wide variety of shapes for structures. Two forms commonly used are the three-hinged arch (Figure 12.2*a*) and the gabled bent (Figure 12.2*b*). A critical consideration is that of the radius of curvature of the member, which must be limited to what the wood species and the laminate thickness can tolerate. For very large elements this is not a problem, but for smaller structures the curvature limits of 2 in. nominal lumber may be critical.

Manufacturers of laminated products usually produce the arch and gabled elements as standard forms. Structural design of the products is usually done by the manufacturer's engineers. Form limits, size range, connection details, and other considerations for these products should be investigated with individual manufacturers.

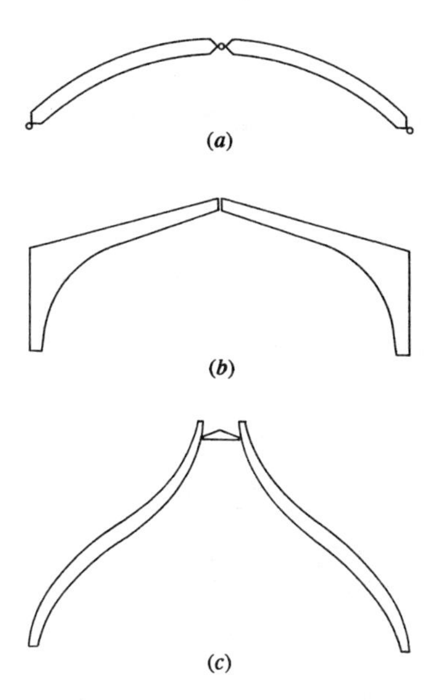

Figure 12.2 Nonstraight laminated elements for spanning structures: (a) arch segments, (b) gabled frame bent, (c) doubly curved elements.

Custom shapes can be produced, such as those with double curvature, as shown in Figure 12.2*c*. Many imaginative structures have been designed using the form variation potential of this process.

Laminated Timber Columns

Columns may be produced with 1.5 in. laminations, presenting the same advantages as those described for beams: higher strength and dimensional stability being most critical. It is also possible to produce glued-laminated columns of greater length than that obtainable with solid-sawn members. In general, laminated columns are used less frequently than beams or girders, and are mostly chosen only when some of the inherent limitations of other options are restrictive.

12.3 STRUCTURAL COMPOSITE LUMBER

Structural composite members are manufactured using wood elements that are bonded together by gluing. Although this classification theoretically includes glued-laminated timber members, this term is used primarily to describe other elements, such as:

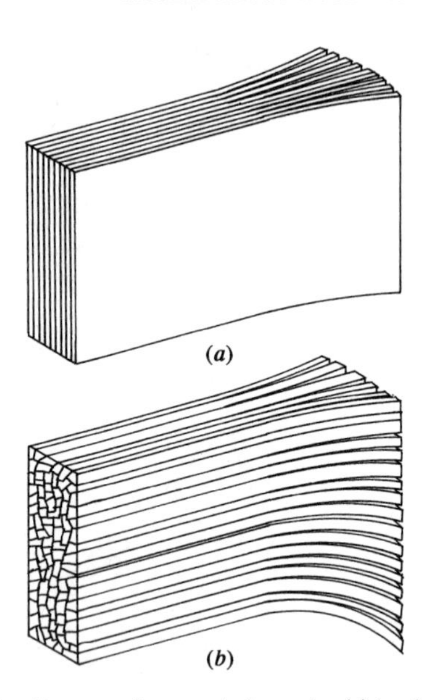

Figure 12.3 Structural composite wood elements: (a) laminated veneer lumber (LVL), (b) parallel strand lumber (PSL).

Laminated Veneer Lumber (LVL). These consist of laminated members using thin veneers similar to those used for plywood, except that here the grain of all the laminations is parallel to the length of the finished members. See Figure 12.3*a*.

Parallel Strand Lumber (PSL). These consist of long, thin strands of wood placed with their orientation parallel to the length of members and bonded together by a combination of resin and compression. See Figure 12.3*b*.

Both of these processes are used to produce linear lumber pieces similar in form and intended function to those produced as solid-sawn elements. Advantages include higher strength, dimensional stability, and virtually unlimited length of pieces. In addition, the raw materials used—especially for PSL—can be obtained from smaller, faster-growth trees. The forest resource is thus similar to that for paper products, which is a vast cultivated system.

A major use for both of these products is for chord members of trusses and I-joists. All of the advantages described are significant for this application.

PSL members were originally introduced as replacements for small lumber pieces such as the 2 by 4. However, they are now produced in a considerable size range and are generally competitive with sawn products in many situations. Only glued-laminated timbers can presently achieve larger cross-sectional sizes and member lengths, but time will tell.

Design of these products is now done with code-approved processes and data. Nevertheless, it is advisable to obtain information from regional suppliers for specific products.

12.4 WOOD STRUCTURAL PANELS

In earlier times, surfacing of walls, roofs, and floors was mostly achieved with solid-sawn wood boards. By the mid-twentieth century, this surfacing was mostly achieved with plywood panels. At present, plywood is still mostly used for floor decks and for shear walls, but other panel materials are increasingly used for walls and roofs. Plywood is discussed in Section 12.5. The following discussion deals with the other two types of panels currently used for structural applications.

Composite panels

Composite panels are produced in a manner similar to that used for plywood. Thin sheet veneers are glued together in layers, called plies, to build up the panel thickness. As with plywood, the two outside veneers are usually of wood with the grain of both veneers in the same direction. The primary difference here is the use of wood fiber materials for the inner plies. Use of fiber materials for the major portion of the bulk of the panel is a cost-saving feature, as well as a utilization of less critical resources. A principal feature of these panels is their close resemblance to plywood in terms of viewed surface. Where the viewed surface is a concern, the cost reduction may be significant.

Oriented Strand Board (OSB)

OSB is produced with thin chips or wafers of wood and a bonding resin. The wafers are placed in layers with a switch in direction of the grain of the wafers in alternating layers, producing a structural similarity to plywood in terms of two-way bending resistance. This product is now widely used for wall and roof sheathing in place of the more expensive and resource-depleting plywood. However, *all* of its properties should be considered for each application, including structural properties, but also moisture resistance, thermal expansion, and fastener holding.

12.5 PLYWOOD

Plywood is the term used to designate structural wood panels made by gluing together multiple layers of thin wood veneer (called *plies*) with alternate layers having their grain direction at right angles. The outside layers are called the *faces* and the others *inner plies*. Inner plies with the grain perpendicular to the faces are called *crossbands*. There is usually an odd number of plies so that the faces have the grain in the same direction. For structural applications in building construction, the common range of panel thickness is from 5/16 to 1-1/8 in.

The alternating grain direction of the plies gives the panels considerable resistance to splitting, and as the number of plies increases, the panels become approximately equal in strength in both directions. Thin panels may have only three plies, but for most structural applications plies will number from 5 to 9.

Types and Grades of Plywood

Many different kinds of panels are produced. For structural applications, the principal distinctions other than panel thickness are the following:

1. *Exposure Classification.* Panels described as being *exterior* are for use where high moisture conditions are enduring, such as outdoor uses and bathrooms, laundry rooms, and other high-moisture interior spaces. A classification of *exterior 1* is for panels whose end usage is for interior conditions, but the panels may be exposed to the weather during construction.
2. *Structural Rating.* Code-approved rated sheathing is identified as to class for purpose of establishing reference design values. Identification is established by marking of panels with an indelible stamp. Information in the stamp designates several properties of the panel, including basic structural capabilities.

Design Usage Data for Plywood

Data for structural design of plywood may be obtained from industry publications or from individual plywood manufacturers. Data is also provided in most building codes. Tables 12.1 and 12.2 are reproductions of tables in the *International Building Code* (Ref. 4). These provide data for the loading and span capabilities of rated plywood panels. Table 12.1 treats panels with the panel face grain perpendicular to the supports, and Table 12.2 treats panels with the face grain parallel to the supports. Footnotes to these tables present various qualifications, including some of the loading and deflection criteria.

Plywood Diaphragms

Plywood deck and wall sheathing is frequently utilized to develop diaphragm actions for resistance to lateral loads from wind or earthquakes. Considerations for design of both horizontal deck diaphragms and vertical wall diaphragms (shear walls) are discussed in Chapter 13. Where both gravity loading and lateral loading must be considered, choices for the construction must relate to both problems.

Usage Considerations for Plywood

The following are some of the principal usage considerations for ordinary applications of structural plywood panels.

1. *Choice of Thickness and Grade.* This is largely a matter of common usage and building code acceptability. For economy the thinnest, lowest-grade panels will always be used unless various concerns require otherwise. In addition to structural spanning capabilities, concerns may include moisture resistance, appearance of face plies, and fastener holding capability.
2. *Modular Supports.* With the usual common panel size of 4 ft × 8 ft, logical spacing for studs, rafters, and joists becomes even number divisions of the 48 or 96 in. dimensions: 12, 16, 24, 32, or 48. However, spacing of framing most often relates to what is attached on the other side of a wall or to a directly attached ceiling.
3. *Panel Edge Supports.* Panel edges not falling on a support may need some provision for nailing, especially for roof and floor decks. Solid blocking is the common answer, although thick deck panels may have tongue-and-groove edges.
4. *Attachment to Supports.* For reference design values for shear loads in diaphragms, attachment is usually considered to be achieved with common wire nails. Required nail size and spacing

TABLE 12.1 Data for Plywood Roof and Floor Deck, Face Grain Perpendicular to Supports

ALLOWABLE SPANS AND LOADS FOR WOOD STRUCTURAL PANEL SHEATHING AND SINGLE-FLOOR GRADES CONTINUOUS OVER TWO OR MORE SPANS WITH STRENGTH AXIS PERPENDICULAR TO SUPPORTS[a,b]

SHEATHING GRADES		ROOF[c]				FLOOR[d]
		Maximum span (inches)		Load[e] (psf)		
Panel span rating roof/floor span	Panel thickness (inches)	With edge support[f]	Without edge support	Total load	Live load	Maximum span (inches)
12/0	$^5/_{16}$	12	12	40	30	0
16/0	$^5/_{16}$, $^3/_8$	16	16	40	30	0
20/0	$^5/_{16}$, $^3/_8$	20	20	40	30	0
24/0	$^3/_8$, $^7/_{16}$, $^1/_2$	24	20[g]	40	30	0
24/16	$^7/_{16}$, $^1/_2$	24	24	50	40	16
32/16	$^{15}/_{32}$, $^1/_2$, $^5/_8$	32	28	40	30	16[h]
40/20	$^{19}/_{32}$, $^5/_8$, $^3/_4$, $^7/_8$	40	32	40	30	20[h,i]
48/24	$^{23}/_{32}$, $^3/_4$, $^7/_8$	48	36	45	35	24
54/32	$^7/_8$, 1	54	40	45	35	32
60/32	$^7/_8$, 1 $^1/_8$	60	48	45	35	32

SINGLE FLOOR GRADES		ROOF[c]				FLOOR[d]
		Maximum span (inches)		Load[e] (psf)		
Panel span rating	Panel thickness (inches)	With edge support[f]	Without edge support	Total load	Live load	Maximum span (inches)
16 o.c.	$^1/_2$, $^{19}/_{32}$, $^5/_8$	24	24	50	40	16[h]
20 o.c.	$^{19}/_{32}$, $^5/_8$, $^3/_4$	32	32	40	30	20[h,i]
24 o.c.	$^{23}/_{32}$, $^3/_4$	48	36	35	25	24
32 o.c.	$^7/_8$, 1	48	40	50	40	32
48 o.c.	$1^3/_{32}$, $1^1/_8$	60	48	50	40	48

For SI: 1 inch = 25.4 mm, 1 pound per square foot = 0.0479 kN/m^2.

a. Applies to panels 24 inches or wider.

b. Floor and roof sheathing conforming with this table shall be deemed to meet the design criteria of Section 2304.7.

c. Uniform load deflection limitations $^1/_{180}$ of span under live load plus dead load, $^1/_{240}$ under live load only.

d. Panel edges shall have approved tongue-and-groove joints or shall be supported with blocking unless $^1/_4$-inch minimum thickness underlayment or 1 $^1/_2$ inches of approved cellular or lightweight concrete is placed over the subfloor, or finish floor is $^3/_4$-inch wood strip. Allowable uniform load based on deflection of $^1/_{360}$ of span is 100 pounds per square foot except the span rating of 48 inches on center is based on a total load of 65 pounds per square foot.

e. Allowable load at maximum span.

f. Tongue-and-groove edges, panel edge clips (one midway between each support, except two equally spaced between supports 48 inches on center), lumber blocking or other. Only lumber blocking shall satisfy blocked diaphragm requirements.

g. For $^1/_2$-inch panel, maximum span shall be 24 inches.

h. Span is permitted to be 24 inches on center where $^3/_4$-inch wood strip flooring is installed at right angles to joist.

i. Span is permitted to be 24 inches on center for floors where $1^1/_2$ inches of cellular or lightweight concrete is applied over the panels.

Source: Reproduced from *International Building Code* (Ref. 4), with permission of the publishers, International Code Council, Inc.

TABLE 12.2 Data for Plywood Roof Deck, Face Grain Parallel to Supports

ALLOWABLE LOAD (PSF) FOR WOOD STRUCTURAL PANEL ROOF SHEATHING CONTINUOUS OVER TWO OR MORE SPANS AND STRENGTH AXIS PARALLEL TO SUPPORTS (Plywood Structural Panels Are Five-Ply, Five-Layer Unless Otherwise Noted)[a,b]

PANEL GRADE	THICKNESS (inch)	MAXIMUM SPAN (inches)	LOAD AT MAXIMUM SPAN (psf)	
			Live	Total
Structural I sheathing	$^{7}/_{16}$	24	20	30
	$^{15}/_{32}$	24	35[c]	45[c]
	$^{1}/_{2}$	24	40[c]	50[c]
	$^{19}/_{32}$, $^{5}/_{8}$	24	70	80
	$^{23}/_{32}$, $^{3}/_{4}$	24	90	100
Sheathing, other grades covered in DOC PS 1 or DOC PS 2	$^{7}/_{16}$	16	40	50
	$^{15}/_{32}$	24	20	25
	$^{1}/_{2}$	24	25	30
	$^{19}/_{32}$	24	40[c]	50[c]
	$^{5}/_{8}$	24	45[c]	55[c]
	$^{23}/_{32}$, $^{3}/_{4}$	24	60[c]	65[c]

For SI: 1 inch = 25.4 mm, 1 pound per square foot = 0.0479 kN/m^2.

a. Roof sheathing conforming with this table shall be deemed to meet the design criteria of Section 2304.7.

b. Uniform load deflection limitations $^{1}/_{180}$ of span under live load plus dead load, $^{1}/_{240}$ under live load only. Edges shall be blocked with lumber or other approved type of edge supports.

c. For composite and four-ply plywood structural panel, load shall be reduced by 15 pounds per square foot.

Source: Reproduced from *International Building Code* (Ref. 4), with permission of the publishers, International Code Council, Inc.

relate to panel thickness and code minimums, as well as to shear capacities in diaphragms. Attachment is now mostly achieved with mechanically driven fasteners rather than old-fashioned pounding with a hand-held hammer. These means of attachment and the actual fasteners used are usually rated for capacity in terms of equivalency to ordinary nailing.

12.6 PREFABRICATED WOOD I-JOISTS

Wood I-shaped joists are formed with chords (top and bottom elements) of sawn lumber or laminated wood veneers. (See Figure 12.4.) Grooves are cut in the chord elements, and a web of plywood or OSB is glued into the grooves. Criteria for design of these structural products are now included in codes and industry standards, but these are essentially the products of individual manufacturers, and specific design data should be obtained from the producers. These products were originally developed to extend the span range slightly beyond the limit represented by 2 × 12 lumber joists and rafters. While this is still the case, smaller I-joists are now competitive with sawn lumber in smaller sizes as well. Advantages include more dimensional stability (resistance to warping, curling, etc.)

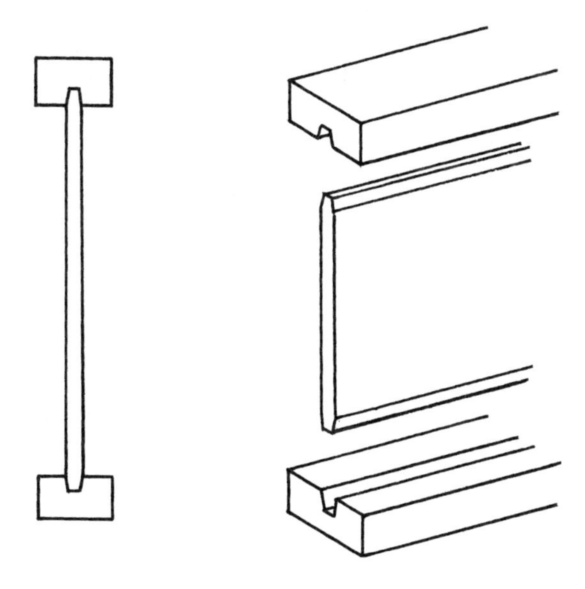

Figure 12.4 Wood I-joists.

and better resistance to lateral and torsional buckling. Elimination of bridging and blocking may be significant.

Not a small consideration today is the issue of conservation of resources. Use of OSB webs and laminated veneer chords represents a reduction of demand for high-quality logs for sawn lumber. A good 2 × 12 cannot be obtained from a small, fast-growth tree.

12.7 BUILT-UP PANEL AND LUMBER BEAMS

Structural wood panels and sawn wood lumber can be combined in various ways. The I-joist is one such product. Another form is the so-called *sandwich panel* or *stressed-skin panel*, which consists of two large panels attached to and separated by a frame of sawn wood elements. (See Figure 12.5*a*.) In this assembly the panels and sawn wood switch roles from those in the I-joist. Here the panels serve as bending resisters at the top and bottom, and the sawn elements serve as the shear-resisting web. Sandwich panels can be assembled to form prefabricated wall or deck units. This usage is sometimes utilized for demountable structures, but is not common for ordinary construction.

Another structural product of this type is the built-up beam formed with top and bottom elements of sawn lumber and panels that are nailed to the sides of the lumber. (See Figure 12.5*b*.) One application of this structure is formed as part of an ordinary stud wall formation. The most frequent such use is for headers over wide openings in walls. For the header the wall top plate (usually two 2 in. nominal elements) and a similar doubled member at the top edge of the opening are used as the top and bottom elements of the beam. The built-up header is then completed by using wall sheathing as the web of the panel/lumber

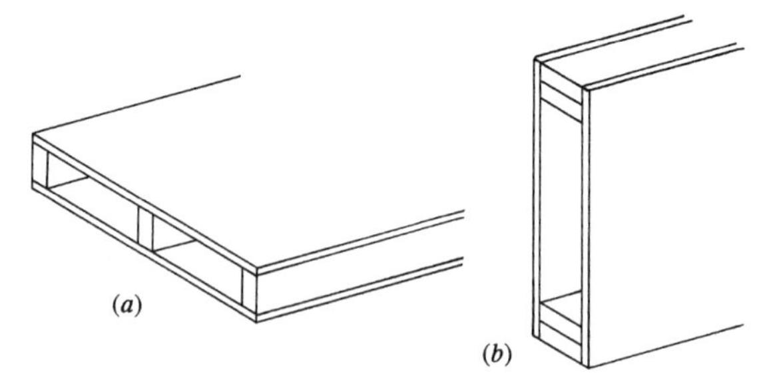

Figure 12.5 Composite, built-up elements with solid-sawn pieces and panels of plywood or wood fiber products.

assemblage. Industry guides and some code criteria are available for design of this structural element.

12.8 FLITCHED BEAMS

In construction using solid-sawn timber beams, there are some occasions when the deflection of a beam is of critical concern. As heavy timber is generally available only in green wood condition, some long-term sag is inevitable, especially when considerable dead load is supported. One means commonly used in this situation is to combine a steel plate with wood members in a *flitched beam*. Two forms for flitched beams are shown in Figure 12.6. The most common form uses a single plate sandwiched between two timber members, with the steel and wood bolted together to ensure interaction, as shown in Figure 12.6*a*. For an even stronger section, two plates may be used with a single timber section, as shown in Figure 12.6*b*.

In most cases, due to the much higher stiffness of the steel, the plates will carry the majority of the load on a flitched beam. A simple design thus consists of considering the plates alone to carry the load and resist deflection. The wood in this case serves primarily to brace the slender plate for lateral and torsional buckling. With the beam used in a general timber structure, the wood is also probably used to achieve necessary connections to other structural elements.

For a true consideration of the composite action it is usually assumed that the wood and steel interact in proportion to their individual stiffnesses under service load conditions. This means that stress levels are assumed to be well below yield stress in the steel or ultimate failure of the wood. Such a procedure may be developed with the stress method, as described in the following example.

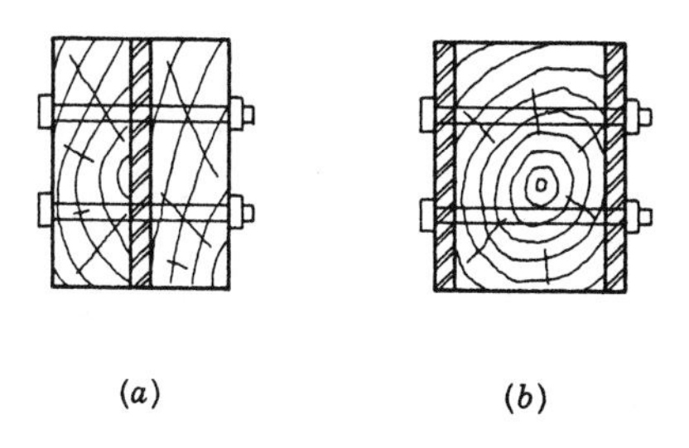

Figure 12.6 Common forms of flitched beams.

If the two materials truly interact, the assumption is made that they deflect (or deform) the same amount. Thus

Δ_1 and Δ_2 = deformation per unit length (strain) of the outermost fibers of the two materials.

f_1 and f_2 = unit bending stresses in the outermost fibers of the two materials.

E_1 and E_2 = moduli of elasticity of the two materials.

Since by definition the modulus of elasticity of a material is equal to the unit stress divided by the accompanying unit strain, then

$$E_1 = \frac{f_1}{\Delta_1} \quad \text{and} \quad E_2 = \frac{f_2}{\Delta_2}$$

and transposing

$$\Delta_1 = \frac{f_1}{E_1} \quad \text{and} \quad \Delta_2 = \frac{f_2}{E_2}$$

Since the two deformations must be equal for interaction,

$$\Delta_1 = \Delta_2 \quad \text{or} \quad \frac{f_1}{E_1} = \frac{f_2}{E_2}$$

and a relationship between the two stresses may thus be expressed as

$$f_2 = f_1 \times \frac{E_2}{E_1}$$

This relationship may be used for investigation or design of a flitched beam by the stress method. The process is demonstrated in the following example.

Example 1. A flitched beam is formed as shown in Figure 12.6*a*, consisting of two 2 × 12 planks of Douglas fir, No. 1 grade, and a steel plate with dimensions of 0.5 × 11.25 in. [13 × 285 mm] and F_y = 36 ksi [248 Mpa]. Determine the allowable uniformly distributed load that this beam can carry on a simple span of 14 ft [4.2 m]. For the steel use E = 29,000 ksi [200 Gpa] and a maximum allowable bending stress of 22 ksi [150 Mpa]. For the wood use E = 1,800 ksi [12.4 Gpa] and a maximum allowable bending stress of 1500 psi [10.3 Mpa]. For the 2 × 12, $S = 31.6\ \text{in.}^3$ [$518 \times 10^3\ \text{mm}^3$].

Solution: A procedure that can be used is to assume the maximum allowable stress in the steel and to find the corresponding stress in the wood. Thus

$$f_w = f_s \times \frac{E_w}{E_s} = (22)\left(\frac{1,800}{29,000}\right) = 1.366 \text{ ksi} \quad \text{or} \quad 1366 \text{ psi}$$

As this produces a stress lower than the limit for the wood, the stress in the steel is the limiting value. That is, if a stress higher than 1366 psi is permitted in the wood, the limiting stress for the steel will be exceeded.

Using the stress limit just determined for the wood, it is now possible to find the share of the load that is carried by the wood members. Calling this total load W_w, the moment on the simple beam due to this load is

$$M_w = \frac{W_w L}{8} = \frac{W_w \times 14}{8} = 1.75 W_w \text{ (in foot units)} \quad \text{or}$$

$$1.75 \times 12 = 21 W_w \text{ (in inch units)}$$

The limit for this moment may be expressed as

$$M = f \times S = 1366 \times (2 \times 31.6) = 86,311 \text{in.-lb}$$

and for determination of W_w

$$M_w = 21 W_w = 86,311 \text{ in.-lb}$$

$$W_w = \frac{86,311}{21} = 4,110 \text{ lb } [18.35 \text{ kN}]$$

For the steel plate

$$S = \frac{bd^2}{6} = \frac{(0.5)(11.25)^2}{6} = 10.55 \text{ in.}^3 \; [176 \times 10^3 \text{ mm}^3]$$

Then

$$M_s = 21 W_s = f_s \times S = (22,000)(10.55) = 232,100 \text{ in.-lb}$$

$$W_s = \frac{232,100}{21} = 11,052 \text{ lb}$$

and the total capacity of the composite beam is

$$W = W_w + W_s = 4{,}110 + 11{,}052 = 15{,}162\, lb\, [68.6\, kN]$$

Although the load-carrying capacity of the wood elements is usually reduced in the flitched beam, the resulting total capacity is substantially greater than that with the wood elements alone. This major increase in strength is achieved with only a small increase in the size of the beam. In addition, there is a significant reduction in the deflection in most applications—most significantly in the reduction of sag over time.

Sag over time is a natural phenomenon with timber, which is normally sold in an assumed green wood condition. One effect of this sag in the flitched beam is the steady shift of load from the wood to the steel plates. Thus, in due time, the wood will in fact carry something less than the load just computed, and it is possible that the result can produce an overload on the steel plate. For this reason, it is advisable that the actual permitted load be something less than the total capacity previously computed.

Design of a flitched beam by the stress method is usually a hit-or-miss situation, requiring some assumptions of properties of the members, a subsequent investigation, and—if necessary—some readjustment of dimensions. It is also necessary to assume some specific amount of reduction of the wood capacity over time. A much simplified design procedure, although somewhat conservative, is to design the steel plate to carry the entire load, relying on the wood members only for lateral and torsional bracing of the plate. This process is demonstrated in the following example, using the strength method for the plate design.

Example 2. Design a flitched beam of the form shown in Figure 12.6*a* with the same wood members as in the preceding example. Assume the steel plate to carry the entire load and determine the minimum thickness required for the plate in this case. For the design load, assume the service load determined as the total capacity of the beam in Example 1 with the loading being 50% dead load and 50% live load.

Solution: The required ultimate load is determined with an average load factor of 1.4, thus

$$W_u = 1.2\left(\frac{15{,}162}{2}\right) + 1.6\left(\frac{15{,}162}{2}\right) = 9{,}097 + 12{,}130 = 21{,}227\ \text{lb}$$

and the required ultimate bending moment is thus

$$M_u = \frac{W_u L}{8} = \frac{(21{,}227)(14)}{8} = 37{,}147 \text{ ft-lb} \quad \text{or} \quad 445{,}767 \text{ in-lb}$$

Using a resistance factor of 0.9, the required resisting moment of the plate is thus

$$M_r = \frac{445{,}767}{0.9} = 495{,}297 \text{ in.-lb}$$

For the plate, the resisting moment is determined as

$$M_r = F_y \times Z$$

For the rectangular form of the plate, Z is determined as

$$Z = \frac{bd^2}{4} = \frac{b(11.25)^2}{4} = 31.64b \text{ in.}^3$$

Thus the thickness of the plate, b, is determined as

$$M_r = 495{,}297 = (36{,}000)(31.64b) = 1{,}139{,}040b$$

$$b = \frac{495{,}297}{1{,}139{,}040} = 0.435 \text{ in.}$$

This indicates that the plate dimensions given in Example 1 are actually sufficient for the load capacity found by the stress method relying completely on the plate.

Problem 12.8.A. A flitched beam consists of a single 10 × 14 of Douglas fir, select structural grade, and two A36 steel plates, each 0.5 × 13.5 in. [13 × 343 mm]. (See Figure 12.6*b*.) Using stress methods, find the magnitude of the single concentrated load this beam can safely support at the center of a 16 ft [4.8 m] simple span. Neglect the weight of the beam. Use a value of 22 ksi [152 MPa] for the limiting bending stress in the plates.

Problem 12.8.B. Using the strength method, find the thickness required for the plates in Problem 12.8.A if the entire load is carried by the plates. Assume the load determined in Problem 12.8.A with an average load factor of 1.4.

12.9 POLE STRUCTURES

A type of construction used extensively in ancient times and still used today in some regions is that which employs wood poles as vertical structural elements. Although processed poles, cut to have a constant diameter, are obtainable, most poles are simply tree trunks with branches and soft outer layers of material stripped away. There are generally three ways in which such poles may be used: as timber piles, driven into the ground; as vertical building columns in a frame structure; or as buried-end poles, extending slightly below grade and slightly above grade (such as fence posts or electric wire transmission poles). The following discussion is limited to consideration of buried-end poles.

As a foundation element, the buried-end pole is typically used to raise a building above the ground. Driven timber piles may be used in this manner, but control of the precise location of the pole top makes them less usable for extended members of the building framework. With regard to the building construction, the two chief means of using poles are for *pole-frame buildings* and *pole-platform buildings*, as shown in Figure 12.7. For a pole-frame building (Figure 12.7*a*), the buried poles are extended above grade to function as building columns. For a pole-platform (Figure 12.7*b*), the poles are cut off at some level above grade, and a flat structure (the platform) is built on top of them, possibly providing support for a conventional wood frame structure.

Pole foundations must usually provide both vertical and lateral support for a building. For vertical loads, the pole end simply transfers vertical load by direct bearing. The three common forms for buried-end

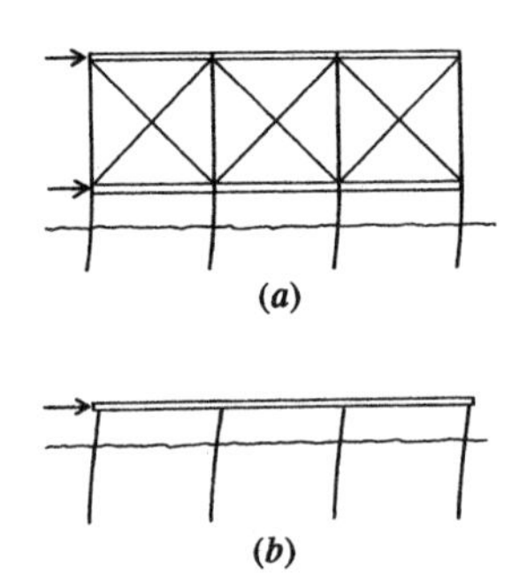

Figure 12.7 Response to lateral forces in pole-frame and pole-platform structures.

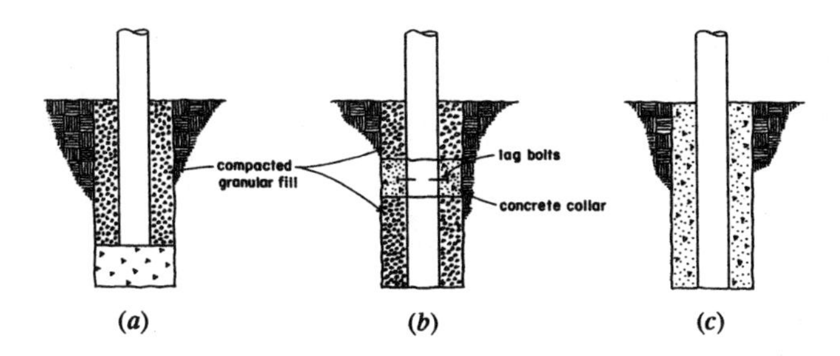

Figure 12.8 Alternate details for pole foundations.

pole foundations are shown in Figure 12.8. In Figure 12.8*a* the bottom of the hole is filled with concrete to provide a footing—a preferred method when the soil at the bottom of the hole is very compressible. In Figure 12.8*b* the pole bears directly on the bottom of the hole, and a stabilizing method sometimes used is to pour a concrete collar near the top to help with lateral loads. If lag bolts are used to anchor the collar, the collar will also assist with vertical support. Of course, both the footing and collar can be used for extra strong support. In Figure 12.8*c* the hole is completely backfilled with concrete, which is the most positive means of support for both vertical and lateral loads.

Pole building is frequently used for utility buildings in regions where poles are readily available. Local codes and common practices—based mostly on experience—often determine details of the construction. Need for depth of embedment is an issue of particular concern, and is often based more on experience than science.

For lateral loads, the situation is usually quite different for pole-frame and pole-platform buildings. For the pole-frame construction, lateral bracing is usually provided in a conventional manner, as for other wood frame construction. The buried pole ends must resist lateral force, but there is less concern for rotation of the pole at grade level. For the pole platform, while the structure built on top of the platform may be conventionally braced, the platform itself depends heavily on the resistance to lateral movement of the poles at grade level. For the platform there should be either some constraint of the pole at ground level or some additional depth of penetration into the ground—or both.

13

WOOD STRUCTURES FOR LATERAL BRACING

This chapter presents materials relating to the development of resistance to the horizontal force effects of wind and earthquakes. Structures with wood frames are most often braced with planar elements that create a boxlike, three-dimensional form. These elements consist of the walls and the roof and floor decks that are formed with wood framing and some surfacing material. When surfacing is not present, it may be possible to use trussing or rigid frame action to resist the horizontal forces. In some cases, however, other rigid construction may be used to brace the wood frame and thus reduce its functions primarily to that of resisting gravity forces. Examples of construction used in this way are masonry or concrete walls and trussed or rigid-jointed steel frames. For a more extensive discussion of design for lateral loads, the reader is referred to *Simplified Building Design for Wind and Earthquake Forces* (Ref. 11), from which most of the materials in this chapter are derived.

13.1 APPLICATION OF WIND AND EARTHQUAKE FORCES

To understand how a building resists the lateral force effects of wind and earthquake forces it is necessary to consider the manner of application of the forces and then to visualize how the forces are transferred through the lateral resistive structural system and into the ground.

Wind Forces

The application of wind forces to a closed building is treated in the form of pressures applied normal to the exterior surfaces of the building and shearing friction effects on surfaces parallel to the wind direction. In one design method the total effect on the building is determined by considering the vertical profile, or silhouette, of the building as a vertical plane surface at right angles to the wind direction. A direct horizontal pressure is assumed to act on this plane.

Figure 13.1 shows a simple rectangular building under the effect of wind normal to one of its flat sides. The lateral resistive structure that responds to this loading consists of the following:

Wall surface elements on the windward side. These are assumed to take the total wind pressure and are typically designed to span vertically between the roof and floor structures.

Roof and floor decks. These are considered as rigid planes (called *diaphragms*) and are assumed to receive the edge loadings from the windward walls and to distribute the loads to the vertical bracing elements.

Vertical frames or shear walls. These act as vertical cantilevers to receive loads from the horizontal diaphragms and to transfer them to supporting structures that consist of lower-story elements or the building foundations.

The building foundations. These constitute the final anchoring construction and transmit the loads to the ground.

The propagation of the loads through the structure is shown in the left part of Figure 13.1, and the functions of the major elements of the lateral bracing system are shown in the right part of the figure. The windward exterior wall functions as a simple spanning element loaded by a uniformly distributed pressure normal to its surface and delivering a reaction force to its supports. In most cases, even though the wall may be continuous through several stories, it is considered as a simple span

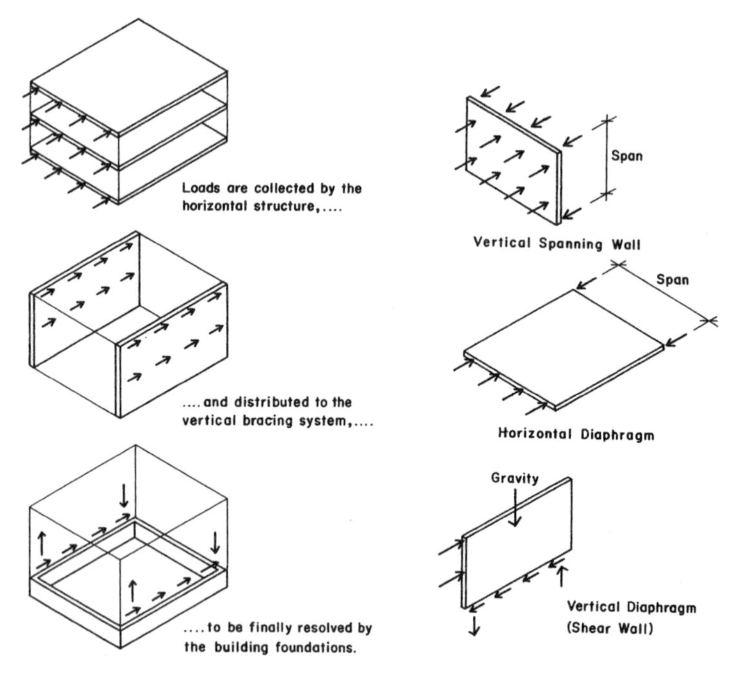

Figure 13.1 Propagation of wind forces and functions of bracing elements in a box system.

at each story level, thus delivering half of its total load to each support. Referring to Figure 13.1, this means that the upper wall delivers half of its load to the roof edge and half to the edge of the second floor. The lower wall delivers half of its load to the second floor and half to the first floor.

This may be a somewhat simplistic view of the function of the walls themselves, depending on their construction. If they are framed walls with windows and doors, there may be many internal load transfers within the wall. Usually, however, the external load delivery to the horizontal structure will be as described.

The roof and second-floor diaphragms function as spanning elements loaded by the edge forces from the exterior walls and spanning between the end shear walls. They are thus subjected to bending that develops tension on the leeward edge and compression on the windward edge. It also produces shear in the plane of the diaphragm that becomes a maximum at the end shear walls. In most cases the shear is assumed to be taken by the deck diaphragm, but the tension and compression forces due to bending are transferred to the framing at the edges. The means of achieving this transfer depends on the materials and details of the construction.

The end shear walls act as vertical cantilevers that also develop internal shear and bending. The total shear force in the upper story is equal to the edge load from the spanning roof diaphragm. The total shear in the lower story is the combination of the loads from the roof and second floor. The total shear in the wall is delivered at its base in the form of a sliding friction between the wall and its support. The bending caused by the lateral load on the wall produces an overturning (toppling) effect at the base of the wall as well as the tension and compression forces at the edges of the wall. The overturning effect is resisted by the stabilizing effect of the dead load on the wall. If this stabilizing moment is not sufficient, a tension tie must be made between the end of the wall and its support.

If the first floor is attached directly to the foundations, it may not actually function as a horizontal diaphragm but rather will push its edge load directly to the foundation wall. In any event, it may be seen in this example that only three quarters of the total wind load on the building is delivered through the upper-level diaphragms to the end shear walls.

This simple example illustrates the basic nature of the propagation of wind forces through the building structure. There are, however, many other possible variations with more complex building forms or with other types of lateral resistive structural systems.

Earthquake (Seismic) Forces

Seismic loads are actually generated by the dead weight of the building and its contents. In visualizing the application of seismic forces, we look at each part of the building and consider its weight as a horizontal force. The forces generated by the building parts will actually be distributed throughout the building, although they will be directed through the parts of the lateral bracing system. Horizontal and vertical diaphragms will generally be loaded as illustrated for the wind loads in Figure 13.1.

For determination of the seismic loads, it is necessary to consider all elements that are permanently attached to the structure. Ductwork, lighting and plumbing fixtures, equipment, signs, and so on, will add to the weight that must be considered for the seismic load. The weight of building contents should also be determined, especially for storage warehouses, parking garages, and other buildings with large amounts of heavy contents.

13.2 HORIZONTAL DIAPHRAGMS

Most lateral resistive structural systems for buildings consist of combinations of vertical and horizontal elements. The horizontal elements are usually the roof and floor framing and decks. When a deck is of sufficient strength and stiffness to be developed as a rigid plane, it is called a *horizontal diaphragm.*

General Behavior

As illustrated in Figure 13.1, a horizontal diaphragm typically functions by collecting the lateral forces at a particular level of the building and then distributing them to the vertical elements of the bracing system. The following are some considerations for the behavior of a horizontal diaphragm and its relationship to the behavior of the general bracing system.

Relative Stiffness of the Horizontal Diaphragm. If the horizontal diaphragm is relatively flexible, it may deflect so much that its continuity is negligible and the distribution of load to the vertical bracing elements is essentially on a *load periphery* basis. If the deck is quite rigid, however, the distribution to vertical elements will be on the basis of the relative stiffness of the vertical elements. These two possibilities are illustrated for a simple box system in Figure 13.2.

Torsional Effects. If the centroid of the lateral forces in the horizontal diaphragm does not coincide with the centroid of the stiffness of the vertical elements, there will be a twisting action (called *rotational effect* or *torsional effect*) on the structure. This effect is in addition to the direct force effect. Figure 13.3 shows a structure in which this effect occurs because of a lack of symmetry of the structure. Torsion is usually of significance only if the horizontal diaphragm is quite stiff. Stiffness in this regard is a matter of the materials of the construction as well as the depth-to-span ratio (that is, the width to length ratio) of the spanning horizontal diaphragm. In general, wood and sheet metal decks are relatively flexible, whereas sitecast concrete decks are very stiff.

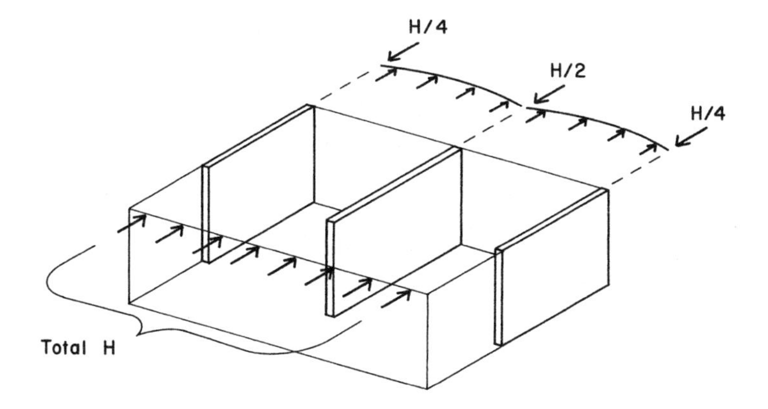

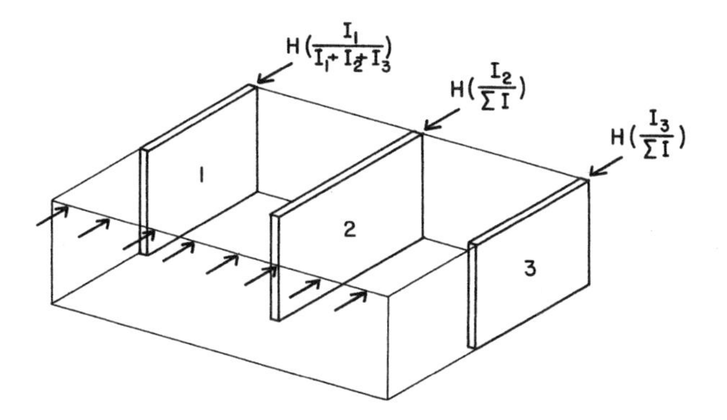

Figure 13.2 Distribution of forces from a horizontal diaphragm to vertical elements of the bracing system.

Relative Stiffness of the Vertical Elements. When vertical elements share load from a rigid horizontal diaphragm, as shown in the lower part in Figure 13.2, their relative stiffness must be determined in order to establish the manner of the load sharing. The determination is comparatively simple when the elements are all similar in type and materials, such as all plywood shear walls. When the vertical elements are different, such as a mix of plywood and masonry shear walls or of some shear walls and some braced frames, their actual deflections must be determined in order to establish the distribution.

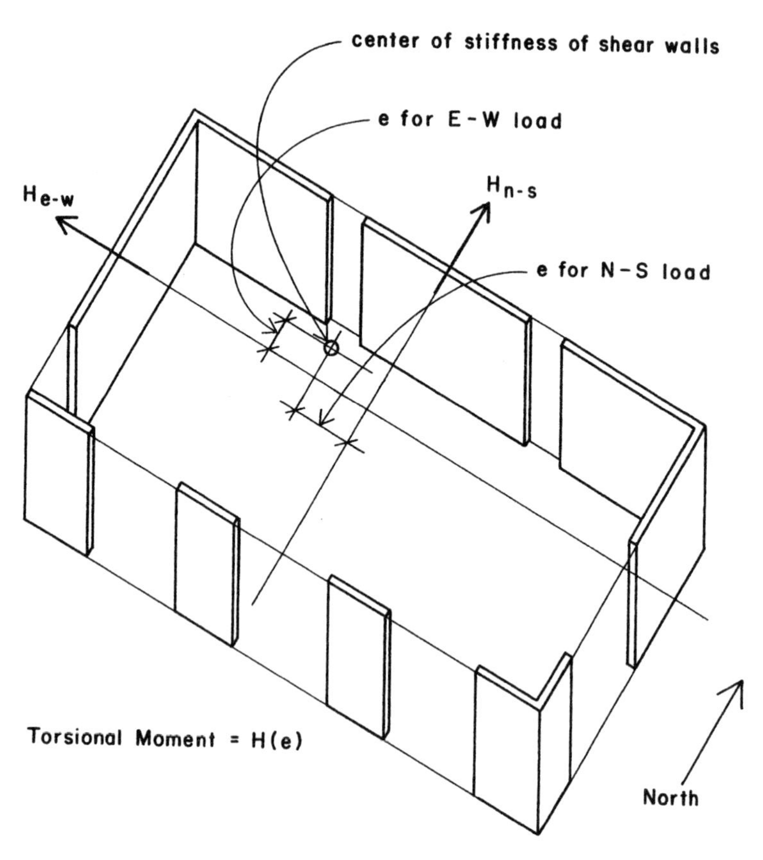

Figure 13.3 Torsional (rotational) moment due to lateral forces on an unsymmetrical structure.

Use of Control Joints. The usual approach in design of lateral load resisting systems is to tie the whole structure together to ensure its overall continuity of movement. Sometimes, however, because of the irregular form or large size of a building, it may be desirable to control its responses to lateral loads by the use of structural control (usually meaning separation) joints. In some cases these joints function to create total separation, thus allowing for completely independent motion of the separated parts of the building. In other cases, the joints may control the movements in a single direction (usually horizontal) while retaining load transfer capabilities in other directions (usually for gravity load transfers).

Design and Usage Considerations. In performing their basic tasks, horizontal diaphragms have a number of potential problems. A major consideration is that of the shear stress in the plane of the diaphragm caused by the spanning action of the diaphragm, as shown in Figure 13.4. This spanning action results in shear stress in the material as well as forces that must be transferred across joints in the deck when it is composed of separate elements such as sheets of plywood or units of sheet metal. Figure 13.5 shows a typical framing detail at

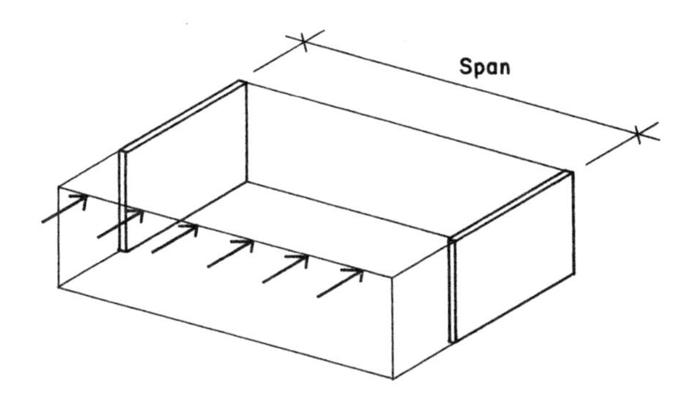

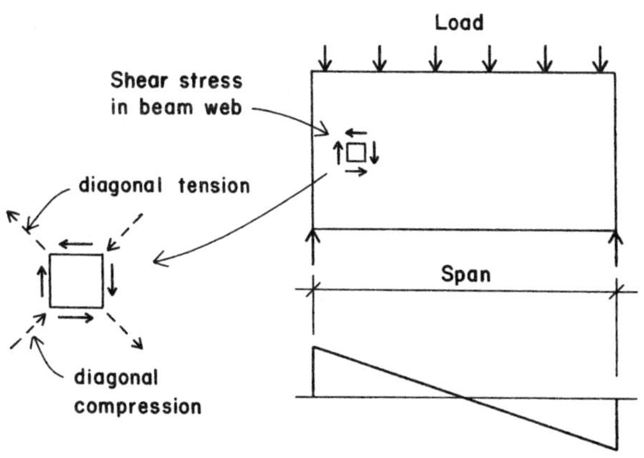

Figure 13.4 Spanning beam functions of a horizontal diaphragm.

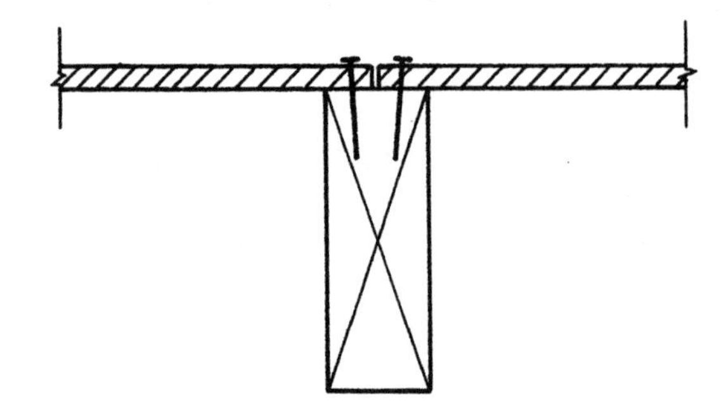

Figure 13.5 Typical nailed joint in a wood-framed diaphragm.

the joint between two plywood panels. The stress in the deck at this location must be passed from one panel through the edge nails to the framing member and then back through the other nails to the adjacent panel. This is a tight situation when the framing members are only 1.5 in. thick.

As is the usual case with shear stress, both diagonal tension and diagonal compression are induced simultaneously with the shear stress (see Figure 13.4). The diagonal tension is critical in concrete with its tension-weak nature. The diagonal compression is a potential source of buckling in decks composed of thin sheets of plywood or metal. In plywood decks the thickness of the plywood relative to the spacing of the supporting framing members must be considered, and it is also why the plywood is nailed to intermediate framing members as well as to those at the edges of the panels.

Diaphragms with continuous deck surfaces are usually designed in a manner similar to that for I-shape steel beams. The deck (web) is designed for the shear, and the flanges (edge framing) are designed for the moment, as shown in Figure 13.6. The edge members are called *chords*, and they are designed for the tension and compression forces at the edges. With diaphragm edges of some length, this usually means that some splicing of the edge framing elements is required. Special edge framing may be provided for this task, but often elements of the ordinary framing system can be utilized.

Typical Construction

The most common horizontal diaphragm is the wood-framed deck sheathed with plywood or OSB panels. For roofs the deck may use

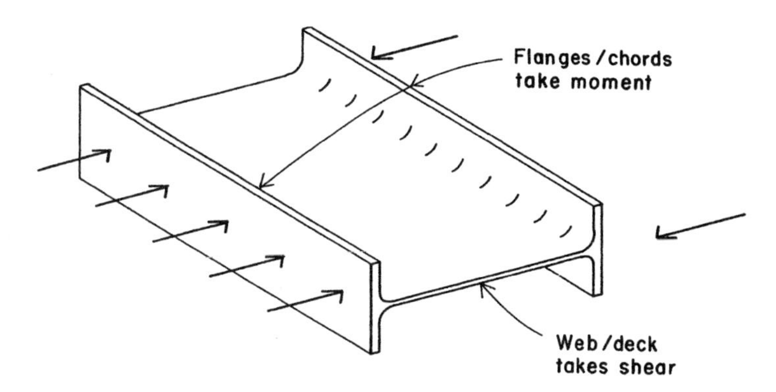

Figure 13.6 Flanged and webbed beam analogy for a horizontal wood-framed diaphragm.

OSB panels or plywood as thin as $^3/_8$ in. However, for roofs with waterproof roofing membranes a thickness of at least $^1/_2$ in. is usually required. The most common floor deck is $^3/_4$-in-thick plywood.

Attachment of sheathing to supporting framing is usually by nailing—by hand or with mechanically driven devices. This form of attachment is also used at the diaphragm boundaries where load transfers must be made to wood framing consisting of the top plates of stud walls or to ledgers attached to the face of supporting walls. Code-acceptable load ratings are based on the plywood type and thickness, the nail size and spacing, and features such as size and spacing of framing members and use of blocking. Load capacities are given in tables such as that shown in Table 13.1. These tables are industry standards and are often incorporated directly in building codes.

In general, wood-framed diaphragms are quite flexible and should be investigated for deflection when spans (distance between bracing) are large or span-depth ratios are high (3 or more). In situations where a deck is continuous over more than two supports, distribution to bracing is ordinarily assumed to be on a peripheral basis, as shown in the upper figure in Figure 13.2.

In some cases the collection of forces into the diaphragm or the distribution of loads to vertical elements may include a stress beyond the capacity of the deck alone. Figure 13.7 shows a building in which a continuous roof deck is connected to a series of shear walls. Load collections and transfers require that some force be dragged along the

TABLE 13.1 Load Values for Plywood Diaphragms

DIAPHRAGMS: RECOMMENDED SHEAR (POUNDS PER FOOT) FOR HORIZONTAL APA PANEL DIAPHRAGMS WITH FRAMING OF DOUGLAS-FIR, LARCH OR SOUTHERN PINE[a] FOR WIND OR SEISMIC LOADING

Panel Grade	Common Nail Size	Minimum Nail Penetration in Framing (inches)	Minimum Nominal Panel Thickness (inch)	Minimum Nominal Width of Framing Member (inches)	Blocked Diaphragms: Nail Spacing (in.) at diaphragm boundaries (all cases), at continuous panel edges parallel to load (Cases 3 & 4), and at all panel edges (Cases 5 & 6)[b]				Unblocked Diaphragms: Nails Spaced 6" max. at Supported Edges[b]	
					6	4	2-1/2[c]	2[c]	Case 1 (No unblocked edges or continuous joints parallel to load)	All other configurations (Cases 2, 3, 4, 5 & 6)
					Nail Spacing (in.) at other panel edges (Cases 1, 2, 3 & 4)					
					6	6	4	3		
APA STRUCTURAL I grades	6d	1-1/4	5/16	2	185	250	375	420	165	125
				3	210	280	420	475	185	140
	8d	1-3/8	3/8	2	270	360	530	600	240	180
				3	300	400	600	675	265	200
	10d[d]	1-1/2	15/32	2	320	425	640	730	285	215
				3	360	480	720	820	320	240
	6d[e]	1-1/4	5/16	2	170	225	335	380	150	110
				3	190	250	380	430	170	125
			3/8	2	185	250	375	420	165	125

APA RATED SHEATHING, APA RATED STURD-I-FLOOR and other APA grades except Species Group 5	8d	1-3/8	3/8	2	240	320	480	545	215	160
				3	270	360	540	610	240	180
			7/16	2	255	340	505	575	230	170
				3	285	380	570	645	255	190
			15/32	2	270	360	530	600	240	180
				3	300	400	600	675	265	200
	10d[(d)]	1-1/2	15/32	2	290	385	575	655	255	190
				3	325	430	650	735	290	215
			19/32	2	320	425	640	730	285	215
				3	360	480	720	820	320	240

(a) For framing of other species: (1) Find specific gravity for species of lumber in the AFPA National Design Specification. (2) Find shear value from table above for nail size for actual grade. (3) Multiply value by the following adjustment factor: Specific Gravity Adjustment Factor = [1 – (0.5 – SG)], where SG = specific gravity of the framing. This adjustment shall not be greater than 1.

(b) Space nails maximum 12 in. o.c. along intermediate framing members (6 in. o.c. when supports are spaced 48 in. o.c.).

(c) Framing at adjoining panel edges shall be 3-in. nominal or wider, and nails shall be staggered where nails are spaced 2 inches o.c. or 2-1/2 inches o.c.

(d) Framing at adjoining panel edges shall be 3-in. nominal or wider, and nails shall be staggered where 10d nails having penetration into framing of more than 1-5/8 inches are spaced 3 inches o.c.

(e) 8d is recommended minimum for roofs due to negative pressures of high winds.

Notes: Design for diaphragm stresses depends on direction of continuous panel joints with reference to load, not on direction of long dimension of sheet. Continuous framing may be in either direction for blocked diaphragms.

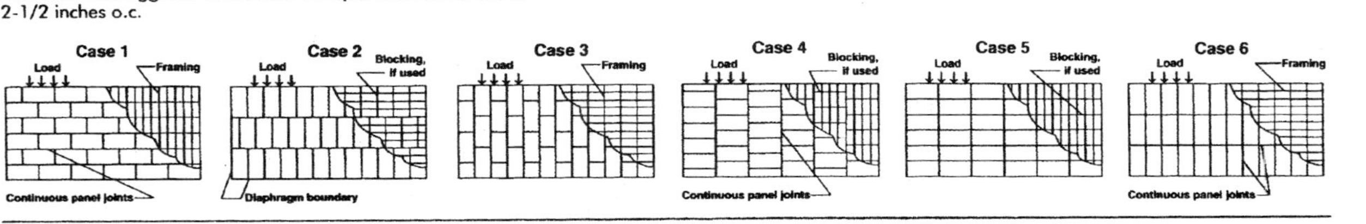

Source: Reproduced from *Introduction to Laternal Design* (Ref. 8), with permission of the publisher APA – The Engineered Wood Association.

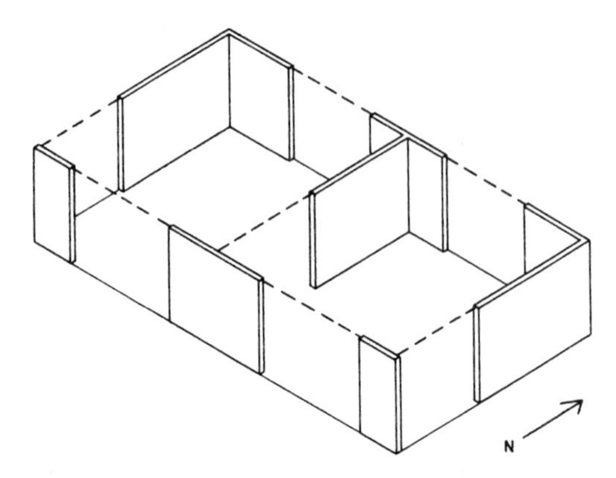

Figure 13.7 Collection and distribution elements in a building braced with shear walls.

dotted lines shown in the figure. For the outside walls it is possible that the edge framing used for the diaphragm chords can do double service for this purpose. For the interior shear wall, and possibly for the exterior walls, if the roof is cantilevered past the walls, some other framing elements may be necessary to reinforce the deck.

Many other types of roof deck construction may function adequately for diaphragm action, especially when the required shear stress resistance is low. Acceptability by local building codes should be determined if any construction other than that described is used.

Shear Capacity of Plywood Decks

Table 13.1 yields the capacity of plywood decks in units of pounds per foot of deck width. The table incorporates a number of variables as follows:

1. *Arrangement of the Plywood Panels.* The table footnotes indicate alternate arrangements of the panels with respect to the supporting framing and the load direction. Data in the table is limited in some cases based on these arrangements.
2. *Provision of Blocking.* This refers to the use of filler pieces inserted between the framing members to provide for support and nailing of the panel edges. The blocking will also function as lateral bracing for the framing. Omission of blocking is permitted only with certain panel arrangements, and in general it results in lowered shear capacities.

3. *Type of Plywood.* Two general classes of grade of the plywood sheets are indicated in the first column of the table.
4. *Size of Nail.* This is generally related to the thickness of the plywood. The table is based on use of common wire nails, although it is common to use shorter nails made specially for nailing plywood sheathing on walls and decks.
5. *Thickness of Plywood.* Table values are given for thickness ranging from 5/16 to 19/32 inch.
6. *Size of framing.* Loads are given for nominal 2 in. and 3 in. framing.
7. *Special Considerations.* Table footnotes provide for additional modifications in some cases.

Building code provisions usually provide for a maximum nail spacing of 6 in. at panel edges and 12 in. at center points of the panels (called the *field* of the panel). The following examples illustrate the use of the table data for ASD work.

Example 1. Determine the maximum shear capacity for a horizontal blocked diaphragm consisting of plywood panels nailed to 2 in. nominal framing. Data for the construction is as follows:

Panels are Structural I, 3/8 in. thick, arranged in the Case 2 pattern and attached to blocked framing.
Nails are 8d common spaced 4 in. at diaphragm boundaries and 6 in. at all other panel edges.

Solution: For this combination of conditions Table 13.1 yields a value of 360 lb/ft.

Example 2. A roof deck made with 15/32 in. Structural I plywood is required to resist a diaphragm shear loading that results in a maximum stress of 450 lb/ft in the deck. The blocked diaphragm has the plywood panels arranged as shown for layout Case 1. Find the nail size and spacing required for (1) 2 in. nominal framing and (2) 3 in. nominal framing.

Solution: (1) For the 2 in. nominal framing, the required spacing for 10d nails is 2.5 in. at the diaphragm boundaries and 4 in. at other panel edges. However, reference to the footnote in Table 13.1 indicates that for the 2.5 in. spacing the framing must be a minimum of 3 in. nominal size. Therefore the 2 in. nominal framing cannot be used to achieve the required shear capacity for this example.

(2) If the 3 in. framing is used, the table indicates a shear capacity of 480 lb/ft with 10d nails at 4 in. at the boundaries and at 6 in. at other panel edges.

For spanning diaphragms that function as uniformly loaded simple beams, the highest shear stress condition occurs only at the ends of the span, as shown for Case 2 in Figure 5.25. When the stress at the ends requires extra diaphragm strength over that provided with minimal construction, it is possible to reduce this construction for the interior portions of the diaphragm. The following example illustrates the procedure for this situation.

Example 3. A one-story wood-framed building has the plan shown in Figure 13.8. The roof consists of wood framing and a plywood deck with 15/32 in. thickness. For lateral load resistance the diaphragm deck must develop a maximum unit shear capacity of 381 lb/ft at the ends of the building. The loading condition produces a uniformly distributed load on the 120-ft span diaphragm. Determine options for the roof construction based on considerations for lateral loading. Assume a Case 1 panel layout as shown in Table 13.1.

Solution: For the maximum shear, options are:

15/32 in. Structural I plywood with 3 in. framing, blocking, and 10d nails (requiring the 3 in. framing) at 4 in. at boundaries and 6 in. at other panel edges. (Capacity 480 lb/ft.) 15/32 in. APA Rated Sheathing with 3 in. framing and 8d nails at 4 in. boundaries and 6 in. at other panel edges. (Capacity 400 lb/ft.)

This construction is required only at the ends of the roof. As the shear drops off toward the middle of the span, a reduction is possible, with a lower limit being the minimum required nailing of 8d nails at 6 in. at all edges and 2 in. framing. On the shear stress diagram in Figure 13.8 the capacity for APA Rated Sheathing with 3 in. framing is chosen for the maximum shear at the diaphragm end and is used out to a distance where a lower capacity is required. Reduction is shown in two stages, the first being one with 2 in. framing, 8d nails at 4 in. spacing at boundaries, and at 6 in. at other edges. Finally a drop is made to the

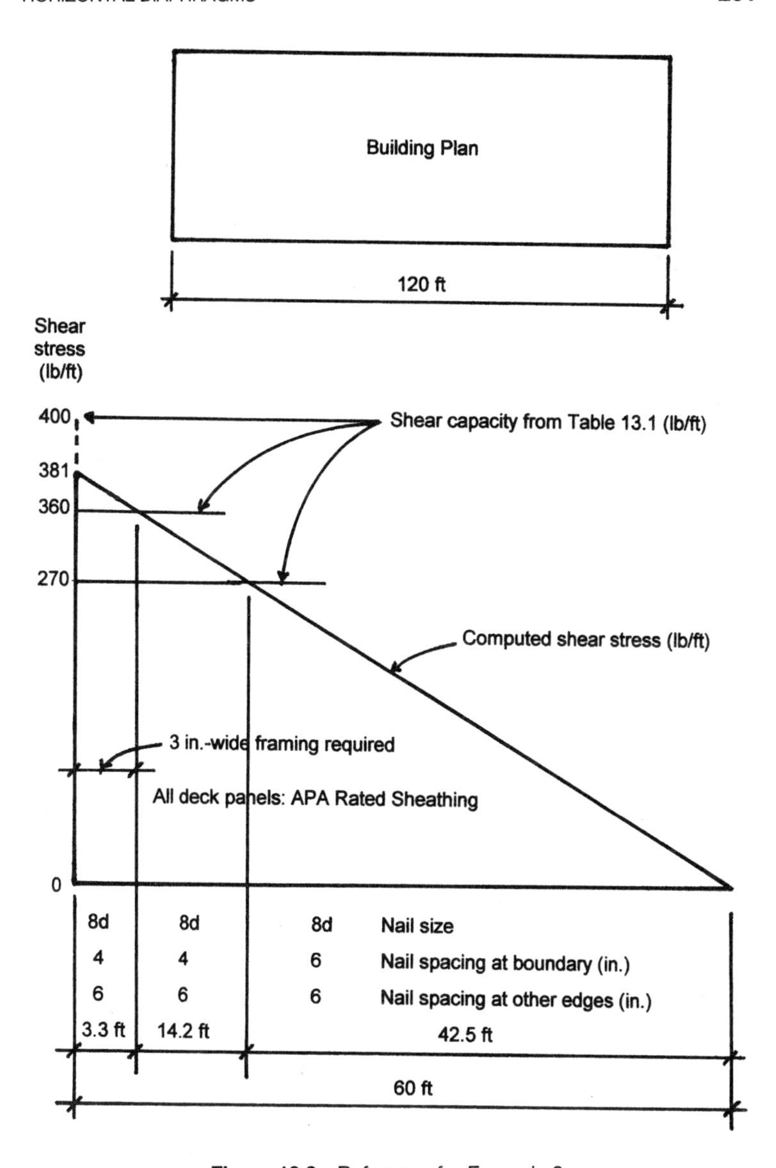

Figure 13.8 Reference for Example 3.

minimum nailing of 6 in. at all panel edges. The distances relating to these variations are shown at the bottom of the shear diagram. Note that the majority of the roof is able to use the minimum construction.

Design of Diaphragms by the LRFD Method

Use can be made of the LRFD method for the design work illustrated previously in this section. As with most other situations, the primary differences consist of determining the design factored loads and converting the unit resistances of sheathing, anchorage devices, and sill bolts to an adjusted value for the LRFD method—a process discussed for various situations in previous chapters of this book. This process is not shown here but is discussed for some of the design example cases in Chapter 15.

Problem 13.2.A. Determine the maximum shear capacity for a horizontal diaphragm consisting of the following: 3 in. nominal framing, 10d nails at 6 in. at all panel edges, 15/32 in. APA Rated Sheathing, blocked diaphragm with Case 2 panel arrangement.

Problem 13.2.B. Same as Problem 13.2.A, except diaphragm is unblocked, panel arrangement is Case 1.

Problem 13.2.C. A roof deck with 3/8 in. APA Rated Sheathing on 2 in. nominal framing is used to form a blocked diaphragm with panel arrangement Case 3. Find the nail size and spacing required for a maximum shear stress of 300 lb/ft.

Problem 13.2.D. Same as Problem 13.2.C, except plywood thickness is 15/32 in. and maximum shear stress is 375 lb/ft.

13.3 VERTICAL DIAPHRAGMS (SHEAR WALLS)

Vertical diaphragms are usually the walls of buildings and are more commonly referred to as *shear walls*. In addition to their shear wall functions, they must fulfill various architectural functions and may also be required to serve structural functions, such as gravity load bearing or receiving of wind pressures on the building exterior. The location of walls, the materials used, and some of the details of their construction must be developed with all of these functions in mind.

The most common shear wall constructions are those of sitecast concrete, masonry, and surfaced stud wall construction on frames of wood or light gauge steel. The framed wall can be made rigid with diagonal bracing but is most often developed with surfacing of sufficient strength for the shear stress capacity required. Choice of the type of construction may be limited by the magnitude of shear caused by the lateral loads, but will also be influenced by fire code requirements and the satisfaction of the various other wall functions, as previously described.

General Behavior

Some of the structural functions usually required of vertical diaphragms are the following (see Figure 13.9):

1. *Direct Shear Resistance.* This usually consists of the transfer of a lateral force in the plane of the wall from some upper level of

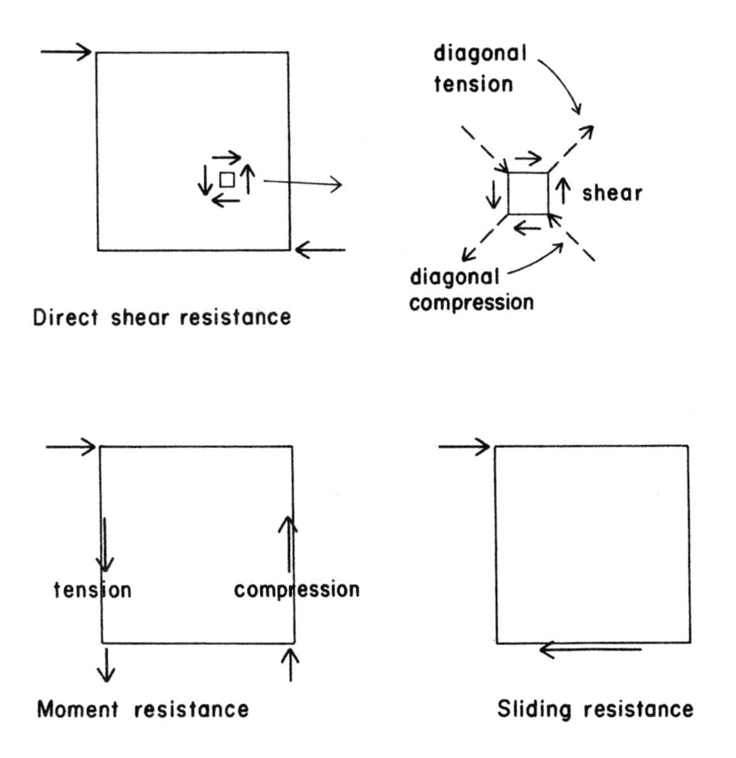

Figure 13.9 Structural functions of a shear wall.

the wall to a lower level in the wall. This results in the typical situation of shear stress and the accompanying diagonal tension and compression stresses, as discussed for horizontal diaphragms. Diagonal tension will cause cracking in concrete and masonry walls, whereas diagonal compression may cause buckling of thin surfacing on framed walls.

2. *Cantilever Moment Resistance.* Shear walls generally work like vertical cantilevers, developing compression on one side and tension on the other side. The tension and compression forces are usually considered to be concentrated at the wall edges, as with the chords in a horizontal diaphragm. The transfer of the moment at the base of the wall must also be considered for attachment of the wall to its supports.
3. *Horizontal Sliding Resistance.* The direct transfer of the lateral load at the base of a wall produces a tendency for the wall to slide horizontally off its supports. Significant friction resistance may be developed, but for framed walls the sliding is usually developed with anchoring of the wall sill to the supports.

Shear stress development in wall surfacing is usually considered separately from other wall structural functions. As with horizontal diaphragms, shear stresses developed from lateral loads are compared to the rated capacity of the wall construction. For a plywood-surfaced wall with a wood frame the construction as a whole is usually rated for its resistance in units of pounds per foot (lb/ft) of wall length. As with horizontal diaphragms, considerations are made for the type and grade of surfacing, size and spacing of frame members, and type, size, and spacing of fasteners. Also of concern are the arrangements of panels and the presence or absence of blocking. Rated shear capacities are provided for wood-framed walls with surfacing of plywood, particleboard, cement plaster, gypsum drywall, and various proprietary products.

Although the possibility exists for the buckling of walls as a result of the diagonal compression effect, this is usually not critical because other limitations exist to constrain wall slenderness. Slenderness of wood studs is limited by design for gravity loads and by code requirements as a function of the stud size and the wall height. Most stud walls are surfaced on both sides, and the sandwich panel effect is usually sufficient to provide for reasonable stiffness.

As in the case of horizontal diaphragms, the moment effect on the wall is usually considered to be resisted by the two vertical edges of the wall acting as flanges for a beam. In wood-framed walls the end-framing members, usually consisting of a minimum of two studs, are considered to fulfill this function. Where the wall end framing also supports beams, it must be designed for the combination of gravity and lateral loads.

The form of the investigation for overturn of a cantilevered wall is shown in Figure 13.10. The overturning moment is resisted by the stabilizing moment produced by the dead load supported by the wall plus the weight of the wall itself. The safety factor (SF) against overturn is defined as

$$\text{SF} = \frac{\text{Stabilizing Moment}}{\text{Overturning Moment}} = \frac{DL \times a}{H \times h}$$

This should be greater than one, and a preferred minimum is usually 1.5. If the stabilizing moment is not sufficient, a tiedown anchorage is required, with the value for its resistance determined as

$$T = \frac{(1.5 \times H \times h) - (DL \times a)}{L}$$

This analysis is sufficient for wind loads, but for earthquake forces it is usually required to reduce the dead load to account for some loss of its effectiveness created by vertical thrust effects of the earthquake.

Resistance to horizontal sliding at the base of a shear wall will be partly resisted by sliding friction due to the dead load. It is common, however, to ignore friction and to design the anchorage—consisting of the sill bolts for a stud wall—for the full lateral forces.

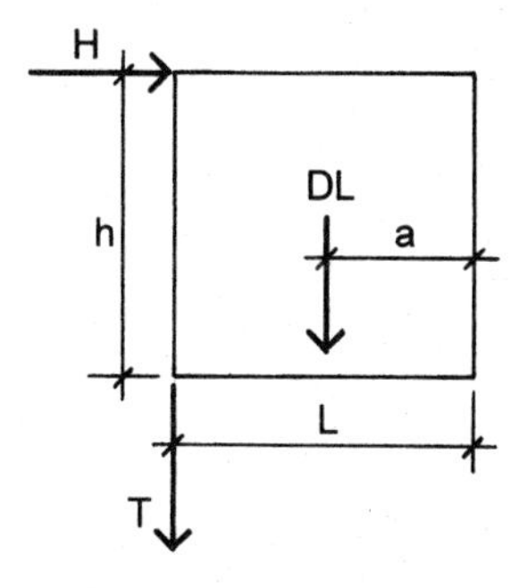

Figure 13.10 Overturn analysis for a shear wall.

Design and Usage Considerations

As discussed for horizontal diaphragms, an important judgment that must often be made in designing for lateral loads is that of the manner of distribution of lateral forces between a number of shear walls that share a load from a horizontal diaphragm. In some cases, the existence of symmetry or of a flexible diaphragm may simplify this consideration. In many cases, however, the relative stiffness of the walls must be determined for this situation.

If considered in terms of static force and elastic stress-strain conditions, the relative stiffness of a wall is inversely proportional to its deflection under a unit load. Figure 13.11 shows the manner of deflection of a shear wall for two assumed conditions. In (*a*) the wall is considered to be fixed at its top and bottom, and to flex in a double curve with an inflection point at midheight. This is the case usually

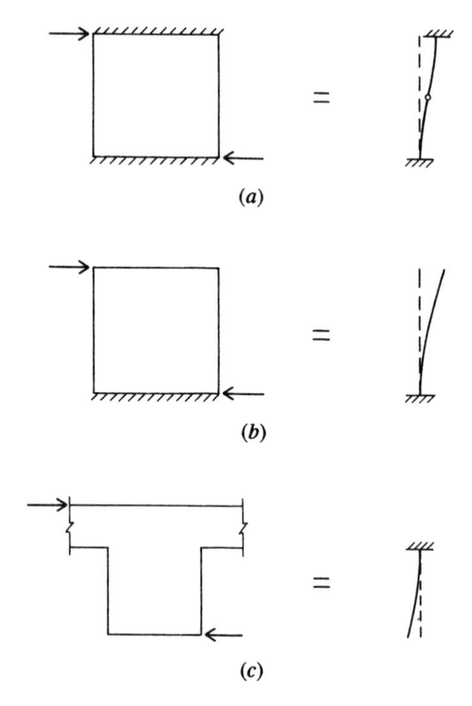

Figure 13.11 Alternative forms of behavior for isolated shear walls (also called individual piers): (a) fixed top and bottom, (b) cantilevered from a fixed bottom, (c) pinned base with fixed or rigidly connected top.

assumed for a continuous wall of concrete or masonry in which a series of individual wall portions (called *piers*) is connected by a continuous upper wall or other construction with considerable stiffness. In (*b*) the wall is considered to be fixed at its base only, functioning as a vertical cantilever. This is the case for independent, freestanding walls or for walls in which the continuous upper structure is relatively flexible. A third possibility is shown in (*c*) in which relatively short piers are assumed to be fixed at their tops only, which essentially produces the same deflection condition as in (*b*).

In some instances the deflection of the wall may result largely from shear distortion rather than from flexural distortion. This may occur because of the wall materials and construction or the proportion of wall height to plan length. Furthermore, stiffness due to dynamic loads (wind gusts and earthquake jolts) is not quite the same as stiffness in resistance to static loads. The following are some recommendations for single-story shear walls.

1. For wood-framed shear walls with height-to-length ratios of 2 or less, assume the stiffness to be proportional to the plan length of the wall.
2. For wood-framed walls with height-to-length ratios over 2, assume the stiffness to be a function of the height-to-length ratio and the method of support (cantilevered or fixed top and bottom—see Figure 13.11).
3. Avoid situations in which walls of significantly great differences in stiffness share loads along a single row. The short walls will tend to receive a small share of the load, especially if stiffness is assumed to be a function of the height-to-length ratio.
4. Avoid mixing shear walls of different construction when they share loads on a deflection basis.

Item 4 in the preceding list can be illustrated by two situations, as shown in Figure 13.12. The first situation is that of a series of shear walls in a single row. If some of these walls are of concrete or masonry and others of wood frame construction, the stiffer concrete or masonry walls will tend to absorb the major portion of the load.

In the second situation shown in Figure 13.12, a set of parallel walls share load from a horizontal diaphragm. Unless the horizontal diaphragm is assumed to be very flexible, the stiffer concrete or masonry construction will tend to attract a disproportionate share of the total diaphragm load.

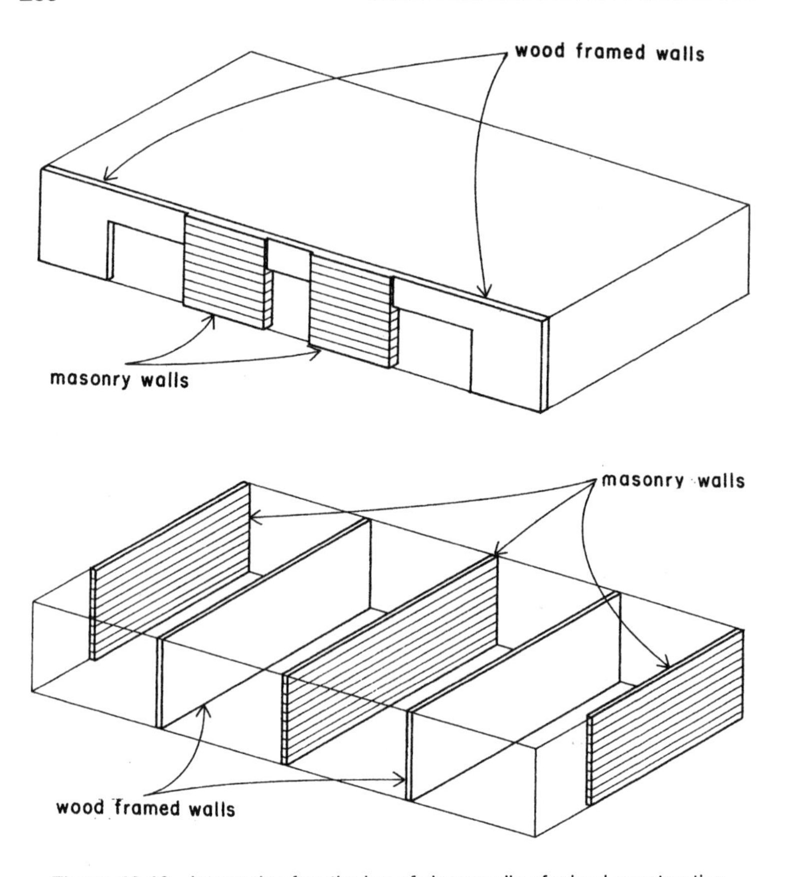

Figure 13.12 Interactive functioning of shear walls of mixed construction.

In addition to the various considerations mentioned for the shear walls themselves, care must be taken to ensure that adequate attachment to the load-distributing diaphragm is made. This may involve some extra framing members or special provisions for splicing continuous members to achieve the collection of forces by the horizontal diaphragm and the transfers to the individual shear walls.

A final consideration for shear walls is that they must be made an integral part of the whole building construction. In long building walls with large door or window openings or other gaps in the wall, shear walls are often considered as entities (isolated, independent piers) for their design. However, the behavior of the entire wall under lateral load should be studied to be sure that the elements not considered to be parts

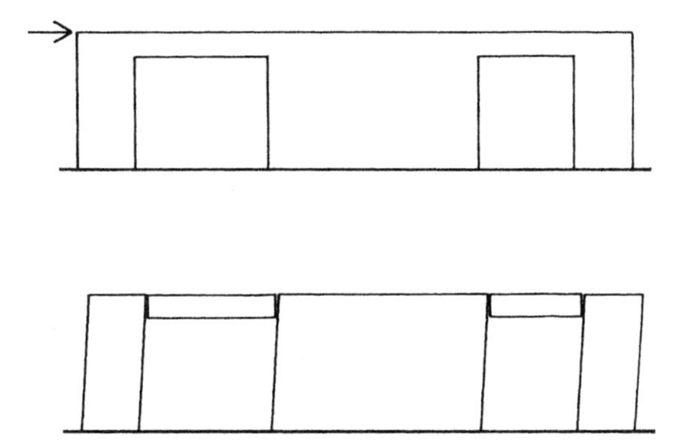

Figure 13.13 Effects of deformation due to lateral loads in continuous wall construction without control joints.

of the lateral bracing system do not suffer damage because of the wall distortions under lateral load.

An example of this situation is shown in Figure 13.13. The relatively long solid portion is assumed to perform the bracing function for the entire wall and would be designed as an isolated pier. However, when the wall deflects, the effect of the movement on the shorter piers, on the headers over openings, and on the door and window framing must be considered. The headers must not be cracked loose from the solid wall portions or pulled off their supports.

Typical Construction

The common forms for shear walls include surfaced wood-frame walls, concrete walls—both sitecast and precast, and masonry walls—usually of reinforced CMU (concrete masonry units) construction. In the past, wood-framed shear walls were mostly surfaced with plywood. Experience and testing have established acceptable shear ratings for various other surfacing, so that plywood is used less, especially when shear stresses are low.

For all types of walls there are various requirements (for good carpentry, fire resistance, proper installation of available products, etc.) that establish a certain minimum construction. In many situations this "minimum" is really adequate for low levels of shear resistance. Obtaining wall strengths above the minimum requires increasing the size or quality of units, adding or strengthening of attachments, developing stronger framing, and so on. There are also upper limits for each type of construction, which means that additional strength can only be obtained by a change in the basic construction.

13.4 INVESTIGATION AND DESIGN OF WOOD-FRAMED SHEAR WALLS

Wood-framed shear walls occur most often as exterior building walls with the shear-resisting surfacing attached to the outside surface of the framing. Shear-resisting sheathing on exterior walls is most often plywood or OSB panels. If stucco is applied to the exterior surface, it will have a stiffness greater than the sheathing, and provision must be made to use the stucco (which has a rated shear capacity) or to provide jointing that prevents the stucco from cracking when the sheathing must take the shear forces.

Interior shear walls are mostly sheathed with gypsum drywall on both surfaces. Although the individual drywall panels are relatively weak, the ability to add the two surfaces for a single wall produces a significant resistance. If this resistance is still not sufficient, some strengthening of the wall is required—possibly by using plywood with the drywall attached as a finish.

Table 13.2 yields the shear capacity of plywood shear walls with ordinary wood stud framing and typical grades of plywood products. The table incorporates a number of variables as follows:

1. *Plywood Grade.* The table contains data for three groups of surfacing, relating to their APA grade.
2. *Thickness of Panels.* A range of standard sizes is given, although the most common are 3/8 and 15/32 in. panels.
3. *Nail Size.* This is associated with the panel thickness; the smallest nails possible are preferred because of ease of installation and less likelihood of splitting of the studs. The table is based on ordinary common wire nails, although various equivalent power-driven fasteners are now commonly used.
4. *Nail Spacing.* Maximum spacing of nails at panel edges is specified elsewhere as 6 in. Data is given here for spacing as close as 2 in., although this close spacing is seldom used because the large number of nails and the possibility of splitting of the studs present installation difficulties. Studs wider than 2 in. nominal are required for close spacing or large nails. As with roof and floor decks, the interior portions of panels are attached to studs with nails at a maximum of 12 in.

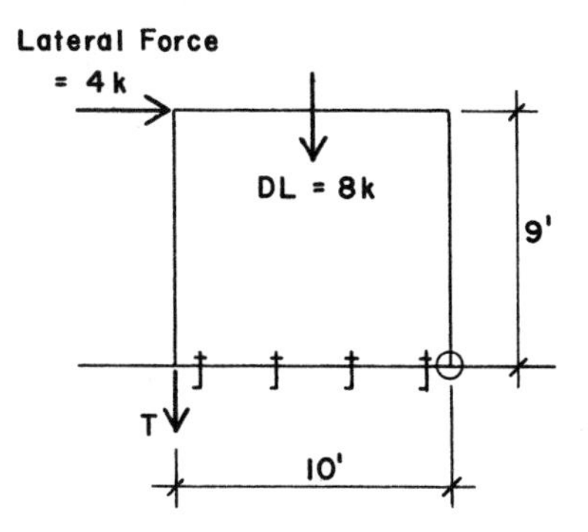

Figure 13.14 Reference for Example 4.

5. *Stud Spacing.* For thin panels the buckling of the plywood may be a problem, which results in some restrictions on stud spacing.
6. *Special Considerations.* Table footnotes give some special restrictions and allowances for special situations.

The following example illustrates the typical procedure for design of a plywood-surfaced shear wall.

Example 4. A plywood shear wall having the form shown in Figure 13.14 is required to resist the lateral force indicated. Assuming the lateral force to be due to wind and the total dead load on the wall to be as shown, design the wall and its framing for the following data:

Plywood: Douglas fir, APA Rated Sheathing
Wall framing: Douglas fir-larch, stud grade
Wall sill-bolted to a concrete base.

Solution: The maximum shear stress in the plywood is

$$v = \frac{\text{Lateral Force}}{\text{Wall Length}} = \frac{4000}{10} = 400 \text{ lb/ft}$$

From Table 13.2 the following options are possible:

3/8 in. plywood, 6d nails at 2 in., $v = 510$ lb/ft (requires 3 in. studs)
3/8 in. plywood, 8d nails at 3 in., $v = 410$ lb/ft
15/32 in. plywood, 10d nails at 4 in., $v = 460$ lb/ft (requires 3 in. studs)

TABLE 13.2 Load Values for Plywood Shear Walls

SHEAR WALLS: RECOMMENDED SHEAR (POUNDS PER FOOT) FOR APA PANEL SHEAR WALLS WITH FRAMING OF DOUGLAS-FIR, LARCH, OR SOUTHERN PINE[a] FOR WIND OR SEISMIC LOADING[b]

Panel Grade	Minimum Nominal Panel Thickness (in.)	Minimum Nail Penetration in Framing (in.)	Panels Applied Direct to Framing					Panels Applied Over 1/2" or 5/8" Gypsum Sheathing				
			Nail Size (common or galvanized box)	Nail Spacing at Panel Edges (in.)				Nail Size (common or galvanized box)	Nail Spacing at Panel Edges (in.)			
				6	4	3	2[e]		6	4	3	2[e]
APA STRUCTURAL I grades	5/16	1-1/4	6d	200	300	390	510	8d	200	300	390	510
	3/8	1-3/8	8d	230[d]	360[d]	460[d]	610[d]	10d[f]	280	430	550	730
	7/16			255[d]	395[d]	505[d]	670[d]					
	15/32			280	430	550	730					
	15/32	1-1/2	10d[f]	340	510	665	870	—	—	—	—	—
APA RATED SHEATHING; APA RATED SIDING[g] and other APA grades except species Group 5	5/16 or 1/4[c]	1-1/4	6d	180	270	350	450	8d	180	270	350	450
	3/8			200	300	390	510		200	300	390	510
	3/8	1-3/8	8d	220[d]	320[d]	410[d]	530[d]	10d[f]	260	380	490	640
	7/16			240[d]	350[d]	450[d]	585[d]					
	15/32			260	380	490	640					
	15/32	1-1/2	10d[f]	310	460	600	770	—	—	—	—	—
	19/32			340	510	665	870	—	—	—	—	—

APA RATED SIDING 303(g) and other APA grades except species Group 5			Nail Size (galvanized casing)					Nail Size (galvanized casing)				
	5/16(c)	1-1/4	6d	140	210	275	360	8d	140	210	275	360
	3/8	1-3/8	8d	160	240	310	410	10d(f)	160	240	310	410

(a) For framing of other species: (1) Find specific gravity for species of lumber in the AFPA National Design Specification. (2) For common or galvanized box nails, find shear value from table above for nail size for actual grade. (3) Multiply value by the following adjustment factor: Specific Gravity Adjustment Factor = [1 – (0.5 – SG)], where SG = specific gravity of the framing. This adjustment shall not be greater than 1.

(b) All panel edges backed with 2-inch nominal or wider framing. Install panels either horizontally or vertically. Space nails maximum 6 inches o.c. along intermediate framing members for 3/8-inch and 7/16-inch panels installed on studs spaced 24 inches o.c. For other conditions and panel thicknesses, space nails maximum 12 inches o.c. on intermediate supports.

(c) 3/8-inch or APA RATED SIDING 16 oc is minimum recommended when applied direct to framing as exterior siding.

(d) Shears may be increased to values shown for 15/32-inch sheathing with same nailing provided (1) studs are spaced a maximum of 16 inches o.c., or (2) if panels are applied with long dimension across studs.

(e) Framing at adjoining panel edges shall be 3-inch nominal or wider, and nails shall be staggered where nails are spaced 2 inches o.c.

(f) Framing at adjoining panel edges shall be 3-inch nominal or wider, and nails shall be staggered where 10d nails having penetration into framing of more than 1-5/8 inches are spaced 3 inches o.c.

(g) Values apply to all-veneer plywood APA RATED SIDING panels only. Other APA RATED SIDING panels may also qualify on a proprietary basis. APA RATED SIDING 16 oc plywood may be 11/32 inch, 3/8 inch or thicker. Thickness at point of nailing on panel edges governs shear values.

Typical Layout for Shear Walls

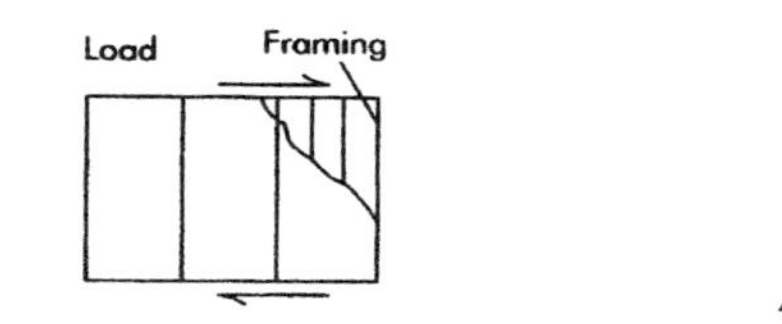

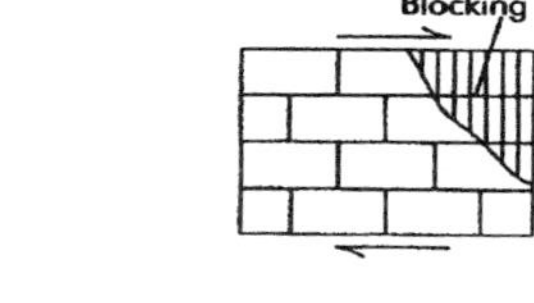

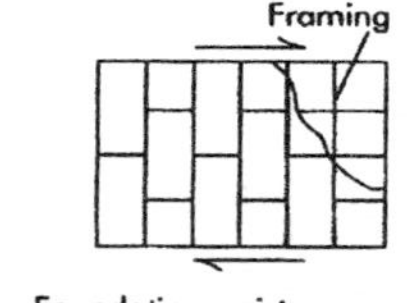

Source: Reproduced from *Introduction to Laternal Design* (Ref. 8), with permission of the publisher APA – The Engineered Wood Association.

Table footnote (d) permits an increase in the value for the 3/8 in. plywood with 8d nails under certain circumstances, although this is not critical for this example. Choice of the plywood and its nailing must be done with consideration of the general construction and other functions of the wall. For the given conditions of this example, any of the three options is adequate. Of course, another option is to use 3/8 in. plywood on both wall surfaces with minimum edge nailing of 6d nails at 6 in. on center, although this is usually not a desirable form of construction in general.

For the overturn analysis, the following is determined:

$$\text{Overturning moment} = (\text{lateral force}) \times (\text{wall height}) \times (\text{safety factor})$$
$$= (4)(9)(1.5) = 54 \text{ kip-ft}$$
$$\text{Restoring moment} = (\text{dead weight}) \times (1/2 \text{ wall length})$$
$$= (8)(10/2) = 40 \text{ kip-ft}$$

Because the overturning effect (with safety factor) prevails, a tiedown force is required at the wall end, with its value equal to

$$T = \frac{54 - 40}{10} = 1.4 \text{ kip}$$

Various patented devices, available as stock hardware items, may be used for this anchorage. This force is well within the capacity of a relatively modest device and not a problem for ordinary framing of the stud wall. Figure 13.15 shows two means commonly employed for anchorage of the end of a shear wall. Figure 13.15*a* shows the general form of a device used for the situation where a wall sits on a concrete foundation. This device is typically attached to the end framing of the shear wall and then to an anchor bolt cast into the supporting concrete. The device itself, the anchor bolt, and the bolts connecting the device to the framing are matched in strength to achieve a rated capacity for the anchorage. A range of sizes and strengths of these fasteners is available. Note that the device does not rest on the sill; this is deliberately done to isolate the function of the device and not use it for horizontal sliding resistance.

Figure 13.15*b* shows a common means for anchoring of the ends of exterior shear walls that are supported by wood framing (usually a shear wall below). This could be achieved with two anchorage devices

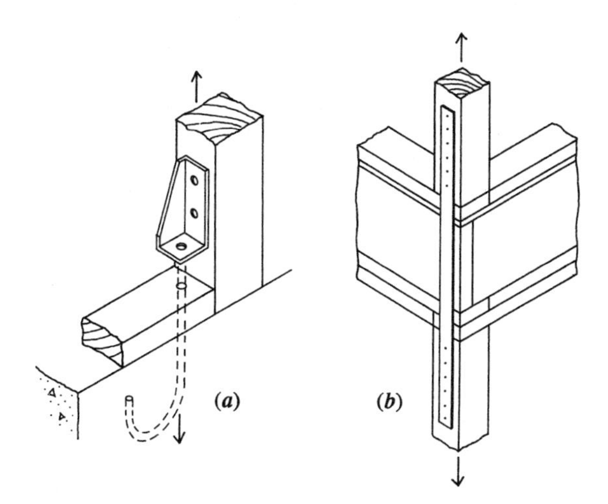

Figure 13.15 Anchorage of shear walls to resist overturning forces: (a) at foundation, (b) between levels in a multistory exterior wall.

like that shown in Figure 13.15*a*, but the cost of the hardware and its installation is much less when the device shown in Figure 13.15*b* is used. This device consists of a heavy-gauge sheet steel strap that is nailed or screwed to the framing above and below the joint. This typically occurs at the outside of an exterior wall, and the strap is usually easily incorporated in the exterior finish materials for the wall.

The usual minimum sill bolting required by building codes is one with $\frac{1}{2}$ in. bolts not over 12 in. from each end of the wall and additional bolts not over 6 ft on center. This would result in a minimum of three bolts for this wall. Based on single shear in a 2 in. nominal thickness sill, Table 10.1 yields a value of 480 lb for one $\frac{1}{2}$ in. bolt. With three bolts and an increase of stress of 1.6 for wind (see Table 4.4), the sliding resistance is thus

$$H = 3 \times 480 \times 1.6 = 2304 \text{ lb}$$

Because this is considerably short of the required 4000 lb, some increase in resistance over the minimum bolting is required. Possible choices include more bolts, larger bolts, a dense-grade sill, or a thicker sill. Again, other factors of the general construction must be considered for a real situation. The least desirable choice is probably for a large number of small bolts because the setting of the bolts in the concrete is a major labor cost. For illustration, try four $3/4$ in. bolts, for which the unit bolt capacity is 720 lb and the total resistance is

$$H = 4 \times 720 \times 1.6 = 4608 \text{ lb}$$

Although the computations will not support the decision, it is probably desirable to increase the sill thickness to a nominal 3 in. with the larger bolts.

As mentioned previously, OSB sheathing is now commonly used for both exterior walls and roof decks. Figure 13.16 shows a residence under construction with an example of this construction. As with any wall sheathing, there is a limit to the material capacity and to the height-to-plan length ratio for the wall. There are two wide panels of wall at the second-story level in Figure 13.16 and they are evidently adequate for the shear at this level. Barely visible in the photo are the straps (as in Figure 13.15*b*) used for anchoring the ends of these two walls. The shorter plan length wall portions are also shown as strapped, but their relative stiffness is too low to make them have any significant contribution to the lateral load on the whole wall. For the shorter wall portions, it is probably a matter of ensuring their anchorage, rather than developing their resistance. The photo in Figure 13.16 also shows part of the front wall of this building, where a lack of sufficient wall plan

Figure 13.16 Use of OSB wall and roof sheathing on a residence. Note anchorage straps at second-floor level.

lengths produces a special problem for lateral bracing in this wall plane. Solutions for this situation are discussed and illustrated in Section 13.6 and illustrated in Figures 13.22 and 13.23.

Problem 13.4.A. A plywood shear wall similar to that shown in Figure 13.14 is to be designed for the following data: lateral force is 6 kips, wall is 10 ft high and 16 ft long, total dead load is 12 kips, plywood is structural I grade. Select the plywood and its nailing. Determine the sill bolting required and investigate the need for tiedown anchorage.

Problem 13.4.B. Same as Problem 13.4.A, except lateral force is 7 kips, wall is 8 ft high and 18 ft long, total dead load is 10 kips.

13.5 TRUSSED BRACING FOR WOOD FRAMES

Incorporation of diagonal members for bracing of otherwise rectangular frameworks dates back to the earliest use of wood frames. In many cases, however, frames achieve lateral stability from the surface materials applied to them, or from other construction elements such as masonry or concrete walls. A common form of bracing used in the past is that of *let-in-bracing*, ordinarily consisting of a thin board recessed (called letting in) on the face of the wall studs. If used today, the primary purpose of this bracing is to temporarily brace the wall to keep it aligned during construction, before the sheathing is applied. If used for bracing now, the diagonal member is more likely to be a metal strap on the wall surface, similar to the anchor straps shown in Figure 13.15*b*.

Post-and-beam structures, consisting of separate vertical and horizontal members, may be inherently stable for resisting gravity loads, but they must be braced in some manner to resist lateral loads. The three basic ways of achieving this bracing are by the use of shear panel surfacing, moment-resistive joints in the frame, or trussing. The trussing, or triangulation, is usually formed by inserting diagonal members in the rectangular spaces of the framing.

If single diagonals are used for trussing, they must serve a dual function: acting in tension for the lateral loads in one direction and in compression when the load direction is reversed (see Figure 13.17*a*). Because long tension members are more efficient than long compression members, frames are often braced with a criss-crossed set of diagonals

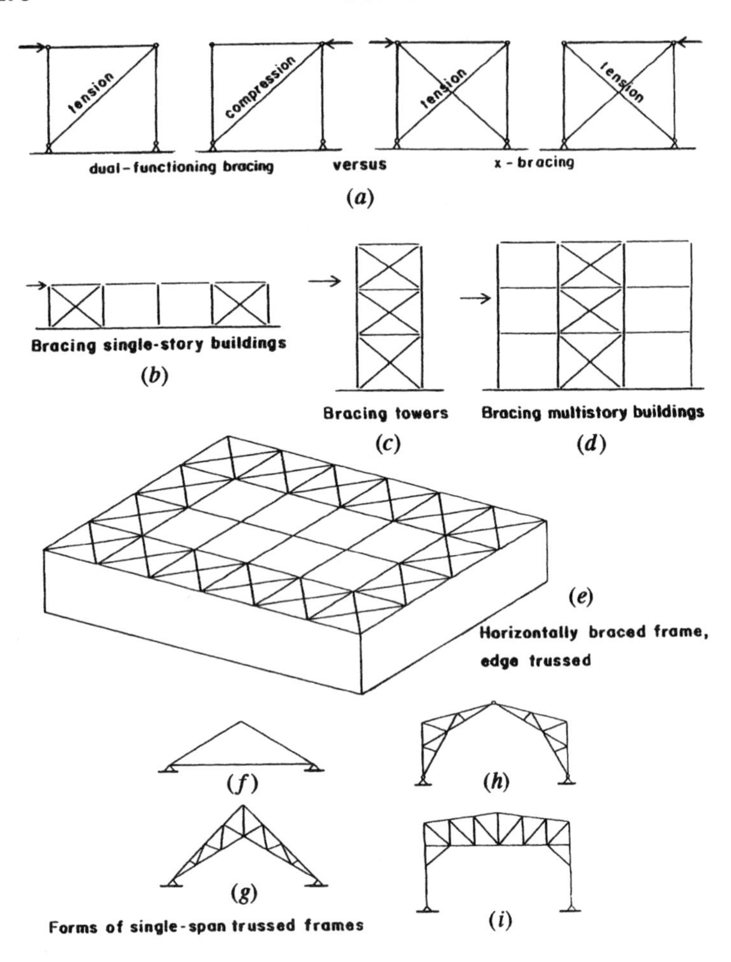

Figure 13.17 Considerations for use of trussed bracing.

(called *x-bracing*) to eliminate the need for the compression members. In any event, the trussing causes the lateral loads to induce only axial forces in the members of the trusses. The trussed frame is generally stiffer than one that is developed as a rigid frame with moment-resisting connections. Unless, of course, the rigid frame is developed with exceptionally stiff members.

Single-story, single-bayed buildings may be braced as shown in Figure 13.17*a*. Single-story, multiple-bayed building may be braced by trussing fewer than all of the bays, as shown in Figure 13.17*b*. The continuity of the frame is used in the latter case to permit the rest of the bays to tag along. Similarly, a single-bayed, multiple-storied, tower-like structure, as shown in Figure 13.17*c*, may have its frame fully braced, whereas the more common multiple-bayed frame may be braced by trussing a single vertical stack in the system, as shown in Figure 13.17*d*.

In general, trussing is inserted only in bays that define solid walls, since the need for openings for doors or windows or the open space required for interior hallways usually conflicts with the truss layout. To a certain extent this is similar to development of a shear wall system, whereas the use of a rigid frame opens up the bays for openings for doors or windows. In many cases this makes the rigid frame bracing system more attractive for architectural planning.

Just about any type of floor construction used for multistoried buildings usually has sufficient capacity for diaphragm action. Roofs, however, often utilize lighter construction or are extensively perforated by openings. Roofs may also be steeply sloped or have curved, arching forms, making them of questionable use for horizontal diaphragms. For such roofs, or for floors with many openings, it may be necessary to use a trussed frame for the horizontal part of a lateral bracing system. Figure 13.17*e* shows a roof for a single-story building in which trussing has been placed in all of the edge bays of the roof framing in order to achieve a horizontal bracing system.

For single-span structures, trussing may be used in a variety of ways for the combined gravity and lateral load resistive system. Figure 13.17*f* shows a typical gabled roof with the rafters tied at their bottom ends by a horizontal member. The tie, in this case, serves the dual function of resisting the outward thrust due to gravity loads and of one of the members of the single-triangle, trussed structure that is resistive to lateral loads. Thus the wind force on the sloping roof surface, or the horizontal seismic force due to the weight of the roof construction, is resisted by the triangular form of the rafter-tie combination.

The horizontal tie shown in Figure 13.17*f* may not be architecturally desirable in all cases. Some other possibilities for the single-span structure are shown in Figures 13.17*g-i*. Figure 13.17*g* shows the so-called *scissors truss*, which can be used to permit more openness on the inside or to produce a ceiling that reflects the form of the roof surface. Figure 13.17*h* shows a trussed bent that is a variation on the three-hinged arch.

The structure in Figure 13.17*i* consists primarily of a single-span truss that rests on top of end columns. If the columns are pin-jointed at the bottom chord of the truss, the structure lacks basic resistance to lateral loads and must be separately braced. If the columns are continuous to the top of the truss, they can be used to develop rigid frame action for resistance to lateral loads. Finally, if the knee-braces shown in the figure are added, the column-to-truss connection is also a rigid one. Of course, development of the rigid frame action means that the columns must also be designed for bending due to gravity loads, which will require much larger columns.

Planning of Bracing

Some of the problems to be considered in using trussed frames are the following:

1. Diagonal members must be placed so as not to interfere with the action of the gravity-resisting structure or with other building functions. If the bracing members are designed to function as axial stress members, they must be located and attached so as to avoid loadings other than those required for their truss functions. They must also be located so as not to interfere with openings for doors and windows, or with ducts, wiring, piping, light fixtures, HVAC equipment, and so on.
2. As mentioned previously, the reversibility of the lateral loads must be considered. As shown in Figure 13.17*a*, such consideration requires that diagonal members be dual-functioning (as single diagonals) or redundant (as x-bracing) with one set of diagonals working in tension for loads in one direction and the other set for reversal loads.
3. Although the diagonal bracing elements usually function only for lateral loading, the vertical and horizontal elements must be considered for the various combinations of gravity and lateral load. Thus the total frame must be analyzed for all the possible loading conditions, and each member must be designed for the particular critical combinations that represent its peak response requirements.
4. Long bracing elements, especially those in x-braced systems, may be quite slender and have significant sag due to their own dead weight. Where this is objectionable, somc provision should be made for their support to reduce this condition.

5. The trussed structure should be "tight." Connections should be made in a manner to ensure that they will be initially free of slack and will not loosen under the load reversals or repeated loading. This means generally avoiding connections that tend to loosen or to progressively deform, such as those that use nails or those with bolts in oversized holes. Use of split rings in bolted wood connections, screws instead of nails with gusset or splice plates, and welding for steel connections are means of producing very tight joints.
6. To avoid possible gravity loading on diagonals, the connections of the diagonals are sometimes made only after the gravity-resisting structure is fully assembled and at least partly loaded by the building dead loads.
7. The deformation of the trussed structure must be considered, and it may relate to its function as a load-distributing element, as in the case of a horizontal diaphragm. It may also relate to its functioning as a member of a set of load-sharing braces, which requires consideration for relative stiffness of the braces.
8. In most cases it is not necessary to brace every individual bay of a multibayed frame, as discussed previously.

The braced frame can be mixed with other bracing systems in some cases. Figure 13.18*a* shows the use of trussed frames for lateral resistance in one direction and a set of shear walls in the other direction. In this example the two systems act independently, except for possible torsion on the building which involves their interaction.

Figure 13.18*b* shows a structure in which the end bays of the roof framing are x-braced. For the loading in the direction shown, these braced bays take the highest shear in the horizontal structure, allowing the roof deck to be designed for lower shear stress.

Although buildings and their structures are often planned and constructed in two-dimensional components (horizontal floor and roof planes and vertical wall or framed bent planes), it must be noted that the building is truly three-dimensional. Bracing against lateral loads is thus unavoidably a three-dimensional problem. Although single horizontal or vertical planes of the structure may be adequately strong, the whole system must interact appropriately. Stability within individual planes may not produce three-dimensional stability; the interaction of—and connections between—individual planes must be considered.

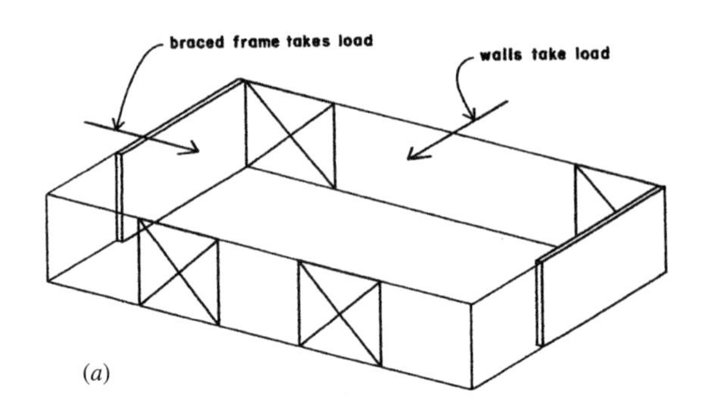

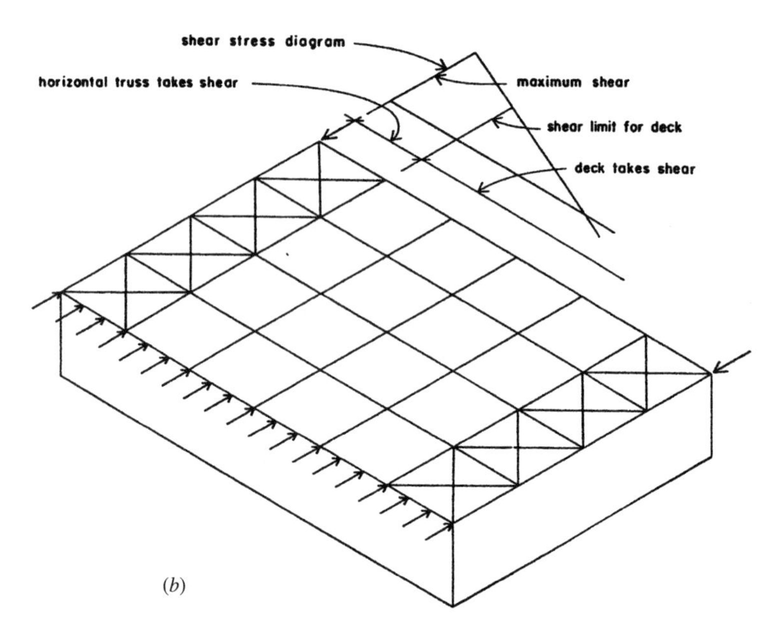

Figure 13.18 Use of trussed bracing in a mixed system with shear diaphragm elements: (a) mixed wall construction, (b) mixed roof construction.

Typical Construction

Development of the details of construction for trussed bracing is in many ways similar to the design of spanning trusses. The materials used (wood or steel), the form of individual truss members, the type of

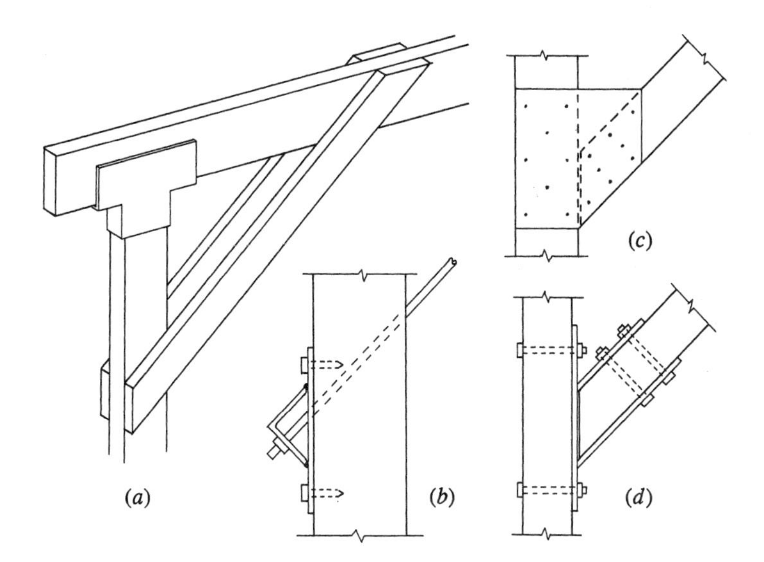

Figure 13.19 Construction details for trussed bracing in wood structures.

fastening (nails, bolts, welds, etc.), and the magnitudes of the forces are all major considerations. Since many of the members of the complete truss serve dual roles for gravity and lateral loads, member selection is seldom based on truss action alone. Quite often trussed bracing is developed by simply adding diagonals (or x-bracing) to a system already determined for gravity loads and for the general development of the building construction.

Figure 13.19 shows some details for wood framing with added diagonal members. Wood framing members are most often rectangular in cross section, and metal devices of various form are commonly used in the assembly of frameworks. Figure 13.19*a* shows a typical beam and column assembly with diagonals consisting of pairs of wood members bolted to the frame. When x-bracing is used, and the diagonals need take only tension forces, slender steel rods may be used; a possible detail for this is shown in Figure 13.19*b*. When single wood diagonals are used, a possible detail is that shown in Figure 13.19*c*, employing gusset plates. Gussets may consist of nailed plywood or various forms of metal devices. An alternative to the gusset plate connection is that shown in Figure 13.19*d*, an advantage being the lack of protrusion of connecting elements beyond the faces of the frame.

13.6 SPECIAL LATERAL BRACING

For various reasons it is sometimes necessary to incorporate special construction into wood-framed structures to achieve adequate lateral bracing. The most common incidence of this is the use of concrete or structural masonry walls, when they can be strategically located to function as major bracing elements. Another solution consists of using steel trussed or rigid frame structures. Figure 13.20 shows a building with a general construction with wood framing, but with elements of a steel trussed frame added for special bracing. In this case the steel structure is developed in the building interior or core, much like the common solution for high-rise buildings. Although x-bracing is used, the diagonals are quite stiff and can function as both tension and compression members. This is now commonly done where seismic forces are high and the light trussed system has not performed well. The light truss is all right for wind, but not for severe seismic loads. Figure 13.21 shows a joint in the trussed frame in Figure 13.20, developed with W-shaped vertical and horizontal members, diagonals consisting of steel channel shapes, and gusset plates in a fully welded joint. This truss is both strong and significantly lacking in deformation.

Figure 13.20 X-braced bents used in a mixed steel and wood bracing system.

Figure 13.21 Detail of connection for the trussed bent in Figure 13.20. Diagonals are stiffened here for seismic resistance.

Another situation requiring special bracing occurs frequently on building exteriors. For wood-framed buildings there is a need for individual wall portions of significant plan length to function as shear wall piers. Figure 13.22 shows a building with a special problem in this regard. In this example the walls on three sides of the building have modest-size openings that leave significant widths of solid wall between the openings. (This is the same building shown in Figure 13.16, which shows the use of the OSB sheathed wall piers.) On the front of this building, however, there is a desire for large openings (because of the view of the ocean), leaving only narrow wall portions at the sides of the windows. The solution here consists of a stiff, two-story, welded steel frame—in this case strong enough to brace the entire front side of the building.

Figure 13.22 Front of the OSB braced building shown in Figure 13.16. Lack of sufficient length of solid wall requires use of special bracing, achieved here with a steel rigid-frame bent.

Figure 13.23 Situation similar to that in Figure 13.22. Vertical bracing is achieved here with special highly-stiffened panels on each side of the large openings.

Another situation involving a building with wide openings is shown in Figure 13.23. In this case bracing is achieved with special panels on each side of the large openings. These panels are a recently developed product, usually consisting of a light steel frame with surfaced faces of heavy-gauge sheet steel.

14

GENERAL CONSIDERATIONS FOR BUILDING STRUCTURES

This chapter contains discussions of some general issues relating to design of building structures. These concerns have mostly not been addressed in the presentations in earlier chapters, but require consideration when dealing with whole building design situations. Application of these materials is illustrated in the design examples in Chapter 15.

14.1 CHOICE OF BUILDING CONSTRUCTION

Materials, methods, and details of building construction vary considerably on a regional basis. There are many factors that affect this situation, including the effects of response to climate and regional availability of construction materials and products. Even in a single region, differences occur between individual buildings, based on styles of architectural design and techniques of builders. Nevertheless, at any given time there are usually a few predominant, popular methods of construction that are employed for most buildings of a given type and size. The construction methods and details shown here are reasonable, but in no way are they intended to illustrate a singular, superior style of building.

14.2 STRUCTURAL DESIGN STANDARDS

Use of methods, procedures, and reference data for structural design is subject to the judgment of the designer. Many guides exist, but some individual selection is often required. Strong influences on choices include:

Building code requirements from the enforceable statutes relating to the location of the building.

Acceptable design standards as published by professional groups, such as the reference from the American Society of Civil Engineers (ASCE) referred to frequently in this book (Ref. 3).

Recommended design standards from industry organizations, such as the AISC and ACI. The body of work from current texts and references produced by respected authors.

Some reference is made to these sources in the book. However, much of the work is also simply presented in a manner familiar to the authors, based on their own experiences. If study of this subject is pursued by readers, they are sure to encounter styles and opinions that differ from those presented here. Making one's own choices in face of those conflicts is part of the progress of professional growth.

14.3 LOADS FOR STRUCTURAL DESIGN

Loads used for structural design must be derived primarily from enforceable building codes. However, the principal concern of codes is public health and safety. Performance of the structure for other concerns may not be adequately represented in the minimum requirements of the building code. Issues sometimes not included in code requirements are:

Effects of deflection of spanning structures on nonstructural elements of the constructions.

Sensations of bounciness of floors by building occupants.

Protection of structural elements from damage due to weather or normal usage.

Damage to nonstructural construction and building services due to movements of the structure during windstorms or earthquakes.

Building codes currently stipulate both the load sources and the form of combinations to be used for design. The following loads are listed in the 2006 edition of the *ASCE Minimum Design Loads for Buildings and Other Structures* (Ref. 3), hereinafter referred to as ASCE 2006.

D = Dead load
E = Earthquake-induced force
L = Live load, except roof load
L_r = Roof live load
S = Snow load
W = Load due to wind pressure

Additional special loads are listed but these are the commonly occurring loads. The following is a description of some of these loads.

14.4 DEAD LOADS

Dead load consists of the weight of the materials of which the building is constructed such as walls, partitions, columns, framing, floors, roofs, and ceilings. In the design of a beam or column, the dead load used must include an allowance for the weight of the structural member itself. Table 14.1, which lists the weights of many construction materials, may be used in the computation of dead loads. Dead loads are due to gravity, and they result in downward vertical forces.

Dead load is generally a permanent load, once the building construction is completed, unless remodeling or rearrangement of the construction occurs. Because of this permanent, longtime, character, the dead load requires certain considerations in design, such as the following:

1. It is always included in design loading combinations, except for investigations of singular effects, such as deflections due to only live load.
2. Its longtime character has some special effects causing sag and requiring reduction of design stresses in wood structures, development of long-term, continuing settlements in some soils, and producing creep effects in concrete structures.
3. It contributes some unique responses, such as the stabilizing effects that resist uplift and overturn due to wind forces.

TABLE 14.1 Weight of Building Construction

	psf[a]	kPa[a]
Roofs		
3-ply ready roofing (roll, composition)	1	0.05
3-ply felt and gravel	5.5	0.26
5-ply felt and gravel	6.5	0.31
Shingles: Wood	2	0.10
Asphalt	2–3	0.10–0.15
Clay tile	9–12	0.43–0.58
Concrete tile	6–10	0.29–0.48
Slate, 3 in.	10	0.48
Insulation: Fiberglass batts	0.5	0.025
Foam plastic, rigid panels	1.5	0.075
Foamed concrete, mineral aggregate	2.5/in.	0.0047/mm
Wood rafters: 2 × 6 at 24 in.	1.0	0.05
2 × 8 at 24 in.	1.4	0.07
2 × 10 at 24 in.	1.7	0.08
2 × 12 at 24 in.	2.1	0.10
Steel deck, painted: 22 gage	1.6	0.08
20 gage	2.0	0.10
Skylights: Steel frame with glass	6–10	0.29–0.48
Aluminum frame with plastic	3–6	0.15–0.29
Plywood or softwood board sheathing	3.0/in.	0.0057/mm
Ceilings		
Suspended steel channels	1	0.05
Lath: Steel mesh	0.5	0.025
Gypsum board, 1/2 in.	2	0.10
Fiber tile	1	0.05
Drywall, gypsum board, 1/2 in.	2.5	0.12
Plaster: Gypsum	5	0.24
Cement	8.5	0.41
Suspended lighting and HVAC, average	3	0.15
Floors		
Hardwood, 1/2 in.	2.5	0.12
Vinyl tile	1.5	0.07
Ceramic tile: 3/4 in.	10	0.48
Thin-set	5	0.24
Fiberboard underlay, 0.625 in.	3	0.15
Carpet and pad, average	3	0.15
Timber deck	2.5/in.	0.0047/mm
Steel deck, stone concrete fill, average	35–40	1.68–1.92
Concrete slab deck, stone aggregate	12.5/in.	0.024/mm
Lightweight concrete fill	8.0/in.	0.015/mm

TABLE 14.1 (*continued*)

	psf[a]	kPa[a]
Floors (Continued)		
Wood joists: 2 × 8 at 16 in.	2.1	0.10
2 × 10 at 16 in.	2.6	0.13
2 × 12 at 16 in.	3.2	0.16
Walls		
2 × 4 studs at 16 in., average	2	0.10
Steel studs at 16 in., average	4	0.20
Lath: Plaster—see *Ceilings*		
Drywall, gypsum board, 1/2 in.	2.5	0.10
Stucco, on paper and wire backup	10	0.48
Windows, average, frame + glazing:		
Small pane, wood or metal frame	5	0.24
Large pane, wood or metal frame	8	0.38
Increase for double glazing	2–3	0.10–0.15
Curtain wall, manufactured units	10–15	0.48–0.72
Brick veneer, 4 in., mortar joints	40	1.92
1/2 in., mastic-adhered	10	0.48
Concrete block:		
Lightweight, unreinforced, 4 in.	20	0.96
6 in.	25	1.20
8 in.	30	1.44
Heavy, reinforced, grouted, 6 in.	45	2.15
8 in.	60	2.87
12 in.	85	4.07

[a] Average weight per square foot of surface, except as noted.
Values given as in. or mm are to be multiplied by actual thickness of material.

Although weights of materials can be reasonably accurately determined, the complexity of most building construction makes the computation of dead loads possible only on an approximate basis. This adds to other factors to make design for structural behaviors a very approximate endeavor. As in other cases, this should not be used as an excuse for sloppiness in computational work, but it should be recognized as a fact to temper concern for high accuracy in design computations.

14.5 BUILDING CODE REQUIREMENTS FOR STRUCTURES

Structural design of buildings is most directly controlled by building codes, which are the general basis for the granting of building permits—the legal permission required for construction. Building codes (and the permit-granting process) are administered by some unit of government: city, county, or state. Most building codes, however, are based on some model code.

Model codes are more similar than different, and are in turn largely derived from the same basic data and standard reference sources, including many industry standards. In the several model codes and many city, county, and state codes, however, there are some items that reflect particular regional concerns. With respect to control of structures, all codes have materials (all essentially the same) that relate to the following issues:

1. *Minimum Required Live Loads.* All building codes have tables that provide required values to be used for live loads.
2. *Wind Loads.* These are highly regional in character with respect to concern for local windstorm conditions. Model codes (see Ref. 4) provide data with variability on the basis of geographic zones.
3. *Seismic (Earthquake) Effects.* These are also regional with special concerns in the western region of the United States. This data and procedures for design are subject to frequent modification, in response to ongoing research and experience.
4. *Load Duration.* Loads or design stresses are often modified on the basis of the time span of the load, varying from the life of the structure for dead load to a few seconds for a wind gust or a single major seismic shock. Safety factors are frequently adjusted on this basis. Some applications are illustrated in the work in the design examples.
5. *Load Combinations.* These were formerly mostly left to the discretion of designers, but are now quite commonly stipulated in codes, mostly because of the increasing use of ultimate strength design and the use of factored loads. (See Section 14.7.)
6. *Design Data for Types of Structures.* These deal with basic materials (wood, steel, concrete, masonry, etc.), specific structures (rigid frames, towers, balconies, pole structures, etc.), and special problems (foundations, retaining walls, stairs, etc.). Industry-wide standards and common practices are generally recognized, but local codes may reflect particular local experience or attitudes. Minimal structural safety is the general basis, and some specified limits may result in questionably adequate performances (bouncy floors, cracked plaster, etc.).

7. *Fire Resistance.* For the structure, there are two basic concerns for fire safety, both of which produce limits for the construction. The first concern is for structural collapse or significant structural loss. The second concern is for containment of the fire to control its spread. These concerns produce limits on the choice of materials (e.g., combustible or noncombustible) and some details of the construction (cover on reinforcement in concrete, fire insulation for steel beams, etc.).

The work in the design examples in Chapter 15 is based largely on criteria from ASCE 2006 (Ref. 3).

14.6 LIVE LOADS

Live loads technically include all the nonpermanent loadings that can occur, in addition to the dead loads. However, the term as commonly used usually refers only to the vertical gravity loadings on roof and floor surfaces. These loads occur in combination with the dead loads, but are generally random in character and must be dealt with as potential contributors to various loading combinations, as discussed in Section 14.7.

Roof Loads

In addition to the dead loads they support, roofs are designed for a uniformly distributed live load. The minimum specified live load accounts for general loadings that occur during construction and maintenance of the roof. For special conditions, such as heavy snowfalls, additional loadings are specified.

The minimum roof live load in psf is specified in ASCE 2006 (Ref. 3) in the form of an equation, as follows:

$$L_r = 20\ R_1\ R_2 \text{ in which } 12 \leq L_r \leq 20$$

In the equation R_1 is a reduction factor based on the tributary area supported by the structural member being designed (designated as A_t and quantified in ft^2) and is determined as follows:

$$\begin{aligned} R_1 &= 1, \text{ for } A_t \leq 200\,\text{ft}^2 \\ &= 1.2 - 0.001\ A_t, \text{ for } 200\ \text{ft}^2 < A_t < 600\ \text{ft}^2 \\ &= 0.6, \text{ for } A_t \geq 600\ \text{ft}^2 \end{aligned}$$

Reduction factor R_2 accounts for the slope of a pitched roof and is determined as follows:

$$\begin{aligned} R_2 &= 1, \text{ for } F \leq 4 \\ &= 1.2 - 0.05\,F, \text{ for } 4 < F < 12 \\ &= 0.6, \text{ for } F \geq 12 \end{aligned}$$

The quantity F in the equations for R_2 is the number of inches of rise per ft for a pitched roof (for example: $F = 12$ indicates a rise of 12 in 12 or an angle of 45°).

The design standard also provides data for roof surfaces that are arched or domed and for special loadings for snow or water accumulation. Roof surfaces must also be designed for wind pressures on the roof surface, both upward and downward. A special situation that must be considered is that of a roof with a low dead load and a significant wind load that exceeds the dead load,

Although the term *flat roof* is often used, there is generally no such thing; all roofs must be designed for some water drainage. The minimum required pitch is usually 1/4 in./ft, or a slope of approximately 1:50. With roof surfaces that are close to flat, a potential problem is that of *ponding*, a phenomenon in which the weight of the water on the surface causes deflection of the supporting structure, which in turn allows for more water accumulation (in a pond), causing more deflection, and so on, resulting in a progressive collapse condition.

Floor Live Loads

The live load on a floor represents the probable effects created by the occupancy. It includes the weights of human occupants, furniture, equipment, stored materials, and so on. All building codes provide minimum live loads to be used in the design of buildings for various occupancies. Since there is a lack of uniformity among different codes in specifying live loads, the local code should always be used. Table 14.2 contains a sample of values for floor live loads as given in ASCE 2006 (Ref. 3) and commonly specified by building codes.

Although expressed as uniform loads, code-required values are usually established large enough to account for ordinary concentrations that occur. For offices, parking garages, and some other occupancies, codes often require the consideration of a specified concentrated load

TABLE 14.2 Minimum Floor Live Loads

Building Occupancy or Use	Uniformly Distributed Load (psf)	Concentrated Load (lbs)
Apartments and Hotels		
Private rooms and corridors serving them	40	
Public rooms and corridors serving them	100	
Dwellings, One and Two-Family		
Uninhabitable attics without storage	10	
Uninhabitable attics with storage	20	
Habitable attics and sleeping rooms	30	
All other areas except stairs and balconies	40	
Office Buildings		
Offices	50	2000
Lobbies and first floor corridors	100	2000
Corridors above first floor	80	2000
Stores		
Retail		
First floor	100	1000
Upper floors	75	1000
Wholesale, all floors	125	1000

Source: Minimum Design Loads for Buildings and Other Structures, ASCE 7-05, (Ref. 3), used with permission of the publishers, American Society of Civil Engineers.

as well as the distributed loading. This required concentrated load is listed in Table 14.2 for the appropriate occupancies.

Where buildings are to contain heavy machinery, stored materials, or other contents of unusual weight, these must be provided for individually in the design of the structure.

When structural framing members support large areas, most codes allow some reduction in the total live load to be used for design. These reductions, in the case of roof loads, are incorporated in the formulas for roof loads given previously. The following is the method given in ASCE 2006 (Ref. 3) for determining the reduction permitted for beams, trusses, or columns that support large floor areas.

The design live load on a member may be reduced in accordance with the formula

$$L = L_0\left(0.25 + \frac{15}{\sqrt{K_{LL}A_T}}\right)$$

where:

L = reduced design live load per square foot of area supported by the member

TABLE 14.3 Live Load Element Factor, K_{LL}

Element	K_{LL}
Interior Columns	4
Exterior Columns without cantilever slabs	4
Edge Columns with cantilever slabs	3
Corner Columns with cantilever slabs	2
Edge Beams without cantilever slabs	2
Interior Beams	2
All Other Members Not Identified Above	1

Source: Minimum Design Loads for Buildings and Other Structures, ASCE 7-05 (Ref. 3), used with permission of the publisher, American Society of Civil Engineers.

L_0 = unreduced live load supported by the member
K_{LL} = live load element factor (see Table 14.3)
A_T = tributary area supported by the member

L shall not be less than 0.50 L_0 for members supporting one floor, and L shall not be less than 0.40 L_0 for members supporting two or more floors.

In office buildings and certain other building types, partitions may not be permanently fixed in location but may be erected or moved from one position to another in accordance with the requirements of the occupants. In order to provide for this flexibility, it is customary to require an allowance of 15 to 20 psf, which is usually added to other dead loads.

14.7 LATERAL LOADS (WIND AND EARTHQUAKE)

As used in building design, the term *lateral load* is usually applied to the effects of wind and earthquakes, as they induce horizontal forces on stationary structures. From experience and research, design criteria and methods in this area are continuously refined, with recommended practices being presented through the various model building codes.

Space limitations do not permit a complete discussion of the topic of lateral loads and design for their resistance. The following discussion summarizes some of the criteria for design in ASCE 2006 (Ref. 3). Examples of application of these criteria are given in the design examples of building structural design in Chapter 15. For a more extensive discussion the reader is referred to *Simplified Building Design for Wind and Earthquake Forces* (Ref. 11).

Wind

Where wind is a regional problem, local codes are often developed in response to local conditions. Complete design for wind effects on buildings includes a large number of both architectural and structural concerns. The following is a discussion of some of the requirements from ASCE 2006 (Ref. 3).

Basic Wind Speed. This is the maximum wind speed (or velocity) to be used for specific locations. It is based on recorded wind histories and adjusted for some statistical likelihood of occurrence. For the United States recommended minimum wind speeds are taken from maps provided in the ASCE standard. As a reference point, the speeds are those recorded at the standard measuring position of 10 m (approximately 33 ft) above the ground surface.

Wind Exposure. This refers to the conditions of the terrain surrounding the building site. The ASCE standard uses three categories, labeled B, C, and D. Qualifications for categories are based on the form and size of wind-shielding objects within specified distances around the building.

Simplified Design Wind Pressure (p_s). This is the basic reference equivalent static pressure based on the critical wind speed and is determined as follows

$$p_s = \lambda\, I p_{S30}$$

where:

λ = adjustment factor for building height and exposure

I = importance factor

P_{S30} = simplified design wind pressure for exposure B, at height of 30 ft, and for $I = 1.0$

The importance factor for ordinary circumstances of building occupancy is 1.0. For other buildings factors are given for facilities that involve hazard to a large number of people, for facilities considered to be essential during emergencies (such as windstorms), and for buildings with hazardous contents.

The design wind pressure may be positive (inward) or negative (outward, suction) on any given surface. Both the sign and the value for the pressure are given in the design standard. Individual building surfaces, or parts thereof, must be designed for these pressures.

Design Methods. Two methods are described in the Code for the application of wind pressures.

Method 1 (Simplified Procedure). This method is permitted to be used for relatively small, low-rise buildings of simple symmetrical shape. It is the method described here and used for the examples in Chapter 15.

Method 2 (Analytical Procedure). This method is much more complex and is prescribed for buildings that do not fit the limitations described for Method 1.

Uplift. Uplift may occur as a general effect, involving the entire roof or even the whole building. It may also occur as a local phenomenon such as that generated by the overturning moment on a single shear wall.

Overturning Moment. Most codes require that the ratio of the dead load resisting moment (called the restoring moment, stabilizing moment, etc.) to the overturning moment be 1.5 or greater. When this is not the case, uplift effects must be resisted by anchorage capable of developing the excess overturning moment. Overturning may be a critical problem for the whole building, as in the case of relatively tall and slender tower structures. For buildings braced by individual shear walls, trussed bents, and rigid-frame bents, overturning is investigated for the individual bracing units.

Drift. Drift refers to the horizontal deflection of the structure due to lateral loads. Code criteria for drift are usually limited to requirements for the drift of a single story (horizontal movement of one level with respect to the next above or below). As in other situations involving structural deformations, effects on the building construction must be considered; thus the detailing of curtain walls or interior partitions may affect limits on drift.

Special Problems. The general design criteria given in most codes are applicable to ordinary buildings. More thorough investigation is recommended (and sometimes required) for special circumstances such as the following:

1. *Tall Buildings.* These are critical with regard to their height dimension as well as the overall size and number of occupants inferred. Local wind speeds and unusual wind phenomena at upper elevations must be considered.
2. *Flexible Structures.* These may be affected in a number of ways, including vibration or flutter as well as simple magnitudes of movement.
3. *Unusual Shapes.* Open buildings, buildings with large overhangs or other projections, and any building with a complex shape should be carefully studied for the special wind effects that may occur.

Earthquakes

During an earthquake a building is shaken up and down and back and forth. The back-and-forth (horizontal) movements are typically more violent and tend to produce major destabilizing effects on buildings; thus structural design for earthquakes is mostly done in terms of considerations for horizontal (called lateral) forces. The lateral forces are actually generated by the weight of the building—or, more specifically, by the mass of the building that represents both an inertial resistance to movement and a source for kinetic energy once the building is actually in motion. In the simplified procedures of the equivalent static force method, the building structure is considered to be loaded by a set of horizontal forces consisting of some fraction of the building weight. An analogy would be to visualize the building as being rotated vertically 90° to form a cantilever beam, with the ground as the fixed end and with a load consisting of the building weight.

In general, design for the horizontal force effects of earthquakes is quite similar to design for the horizontal force effects of wind. The same basic types of lateral bracing (shear walls, trussed bents, rigid frames, etc.) are used to resist both force effects. There are indeed some significant differences, but in the main a system of bracing that is developed for wind resistance will most likely serve reasonably well for earthquake resistance as well.

Because of its considerably more complex criteria and procedures, we have chosen not to illustrate the design for earthquake effects in the examples in this book. Nevertheless, the development of elements and systems for the lateral bracing of the building in the design examples here is quite applicable in general to situations where earthquakes are a predominant concern. For structural investigation, the principal difference is in the determination of the loads and their distribution in the building. Another major difference is in the true dynamic effects, critical wind force being usually represented by a single, major, one-direction punch from a gust, while earthquakes represent rapid back-and-forth, reversing-direction actions. However, once the dynamic effects are translated into equivalent static forces, design concerns for the bracing systems are very similar, involving considerations for shear, overturning, horizontal sliding, and so on.

For a detailed explanation of earthquake effects and illustrations of the investigation by the equivalent static force method, the reader is referred to *Simplified Building Design for Wind and Earthquake Forces* (Ref. 11).

14.8 LOAD COMBINATIONS AND FACTORS

The various types of load sources, as described in the preceding section, must be individually considered for quantification. However, for design work the possible combination of loads must also be considered. Using the appropriate combinations, the design load for individual structural elements must be determined. The first step in finding the design load is to establish the critical combinations of load for the individual element. Using ASCE 2006 (Ref. 3) as a reference, the following combinations are to be considered.

For the ASD method:

Dead load

Dead load + live load

Dead load + 0.75(wind load) + 0.75(live load)

Dead load + 0.70(earthquake load) + 0.75(live load)

For the LRFD method:

1.4(dead load)

1.2(dead load) + 1.6(live load) + 0.5(roof load)

1.2(dead load) + 1.6(roof load) + live load or 0.8(wind load)

1.2(dead load) + 1.6(wind load) + (live load) + 0.5(roof load)

1.2(dead load) + 1.0(earthquake load) + live load + 0.2(snow load)

0.9(dead load) + 1.0(earthquake load) or 1.6(wind load)

14.9 DETERMINATION OF DESIGN LOADS

The load to be carried by each element of a structure is defined by the unit loads for dead load and live load and the *load periphery* for the individual elements. The load periphery for an element is established by the layout and dimensions of the framing system. Any possible live load reduction (as described in Section 14.6) is made for the individual elements based on their load periphery area. An example of determination of peripheral loading is given in Section 3.2. The loads so determined are used in the defined combinations described in Section 14.8. If any of these elements are involved in the development of the lateral bracing structure, the appropriate wind or earthquake loads are also added. Floor live loads may be reduced by the method described in Section 14.6. Reductions are based on the tributary area supported and the number of levels supported by members. Computations of design loads using this process are given for the building design cases in Chapter 15.

14.10 STRUCTURAL PLANNING

Planning a structure requires the ability to perform two major tasks. The first is the logical arranging of the structure itself, regarding its geometric form, its actual dimensions and proportions, and the ordering of the elements for basic stability and reasonable interaction. All of these issues must be faced, whether the building is simple or complex, small or large, of ordinary construction or totally unique. Spanning beams must be supported and have depths adequate for the spans; horizontal thrusts of arches must be resolved; columns above should be centered over columns below; and so on.

The second major task in structural planning is the development of the relationships between the structure and the building in general. The building plan must be "seen" as a structural plan. The two may not be quite the same, but they must fit together. "Seeing" the structural plan (or possibly alternative plans) inherent in a particular architectural plan is a major task for designers of building structures.

Hopefully, architectural planning and structural planning are done interactively, not one after the other. The more the architect knows about the structural problems and the structural designer (if another person) knows about architectural problems, the more likely it is that an interactive design development may occur.

Although each individual building offers a unique situation if all of the variables are considered, the majority of building design problems are highly repetitious. The problems usually have many alternative solutions, each with its own set of pluses and minuses in terms of various points of comparison. Choice of the final design involves the comparative evaluation of known alternatives and the eventual selection of one.

The word *selection* may seem to imply that all the possible solutions are known in advance, not allowing for the possibility of a new solution. The more common the problem, the more this may be virtually true. However, the continual advance of science and technology and the fertile imagination of designers make new solutions an ever-present possibility, even for the most common problems. When the problem is truly a new one in terms of a new building use, a jump in scale, or a new performance situation, there is a real need for innovation. Usually, however, when new solutions to old problems are presented, their merits must be compared to established previous solutions in order to justify them. In its broadest context the selection process includes the consideration of all possible alternatives: those well known, those new and unproven, and those only imagined.

14.11 BUILDING SYSTEMS INTEGRATION

Good structural design requires integration of the structure into the whole physical system of the building. It is necessary to realize the potential influences of structural design decisions on the general architectural design and on the development of the systems for power, lighting, thermal control, ventilation, water supply, waste handling, vertical transportation, firefighting, and so on. The most popular structural systems have become so in many cases largely because of their ability to accommodate the other subsystems of the building and to facilitate popular architectural forms and details.

14.12 ECONOMICS

Dealing with dollar cost is a very difficult, but necessary, part of structural design. For the structure itself, the bottom-line cost is the delivered cost of the finished structure, usually measured in units of dollars per square foot of the building. For individual components, such as a single wall, units may be used in other forms. The individual cost factors or components, such as cost of materials, labor, transportation, installation, testing, and inspection, must be aggregated to produce a single unit cost for the entire structure.

Designing for control of the cost of the structure is only one aspect of the cost problem, however. The more meaningful cost is that for the entire building construction. It is possible that certain cost-saving efforts applied to the structure may result in increases of cost of other parts of the construction. A common example is that of the floor structure for multistory buildings. Efficiency of floor beams occurs with the generous provision of beam depth in proportion to the span. However, adding inches to beam depths with the unchanging need for dimensions required for floor and ceiling construction and installation of ducts and lighting elements means increasing the floor-to-floor distance and the overall height of the building. The resulting increases in cost for the added building skin, interior walls, elevators, piping, ducts, stairs, and so on, may well offset the small savings in cost of the beams. The really effective cost-reducing structure is often one that produces major savings of nonstructural costs, in some cases at the expense of structural efficiency.

Real costs can only be determined by those who deliver the completed construction. Estimates of cost are most reliable in the form of actual offers or bids for the construction work. The farther the cost estimator is from the actual requirement to deliver the goods, the more speculative the estimate. Designers, unless they are in the actual employ of the builder, must base any cost estimates on educated guesswork deriving from some comparison with similar work recently done in the same region. This kind of guessing must be adjusted for the most recent developments in terms of the local markets, competitiveness of builders and suppliers, and the general state of the economy. Then the four best guesses are placed in a hat and one is drawn out.

Serious cost estimating requires training and experience and a source of reliable, timely information. For major projects various sources are available, in the form of publications or computer databases.

The following are some general rules for efforts that can be made in the structural design work in order to have an overall general cost-saving attitude.

1. Reduction of material volume is usually a means of reducing cost. However, unit prices for different grades must be noted. Higher grades of steel or wood may be proportionally more expensive than the higher stress values they represent; more volume of cheaper material may be less expensive.
2. Use of standard, commonly stocked products is usually a cost savings, as special sizes or shapes may be premium priced. Wood

2 × 3 studs may be higher in price than 2 × 4 studs, since the 2 × 4 is so widely used and bought in large quantities.

3. Reduction in the complexity of systems is usually a cost savings. Simplicity in purchasing, handling, managing of inventory, and so on, will be reflected in lower bids as builders anticipate simpler tasks. Use of the fewest number of different grades of materials, sizes of fasteners, and other such variables is as important as the fewest number of different parts. This is especially true for any assemblage done on the building site; large inventories may not be a problem in a factory, but usually are on a restricted site.
4. Cost reduction is usually achieved when materials, products, and construction methods are familiar to local builders and construction workers. If real alternatives exist, choice of the "usual" one is the best course.
5. Do not guess at cost factors; use real experience, yours or others'. Costs vary locally, by job size and over time. Keep up to date with cost information.
6. In general, labor cost is greater than material cost. Labor for building forms, installing reinforcement, pouring, and finishing concrete surfaces is *the* major cost factor for site-poured concrete. Savings in these areas are much more significant than saving of material volume.
7. For buildings of an investment nature, time is money. Speed of construction may be a major advantage. However, getting the structure up fast is not a true advantage unless the other aspects of the construction can take advantage of the time gained. Steel frames often go up quickly, only to stand around and rust while the rest of the work catches up.

15

BUILDING DESIGN EXAMPLES

This chapter presents illustrations of the design of wood structures for four example buildings. Design of the individual elements for these structures is based on the materials presented in earlier chapters. The principal purpose here is to show the broader context of design work by dealing with whole structures and the buildings they serve.

Materials, methods, and details of building construction vary considerably on a regional basis. There are many factors that affect this situation, including the effects of climate and the availability of construction materials and products. Even in a single region, differences occur between individual buildings, based on styles of architectural design and particular techniques of builders. Nevertheless, at any given time there are usually a few predominant, popular methods of construction employed for most buildings of a given type and size. The construction methods and details shown here are reasonable, but in no way are they intended to illustrate a singular, superior style of building.

15.1 BUILDING ONE: SINGLE-STORY LIGHT WOOD FRAME

Figure 15.1 shows the general form, the construction of the basic building shell, and the form of the wind-bracing shear walls for Building One. The drawings show a building profile with a generally flat roof (with minimal slope for drainage) and a short parapet at the roof edge. This structure is generally described as a *light wood frame*, and is the first alternative to be considered for Building One. The following data is used for design:

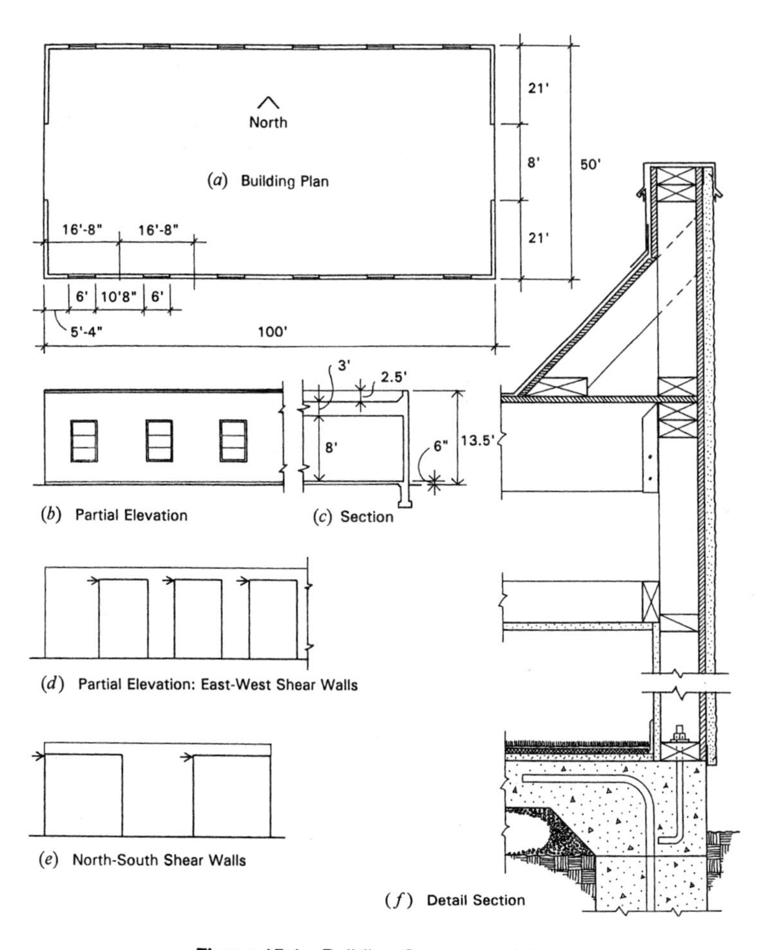

Figure 15.1 Building One, general form.

Roof live load = 20 psf (reducible).

Wind load as determined from the ASCE 2006 (Ref. 3).

Wood framing lumber of Douglas fir-larch.

The general profile of the building is shown in Figure 15.1*c*, which shows a flat roof and ceiling and a short parapet at the exterior walls. The general nature of the construction is shown in the detailed wall section in Figure 15.1*f*. Specific details of the framing will depend on various decisions in the design of the structure as developed in the following discussion. The general form of the exterior shear walls is indicated in Figures 15.1*d* and *e*. Design considerations for lateral loads are presented at the end of this section. Considerations will first be made for the design of the roof structural system for only gravity loads, although it should be kept in mind that the roof must eventually be developed as a horizontal diaphragm and the exterior walls as shear walls.

Design for Gravity Loads

With the construction as shown in Figure 15.1*f*, the roof dead load is determined as follows:

Three-ply felt and gravel roofing	5.5 psf
Glass fiber insulation batts	0.5
2-in. thick plywood roof deck	1.5
Wood rafters and blocking (estimate)	2.0
Ceiling framing	1.0
2-in.-thick drywall ceiling	2.5
Ducts, lights, etc.	3.0
Total roof dead load for design:	16.0 psf

Assuming a partitioning of the interior as shown in Figure 15.2*a*, various possibilities exist for the development of the spanning roof and ceiling framing systems and their supports.

Scheme 1: Interior Bearing Walls

For this scheme, the roof is supported by a series of bearing walls that are 16 ft 8 in. on center. These walls may be continuous across the full building width or may be used to develop the building plan shown in Figure 15.2*a*, with short beams used at the central hall to continue the line of support developed by the walls. Primary structural

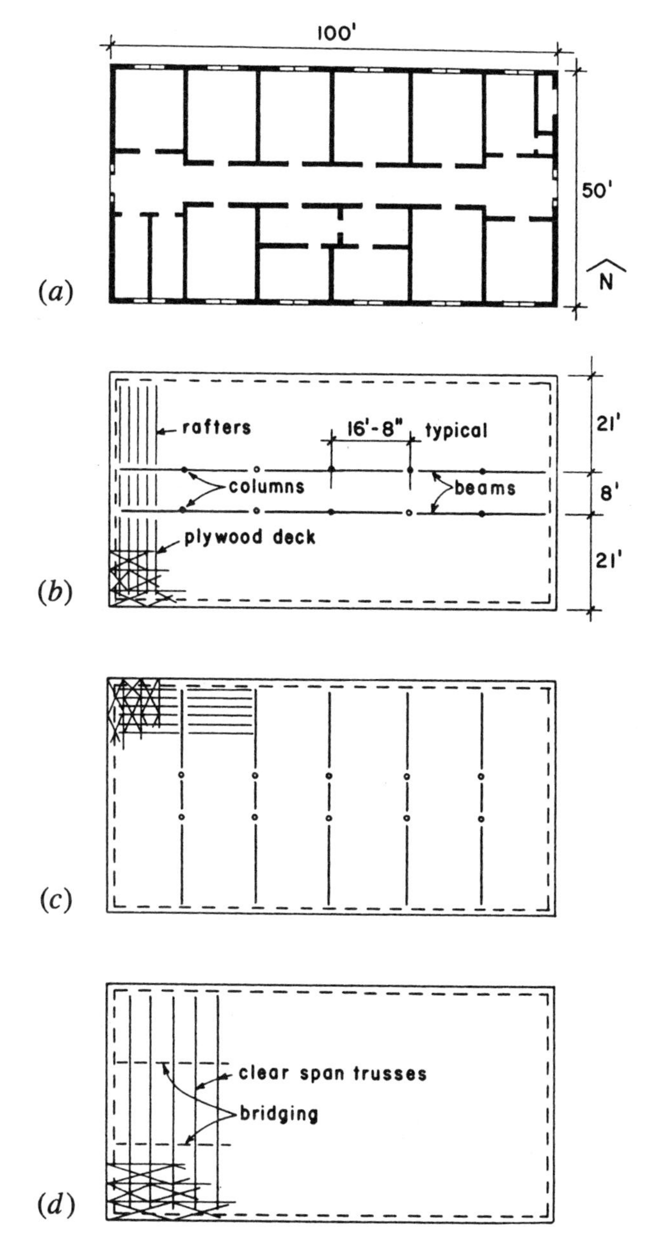

Figure 15.2 Developed plan for interior partitioning and alternatives for the roof framing.

elements for the support of the roof are the roof deck and rafters, the wall studs, and the wall footings. A partial plan for the layout of the roof framing is shown in Figure 15.3. Some special framing is required to achieve the wall openings, which usually involves the use of beams as headers above the openings and columns at the sides of the openings. The simplest form for the headers and columns is a structural member with one dimension equal to the broad dimension of the wall studs. An alternative selection for beams is a multiple of 2 in. lumber members. Columns are most often achieved with doubled studs, a form also used at wall ends.

Selection of roof decking and wall sheathing involves many considerations for the full development of the construction for the roof, exterior

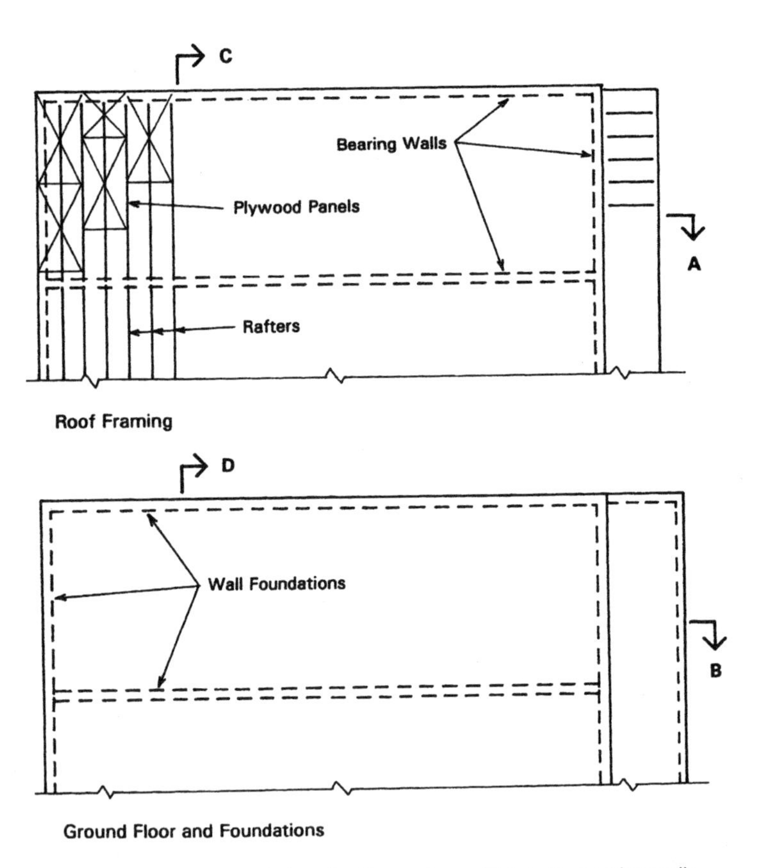

Figure 15.3 Structural plans for the system with interior bearing walls.

walls, and interior walls. All surfaces could be achieved with plywood, but many other surfacing materials are available and are frequently used. A range of possibilities for panel-formed surfacing is discussed in Section 12.4. For illustration here, plywood is generally chosen and properties for structural grades of plywood are given in table form. The required strength for plywood deck depends on the loads and the rafter spacing. For flat roofs, a minimum thickness is usually considered to be 15/32 in., which has to do with the attachment of roofing materials.

An approximate rafter size and spacing combination for the 16 ft 8 in. span can be determined using Table 7.2. In the table maximum spans are given for two load combinations, one with a live load of 20 psf and a dead load of 10 psf and the other with a live load of 20 psf and a dead load of 20 psf. As the estimated dead load for this example is 16 psf, neither table combination is correct for our situation. However, a good guess is possible by averaging the two span limits in the table. For example:

For 2 × 8s at 16 in., span limit is between 18 ft 2 in. and 15 ft 9 in.

For 2 × 10s at 24 in., span limit is between 18 ft 2 in. and 15 ft 8 in.

Either of these choices is probably adequate for the 16 psf dead load, although, if desired, computations may be done to prove the case. Blocking or bridging should be used to provide lateral bracing for the rafters. If bridging is used, it is usually desirable to place the plywood with the 8 ft panel dimension parallel to the rafters, as this requires blocking only every 8 ft for support of the plywood panel edges. This issue must also be considered for the function of the roof deck as a horizontal diaphragm for lateral loads.

Reference to Tables 12.1 and 12.2 will show that the 15/32 (1/2) in. plywood can be used for 24 in. rafter spacing. Although wider rafter spacing is possible with fabricated I-joists, the thicker plywood required may not be cost-effective. At this span, the solid-sawn rafters are still adequate, so fabricated products are less likely to be required.

Reference to Table 9.2 will show that 2 × 4 studs at 24 in. spacing could be used, although 16 in. spacing is more often chosen, depending on wall surfacing materials. Laterally unsupported height here will be less than the 10 ft limit given in Table 9.2. Studs in the exterior walls should also be investigated for the combination of vertical gravity compression and lateral bending due to wind, as illustrated in Example 5 in Section 9.8. The loadings shown for the studs in the example are slightly greater than those encountered here, so the studs for this building should be quite adequate.

Scheme 2: Beam Framing with Interior Columns

Interior walls may be used for supports, but a more desirable situation in commercial uses is sometimes obtained by using interior columns that allow for rearrangement of interior spaces. The roof framing system shown in Figure 15.2*b* is developed with two rows of interior columns placed at the location of the corridor walls. If the partitioning shown in Figure 15.2*a* is used, these columns may be totally out of view (and not intrusive in the building plan) if they are incorporated in the wall construction. Figure 15.2*c* shows a second possibility for the roof framing using the same column layout as in Figure 15.2*b*. There may be various reasons for favoring one of these framing schemes over the other. Problems of installation of ducts, lighting, wiring, roof drains, and fire sprinklers may influence this structural design decision. For this example the scheme shown in Figure 15.2*b* is arbitrarily selected for illustration of the design of the elements of the structure.

As discussed for Scheme 1, installation of a membrane-type roofing ordinarily requires at least a 1/2-in.-thick roof deck. Such a deck is capable of up to 32-in. spans in a direction parallel to the face ply grain (the long direction of ordinary 4 × 8 ft panels). If rafters are not over 24 in. on center—as they are likely to be for the schemes shown in Figure 15.2*b* and *c*—the panels may be placed so that the plywood span is across the face grain. An advantage in the latter arrangement is the reduction in the amount of blocking between rafters that is required at panel edges not falling on a rafter. The reader is referred to the discussion of plywood decks in Section 12.5.

As shown in the framing plan in Figure 15.2*b*, the rafters here must achieve a 21 ft span. As with Scheme 1, an approximation of the required rafters may be obtained from Table 7.2.

Again, an average may be interpolated between the following choices from the table.

For 2 × 10s at 16 in., span limit is between 22 ft 3 in. and 19 ft 3 in.
For 2 × 12s at 19.2 in., span limit is between 23 ft 6 in. and 20 ft 4 in.

The 2 × 10s are very close to the limit, so the 2 × 12s are a better choice. As with Scheme 1, bridging or blocking should be used for lateral bracing, and the use of blocked edges for the deck panels should be investigated with regard to diaphragm action for lateral loads.

A ceiling may be developed by direct attachment to the underside of the rafters. However the construction as shown here indicates a ceiling at some distance below the rafters, allowing for various service elements to be incorporated above the ceiling. Such a ceiling might be framed independently for short spans (such as at a corridor), but is more often developed as a *suspended ceiling*, with hanger elements from the overhead structure used to shorten the span of ceiling framing.

The wood beams as shown in Figure. 15.2*b* are continuous through two spans, with a total length of 33 ft 4 in. and beam spans of 16 ft 8 in. For the two-span beam the maximum bending moment is the same as for a simple span, the principal advantage being a reduction in deflection. The total load area for one span is

$$A = \left(\frac{21 + 8}{2}\right) \times 16.67 = 242 \text{ ft}^2$$

This permits the use of a live load of 16 psf. Thus the unit of the uniformly distributed load on the beam is found as

$$w = (16 \text{ psf } LL\ + 16 \text{ psf } DL) \times \frac{21 + 8}{2} = 464 \text{ lb/ft}$$

Adding a bit for the beam weight, a design for 480 lb/ft is reasonable, for which the maximum bending moment is

$$M = \frac{wL^2}{8} = \frac{480 \times (16.67)^2}{8} = 16{,}673 \text{ ft-lb}$$

A common minimum grade for beams is No. 1. The allowable bending stress depends on the beam size and the load duration. Assuming a 15% increase for load duration, Table 4.1 yields the following:

For a 4 × member: $F_b = 1.15(1000) = 1150$ psi

For a 5 × or larger: $F_b = 1.15(1350) = 1552$ psi

Then, for a 4 × :

$$\text{Required } S = \frac{M}{F_b} = \frac{16{,}673 \times 12}{1150} = 174 \text{ in.}^3$$

From Table A.3, the largest 4-in.-thick member is a 4 × 16 with $S = 135.7\ \text{in.}^3$, which is not adequate. (*Note*: deeper 4 × members are available, but are quite laterally unstable and thus not recommended.) For a thicker member the required S may be determined as

$$S = \frac{1150}{1552} \times 174 = 129\ \text{in.}^3$$

for which possibilities include a 6 × 14 with $S = 167\ \text{in.}^3$ or an 8 × 12 with $S = 165\ \text{in.}^3$

Although the 6 × 14 has the least cross-sectional area and ostensibly the lower cost, various considerations of the development of construction details may affect the beam selection. This beam could also be formed as a built-up member from a number of 2 × members. Where deflection or long-term sag are critical, a wise choice might be to use a glued laminated section or even a steel rolled section. This may also be a consideration if shear is critical, as is often the case with heavily-loaded beams.

A minimum slope of the roof surface for drainage is usually 2%, or approximately 1/4-in. per ft. If drainage is achieved as shown in Fig 15.4*a*, this requires a total slope of 1/4 × 25 = 6.25 in. from the center to the edge of the roof. There are various ways of achieving this sloped surface, including the simple tilting of the rafters.

Figure 15.4*b* shows some possibilities for the details of the construction at the center of the building. As shown here the rafters are kept flat and the roof profile is achieved by attaching cut 2 × members to the tops of the long rafters and using a short profiled rafter at the corridor. Ceiling joists for the corridor are supported directly by the corridor walls. Other ceiling joists are supported at their ends by the walls and at intermediate points by suspension from the rafters.

The typical column at the corridor supports a load approximately equal to the entire spanning load for one beam, or

$$P = 480 \times 16.67 = 8000\ \text{lb}$$

This is a light load, but the column height requires something larger than a 4 × 4 size, possibly a 4 × 8 (see Table 9.1.). If a 6 × 6 is not objectionable, it is adequate in the lower stress grades. However, it is common to use a steel pipe or tubular section, either of which can probably be accommodated in a stud partition wall.

As with Scheme 1, the stud walls should be adequate with 2 × 4 studs at 16 in. spacing.

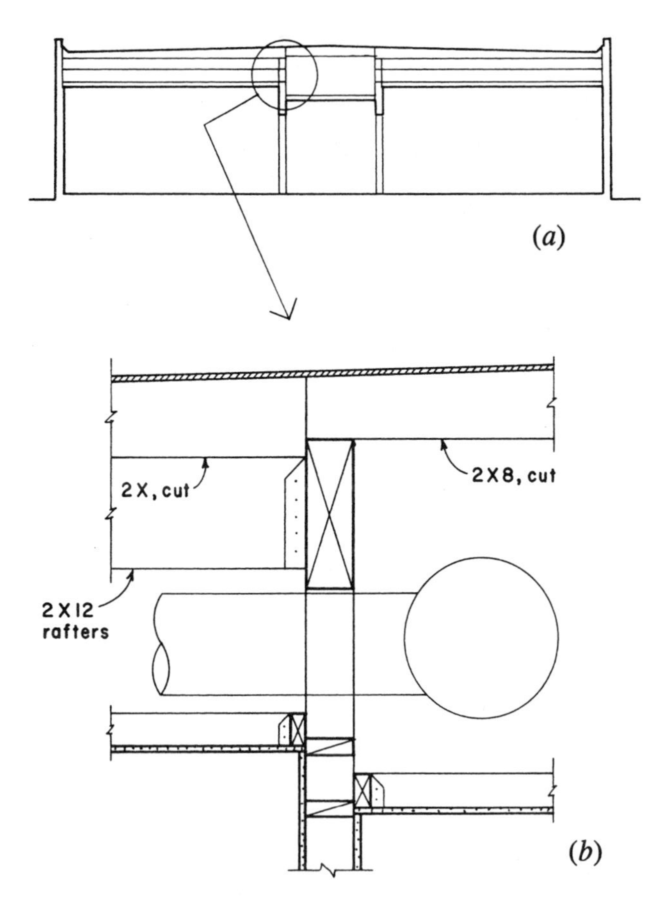

Figure 15.4 Construction details.

Scheme 3: Clear-Spanning Roof Trusses

A partial framing plan for a modified roof structure is shown in Figure 15.5, indicating the use of spanning trusses to achieve a clear span of 50 ft. A possible choice for these trusses is the manufactured truss using wood chords and steel interior members. (See discussion in Section 11.8.) This scheme offers the greatest opportunity for customized development of the building interior and is thus a popular solution.

The framing plan shows the layout for this system, indicating the spacing of the trusses and the arrangement of the 4 ft × 8 ft deck panels.

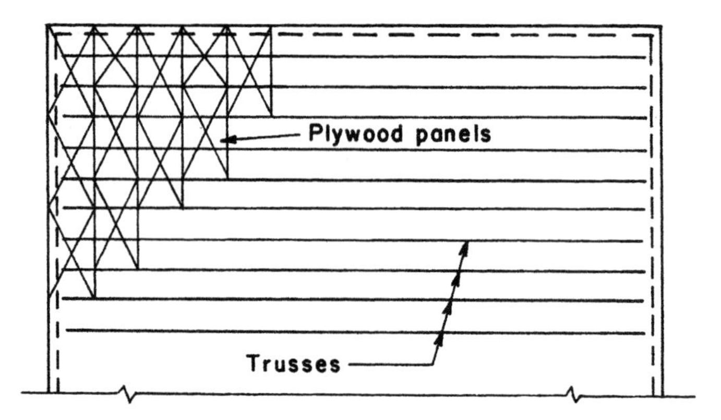

Figure 15.5 Partial plan for the Building One roof with the clear-spanning truss system.

Not shown is the required lateral bracing, which is required by code specifications. For practical purposes, the lateral bracing is also used to ensure the accurate spacing, vertical straightness, and longitudinal straightness of the quite laterally flexible trusses. The truss system is very sturdy when properly erected, but the correct bracing of the trusses is essential to their stability.

These trusses will undoubtedly be the proprietary products of some manufacturer. Safe load tables prepared by the manufacturer can be used for their design.

Selection of truss spacing relates not only to the load capacity of the trusses but also to the development of the roof deck and the ceiling construction. A 2-ft spacing would permit use of ordinary plywood panels for the deck and directly applied gypsum drywall for the ceiling. The truss chords can be fabricated with the bottom flat and the top sloped for roof drainage. This is undoubtedly the simplest and most economical form for the general construction.

However, the 2-ft spacing is a bit close for this span, so the scheme shown here uses a 32-in. spacing (one third of the 96 in. panel length). Use of 7/8-in.-thick plywood with tongue-and-groove edges would eliminate the need for edge blocking between the trusses. As for other situations, this design must also satisfy the need for development of the roof deck as a horizontal diaphragm for lateral loads.

The long-span trusses generate a significant vertical load for the supporting stud walls. The combination of this load with bending induced by wind pressure on the exterior walls requires an investigation of the studs, as was described for Scheme 1. The form of this investigation is illustrated in Example 5 in Section 9.8.

The design of the basic elements for this scheme is summarized as follows:

Roof Deck. 4 ft × 8 ft plywood panels, face grain perpendicular to the trusses, 32 in. span. From Table 12.1, thicknesses of 19/32, 5/8, 3/4, and 7/8 in. are permitted. As discussed previously, 7/8 in. thickness can be obtained with tongue-and-groove edges, providing better support for the edges between trusses.

Truss. 50-ft span, 32 in. on center. Live load = 20 psf × 32/12 = 53.3 lb/ft. Dead load: roofing + insulation + deck + ceiling + equipment (lights, HVAC, fire sprinklers), say 25 psf. 25 × 32/12 = 66.7 lb/ft. Total load = 53.3 + 66.7 + truss = 120 lb/ft + truss weight. Use this load to pick a truss from a manufacturer's load tables.

Bearing Wall. For front wall assume a 13.5 ft high wall + a 5-ft canopy cantilevered from the wall. Roof dead load = (50/2) × 30 psf (including trusses) = 750 lb/ft. Roof live load = 20 psf × (25 + 5) = 600 lb/ft. Wall dead load = 13.5 × 20 psf = 270 lb/ft. Canopy dead load = 100 lb/ft (estimate). Total dead load = 1120 lb/ft. Total load = 1120 + 600 = 1720 lb/ft. With studs at 16 in. spacing, load on one stud is 1720 × (16/12) = 2294 lb. An investigation of the stud for combined gravity and wind loads is given in Example 5, Section 9.8.

Design for Lateral Loads

Design of the building structure for wind includes consideration for the following:

1. Inward and outward pressure on exterior building surfaces, causing bending of the wall studs and an addition to the gravity loads on roofs.
2. Total lateral (horizontal) force on the building, requiring bracing by the roof diaphragm and the shear walls.
3. Uplift on the roof, requiring anchorage of the roof structure to its supports.
4. Total effect of uplift and lateral forces, possibly resulting in overturn (toppling) of the entire building.

Uplift on the roof depends on the roof shape and the height above ground. For this low, flat-roofed building, the ASCE standard (Ref. 3) requires an uplift pressure of 10.7 psf. In this case the uplift pressure does not exceed the roof dead weight of 16 psf, so anchorage of the roof construction is not required. However, common use of metal framing devices for light wood frame construction provides an anchorage with considerable resistance.

Overturning of the building is not likely critical for a building with this squat profile (50 ft wide by only 13.5 ft high). Even if the overturning moment caused by wind exceeds the restoring moment due to the building dead weight, the sill anchor bolts will undoubtedly hold the building down in this case. Overturn of the whole building is usually more critical for tower-like building forms or for extremely light construction. Of separate concern is the overturn of individual bracing elements, in this case the individual shear walls, which will be investigated later.

Wind Force on the Bracing System

The building's bracing system must be investigated for horizontal force in the two principal orientations: east-west and north-south. And, if the building is not symmetrical, in each direction on each building axis: east, west, north, and south.

The horizontal wind force on the north and south walls of the building is shown in Figure 15.6. This force is generated by a combination of positive (direct, inward) pressure on the windward side and negative (suction, outward) on the lee side of the building. The pressures shown as Case 1 in Figure 15.6 are obtained from data in the ASCE 2006 (Ref. 3) chapter on wind loads. (See discussion of wind loads in Chapter 13.) The single pressures shown in the figure are intended to account for the combination of positive and suction pressures. The ASCE standard provides for two zones of pressure—a general one and a small special increased area of pressure at one end. The values shown in Figure 15.6 for these pressures are derived by considering a critical wind velocity of 90 MPH and an exposure condition B, as described in the standard.

The range for the increased pressure in Case 1 is defined by the dimension *a* and the height of the windward wall. The value of *a* is established as 10 percent of the least plan dimension of the building or 40 percent of the wall height, whichever is smaller, but not less than

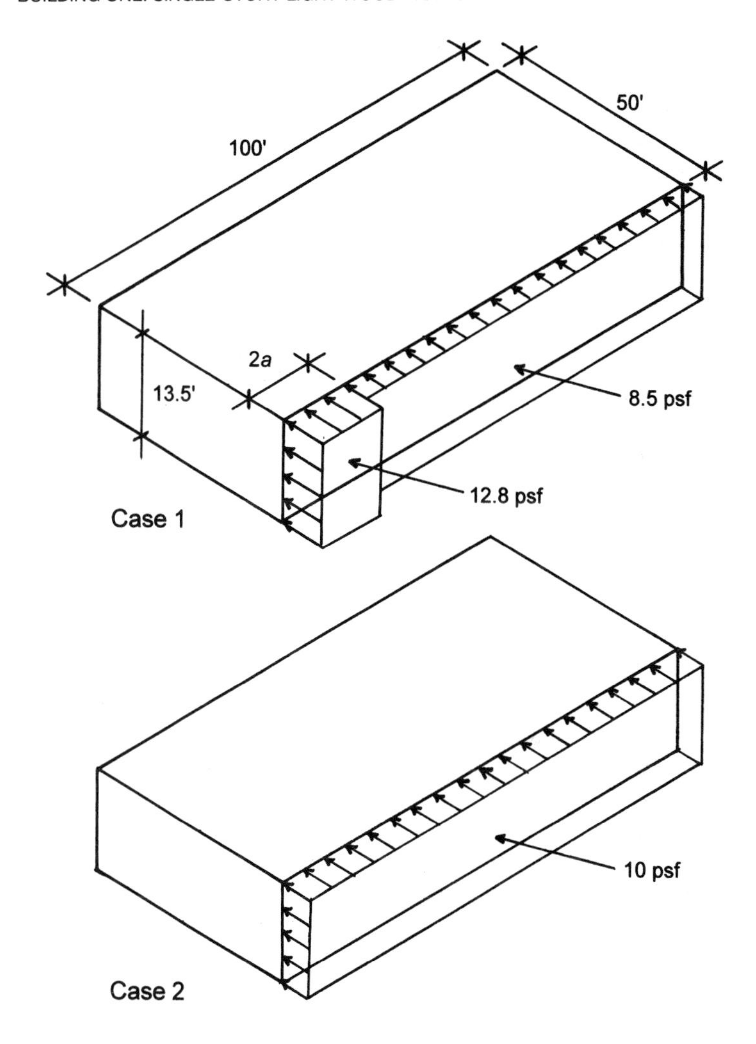

Figure 15.6 Wind pressure on the south wall, ASCE 2006 (Ref. 3).

3 ft. For this example a is determined as

$$a = 0.10 \times 50 = 5\,\text{ft}$$

or

$$a = 0.4 \times 13.5 = 5.4\,\text{ft}$$

The distance for the pressure of 12.8 psf in Case 1 is thus $2(a) = 2(5.0) = 10$ ft.

The design standard also requires that the bracing system be designed for a minimum pressure of 10 psf on the entire area of the wall. This sets up two cases (Case 1 and Case 2 in Figure 15.6) that must be considered. Since the concern for the design is the generation of maximum effect on the roof diaphragm and the end shear walls, the critical conditions may be determined by considering the development of end reaction forces and maximum shear for an analogous beam subjected to the two loadings. This analysis is shown in Figure 15.7, from which it is apparent that the critical concern for the end shear walls and the maximum effect in the roof diaphragm is derived from Case 2 in Figure 15.6.

The actions of the horizontal wind force resisting system in this regard are illustrated in Figure 15.8. The initial force comes from wind pressure on the building's vertical sides. The wall studs span vertically

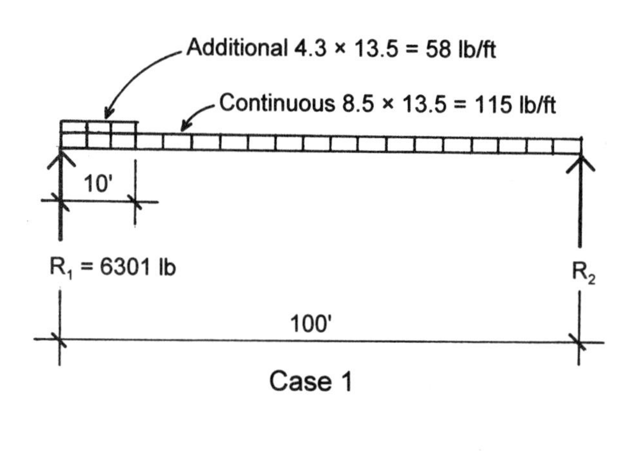

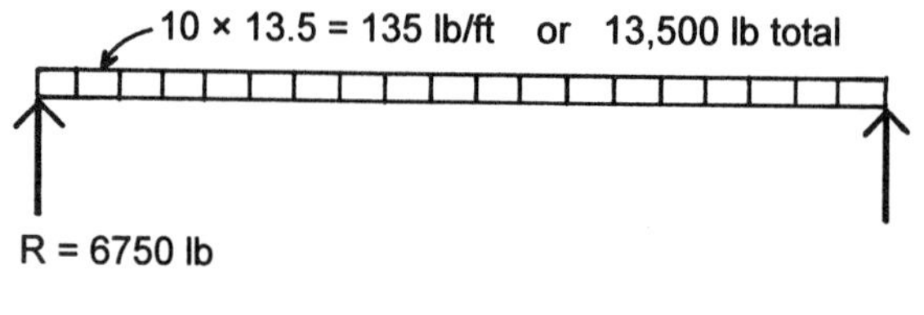

Case 2

Figure 15.7 Resultant wind forces on the end shear walls.

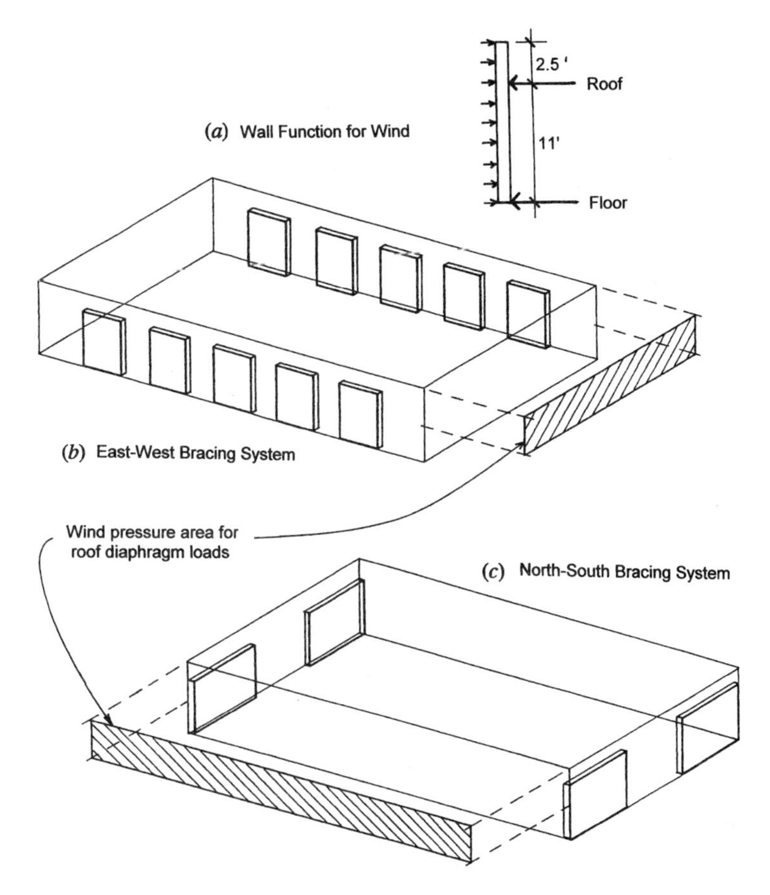

Figure 15.8 Wall functions and wind pressure development.

to resist this uniformly distributed load, as shown in Figure 15.8*a*. Assuming the wall function to be as shown in Figure 15.8*a*, the north-south wind force delivered to the roof edge is determined as

$$\text{Total } W = (10 \text{ psf})(100 \times 13.5) = 13{,}500 \text{ lb}$$

$$\text{Roof Edge } W = 13{,}500 \times \frac{6.75}{11} = 8284 \text{ lb}$$

In resisting this load the roof functions as a spanning member supported by the shear walls at the east and west ends of the building. The investigation of the diaphragm as a 100-ft simple span beam with uniformly distributed loading is shown in Figure 15.9. The end reaction and maximum diaphragm shear force is found as

$$R = V = \frac{8284}{2} = 4142 \text{ lb}$$

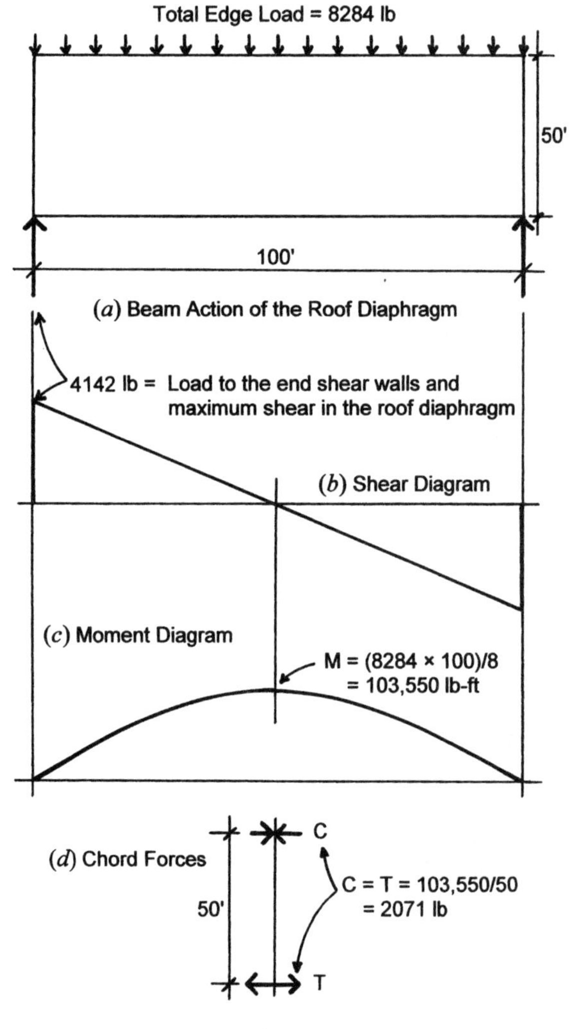

Figure 15.9 Spanning functions of the roof diaphragm.

which produces a maximum unit shear in the 50-ft-wide diaphragm of

$$v = \frac{\text{shear force}}{\text{roof width}} = \frac{4142}{50} = 82.8 \text{ lb/ft}$$

From Table 13.1 a variety of selections is possible. Variables include the class of the plywood, the panel thickness, the width of supporting rafters, the nail size and spacing, the use of blocking, and the layout pattern of the plywood panels. Assuming a minimum plywood thickness for the flat roof at 1/2 in. (given as 15/32 in the table), a possible choice is as follows:

APA Rated sheathing, 15/32-in.-thick, 2 × rafters, 8d nails at 6 in. at all panel edges, unblocked diaphragm, allowable shear = 180 lb/ft.

In this example, if the need for the minimum thickness plywood is accepted, it turns out that the minimum construction is more than adequate for the required lateral force resistance. Had this not been the case, and the required capacity had resulted in considerable nailing beyond the minimum, it would be possible to graduate the nailing spacing from that required at the building ends to minimal nailing in the center portion of the roof. (See the form of the shear variation across the roof width.) (See also Example 3 in Section 13.2.)

The moment diagram shown in Figure 15.9 indicates a maximum value of 104 kip-ft at the center of the span. This moment is used to determine the maximum force in the diaphragm chords at the roof edges. The force must be developed in both compression and tension as the wind direction reverses. With the construction as shown in Figure 15.1*f*, the top plate of the stud wall is the most likely element to be utilized for this function. In this case the chord force of 2071 lb, as shown in Figure 15.9, is quite small, and the doubled 2 × member should be capable of resisting the force. However, the building length requires the use of several pieces to create this continuous plate, so the splices for the double member should be investigated.

The end reaction force for the roof diaphragm, as shown in Figure 15.9, must be developed by the end shear walls. As shown in Figure 15.1, there are two walls at each end, both 21 ft long in plan. Thus the total shear force is resisted by a total of 42 ft of shear wall and the unit shear in the wall is

$$v = \frac{4142}{42} = 98.6 \text{ lb/ft}$$

As with the roof, there are various considerations for the selection of the wall construction. Various materials may be used on both the exterior and interior surfaces of the stud wall. The common form of construction shown in Figure 15.1*f* indicates gypsum drywall on the inside and a combination of plywood and stucco (cement plaster) on the outside of this wall. All three of these materials have rated resistances to shear wall stresses. However, with a combination of materials, it is common practice to consider only the strongest of the materials to be the resisting element. In this case that means the plywood sheathing on the exterior of the wall. From Table 13.2 a possible choice is

APA rated sheathing, 3/8-in.-thick, with 6d nails at 6-in. spacing at all panel edges, capacity = 200 lb/ft.

Again, this is minimal construction. For higher loadings a greater resistance can be obtained by using better plywood, thicker panels, larger nails, closer nail spacing, and—sometimes—wider studs. Unfortunately, the nail spacing cannot be graduated—as it may be for the roof—as the unit shear is a constant value throughout the height of the wall.

Figure 15.10*a* shows the loading condition for investigation of the overturn effect on the end shear wall. Overturn is resisted by the

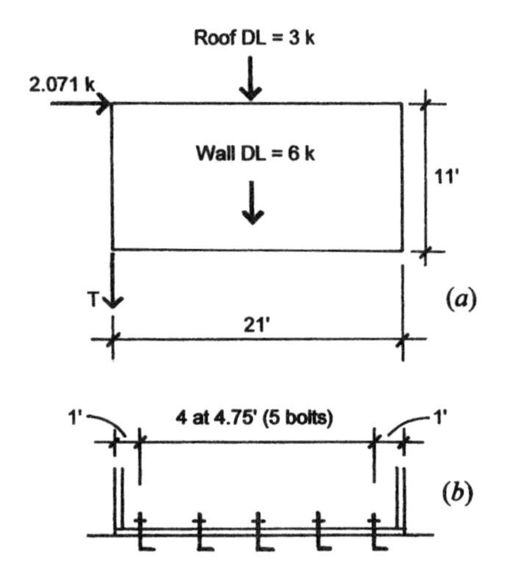

Figure 15.10 Functions of the end shear walls.

so-called restoring moment, due to the dead load on the wall—in this case a combination of the wall weight and the portion of roof dead load supported by the wall. Safety is considered adequate if the restoring moment is at least 1.5 times the overturning moment. A comparison is therefore made between the value of 1.5 times the overturning moment and the restoring moment, as follows:

Overturning moment = (2.071)(11)(1.5) = 34.2 kip-ft
Restoring moment = (3 + 6)(21/2) = 94.5 kip-ft

This indicates that no tiedown force is required at the wall ends (force T as shown in Figure 15.10). Details of the construction and other functions of the wall may provide additional resistances to overturn. However, some designers prefer to use end anchorage devices (called tiedown anchors) at the ends of all shear walls, regardless of loading magnitudes.

Finally, the walls will be bolted to the foundation with code-required sill bolts, which provide some resistance to uplift and overturn effects. At present most codes do not permit sill bolts to be used for computed resistances to these effects due to the cross-grain bending that is developed in the wood sill members.

The sill bolts are used, however, for resistance to the sliding of the wall. The usual minimum bolting is with 1/2-in. bolts, spaced at a maximum of 6 ft centers, with a bolt not less than 12 in. from the wall ends. This results in a bolting for this wall as shown in Figure 15.10*b*. The five bolts shown should be capable of resisting the lateral force, using the values given by the codes.

For buildings with relatively shallow foundations, the effects of shear wall anchorage forces on the foundation elements should also be investigated. For example, the overturning moment is also exerted on the foundations, and may cause undesirable soil stresses or require some structural resistance by foundation elements.

Another area of concern has to do with the transfer of forces from element-to-element in the whole lateral force resisting structural system. A critical point of transfer in this example is at the roof-to-wall joint. The force delivered to the shear walls by the roof diaphragm must actually be passed through this joint, by the attachments of the construction elements. The precise nature of this construction must be determined and must be investigated for these force actions.

Foundations

Foundations for Building One would be quite minimal. For the exterior bearing walls, the construction provided will depend on concerns for frost and the location below ground of suitable bearing material. Options for the wood structure are shown in Figure 15.11.

Where frost is not a problem and suitable bearing can be achieved at a short distance below the finished grade, a common solution is to use the construction shown in Figure 15.11*a*, called a *grade beam*. This is essentially a combined footing and short foundation wall in one. It is typically reinforced with steel bars in the top and bottom to give it some capacity as a continuous beam, capable of spanning over isolated weak spots in the supporting soil.

Where frost *is* a problem, local codes will specify a minimum distance from the finished grade to the bottom of the foundation. To reach this distance it may be more practical to build a separate footing and foundation wall, as shown in Figure 15.11*b*. This short, continuous wall may also be designed for some minimal beamlike action, similar to that for the grade beam.

For either type of foundation, the light loading of the roof and the wood stud wall will require a very minimal width of foundation, if the bearing soil material is at all adequate. If bearing is not adequate, then this type of foundation (shallow bearing footings) must be replaced

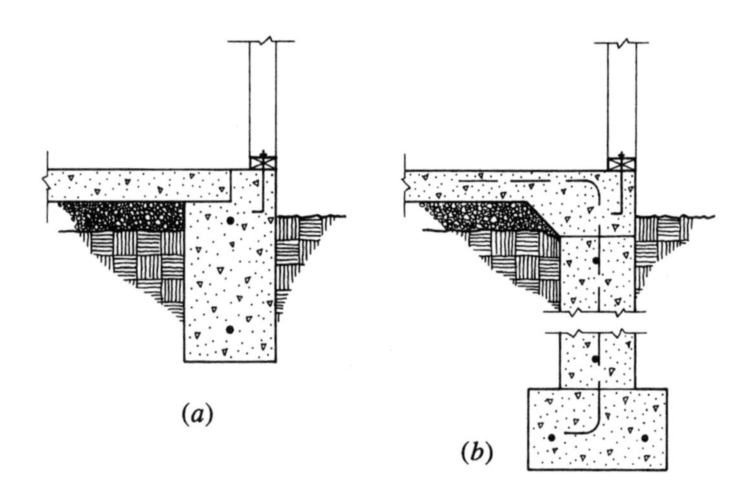

Figure 15.11 Options for the exterior wall foundations.

with some form of deep foundation (piles or caissons), which presents a major structural design problem.

Footings for any interior columns for Building One would also be minimal, due to the light loading from the roof structure and the low value of live load for the roof.

15.2 BUILDING TWO: MULTISTORY LIGHT WOOD FRAME

Figure 15.12 shows a building that consists essentially of stacking the plan for Building One to produce a two-story building. The profile section of the building shows that the structure for the second story is developed in a manner similar to that for the roof structure and walls for Building One. Here, for both the roof and the second floor, the framing option chosen is that shown in Figure 15.2*b*.

Even though the framing layout is similar, the principal difference between the roof and floor structures has to do with the loadings. Both the dead load and live load are greater for the floor. In addition, the deflection of long-spanning floor members is a concern both for the dimension and for the bounciness of the structure.

The two-story building sustains a greater total wind load, although the shear walls for the second story will be basically the same as for Building One. The major effect in this building is the force generated in the first-story shear walls. In addition, there is a second horizontal diaphragm: the second-floor deck.

Some details for the second-floor framing are shown in Figure 15.12*c*. Roof framing details are similar to those shown for Building One in Figure 15.4*b*. As with Building One, an option here is to use a clear-spanning roof structure—most likely with light trusses—which would eliminate the need for the corridor wall columns in the second floor.

Design for Gravity Loads

For design of the second-floor structure, the following construction is assumed. The weight of the ceiling is omitted, assuming it to be supported by the first-story walls.

Carpet and pad	3.0 psf
Fiberboard underlay	3.0
Concrete fill, 1.5 in.	12.0

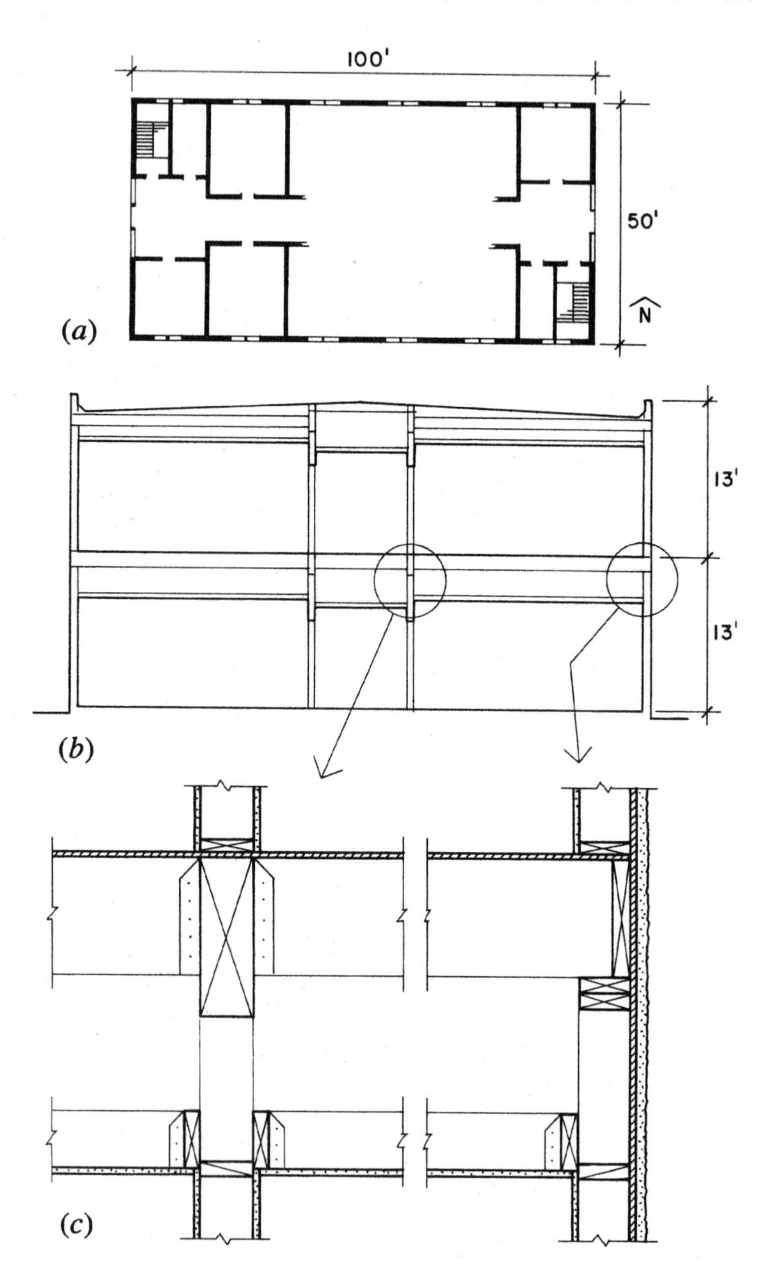

Figure 15.12 Building Two, general form and construction details.

Plywood deck, 3/4 in.	2.5
Ducts, lights, wiring	3.5
Total, without joists	24.0 psf

Minimum live load for office areas is 50 psf. However, the code requires the inclusion of an extra load to account for possible additional partitioning, usually 25 psf. Thus, the full design live load is 75 psf. At the corridor the live load is 100 psf. Many designers would prefer to design the whole floor for a live load of 100 psf, thereby allowing for other arrangements or occupancies in the future. As the added partition load is not required for this live load, it is only an increase of about 20% in the total load. With this consideration, the total design load for the joists is thus 124 psf.

With joists at 16-in. centers, the superimposed uniformly distributed load on a single joist is thus

$$DL = \frac{16}{12}(24) = 32\ \text{lb/ft} + \text{ the joist, say 40 lb/ft}$$

$$LL = \frac{16}{12}(100) = 133\ \text{lb/ft}$$

and the total load is 173 lb/ft. For the 21-ft-span joists the maximum bending moment is

$$M = \frac{wL^2}{8} = \frac{173 \times (21)^2}{8} = 9537\ \text{ft-lb}$$

For Douglas fir-larch joists of select structural grade and 2-in. nominal thickness, F_b from Table 4.1 is 1725 psi for repetitive member use. Thus the required section modulus is

$$S = \frac{M}{F_b} = \frac{9537 \times 12}{1725} = 66.3\ \text{in.}^3$$

Inspection of Table A. 3 shows that there is no 2 × member with this value for section modulus. A possible choice is for a 3 × 14 with $S =$ 73.151 in.^{3}.

This is not a good design, since the 3 × members in select structural grade are very expensive. Furthermore, as inspection of Figure 6.6 reveals, the deflection is considerable—not beyond code limits, but sure to result in some sag and bounciness. A better choice for this span and load is probably for one of the proprietary fabricated I-joists which are available in depths greater than 12 in.

The beams support both the 21-ft joists and the short 8-ft corridor joists. With the closely spaced joists, this amounts to a uniformly distributed load on a strip of floor that has a width of (21/2) + 8/2) = 14.5 ft. The total load periphery carried by one beam is thus 14.5 × 16.67 = 242 ft^2, for which a reduction of 7% is allowed for the live load (see discussion in Sec. 14.6). Dead load is increased to 30 psf to account for the weight of the joists and blocking. Using the same loading for both the corridor and the offices, the beam load is determined as

$$\begin{aligned}
DL = (30)(14.5) &= 435 \text{ lb/ft} \\
+ \text{ beam weight} &= 50 \\
+ \text{ wall above} &= 150 \\
\text{Total } DL &= 635 \text{ lb/ft} \\
LL = (0.93)(100)(14.5) &= 1349 \text{ lb/ft} \\
\text{Total load on beam} &= 1984, \text{ say } 2000 \text{ lb/ft}
\end{aligned}$$

For the uniformly loaded simple beam with a span of 16.67 ft

Total load = W = (2)(16.67) = 33.4 kips
End reaction = maximum beam shear = $W/2$ = 16.7 kips
Maximum bending moment is

$$M = \frac{WL}{8} = \frac{33.4 \times 16.67}{8} = 69.6 \text{ kip-ft}$$

For a Douglas fir-larch, dense No. 1 grade beam, Table 4.1 yields values of F_b = 1550 psi, F_v = 170 psi, and E = 1,700,000 psi. To satisfy the flexural requirement, the required section modulus is

$$S = \frac{M}{F_b} = \frac{69.6 \times 12}{1.550} = 539 \text{ in.}^3$$

From Table A.3 the least weight section is a 10 × 20 or a 12 × 18.

If the 20-in.-deep section is used, its effective bending resistance must be reduced (see discussion in Section 6.2). Thus the actual moment capacity of the 10 × 20 is reduced by the size factor from Table 6.1 and is determined as

$$M = C_F \times F_b \times S = (0.947)(1.550)(602.1)(1/12) = 73.6\,\text{kip-ft}$$

As this still exceeds the requirement, the selection is adequate. Similar investigations will show the other size options to also be acceptable.

If the actual beam depth is 19.5 in., the critical shear force may be reduced to that at a distance of the beam depth from the support. Thus an amount of load equal to the beam depth times the unit load can be subtracted from the maximum shear. The critical shear force is thus

$$V = (\text{actual end shear force}) - (\text{beam depth in ft times unit load})$$
$$= 16.7 - 2.0(19.5/12) = 16.7 - 3.25 = 13.45\ \text{kips}$$

For the 10 × 20 the maximum shear stress is thus

$$f_v = 1.5\frac{V}{A} = 1.5\frac{13,450}{185.25} = 108.9\ \text{psi}$$

This is less than the limiting stress of 170 psi as given in Table 4.1, so the beam is acceptable for shear resistance. However, this is still a really big piece of lumber, and questionably feasible, unless this building is in the heart of a major timber region. It is probably logical to modify the structure to reduce the beam span or to choose a steel beam or a glued-laminated section in place of the solid-sawn timber.

Although deflection is often critical for long spans with light loads, it is seldom critical for the short-span, heavily loaded beam. The reader may verify this by investigating the deflection of this beam, but the computation is not shown here (see Section 6.3).

For the interior column at the first story, the design load is approximately equal to the total load on the second-floor beam plus the load from the roof structure. As the roof loading is about one third of that for the floor, the design load is about 50 kips for the 10-ft-high column. Table 9.1 yields possibilities for an 8 × 10 or 10 × 10 section. For various reasons it may be more practical to use a steel member here—a round pipe or a square tubular section—which may actually be accommodated within a relatively thin stud wall at the corridor.

Columns must also be provided at the ends of the beams in the east and west walls. Separate column members may be provided at these locations, but it is also common to simply build up a column from a number of studs.

Design for Lateral Loads

Lateral resistance for the second story of Building Two is essentially the same as for Building One. Design consideration here will be limited to the diaphragm action of the second-floor deck and the two-story end shear walls.

The wind loading condition for the two-story building is shown in Figure 15.13*a*. For the same design conditions assumed for wind in

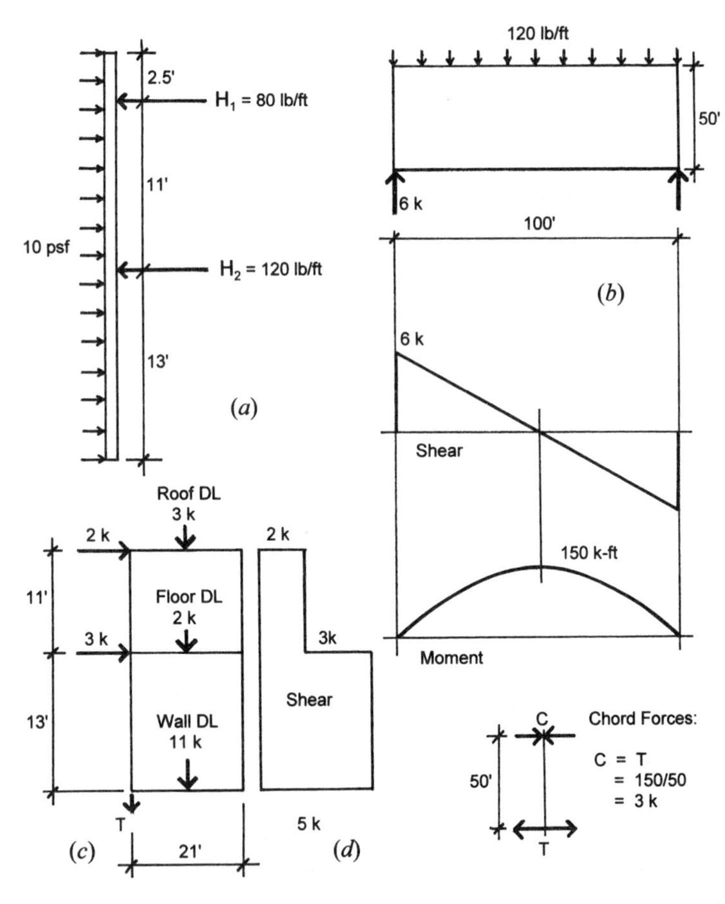

Figure 15.13 Building Two, development of lateral forces due to wind: (a) functions of the exterior wall under wind pressure and bracing from the roof and floors, (b) spanning functions of the second-floor diaphragm, (c) loading of the two-story shear wall, (d) shear diagram for the shear wall.

Chapter 13, the pressure used for horizontal force on the building bracing system is 10 psf for the entire height of the exterior wall. At the second-floor level the wind load delivered to the edge of the diaphragm is 120 lb/ft, resulting in the spanning action of the diaphragm as shown in Figure 15.13*b*. Referring to the building plan in Figure 15.12*a*, it may be observed that the opening required for the stairs creates a void in the floor deck at the ends of the diaphragm. The net width of the diaphragm is thus reduced to approximately 35 ft at this point, and the unit stress for maximum shear is

$$v = \frac{6000}{35} = 171 \text{ lb/ft}$$

From Table 13.1 it may be determined that this requires only minimum nailing for a 19/32-in.-thick plywood deck, which is the usual minimum thickness used for floor decks.

As discussed for the roof diaphragm in Section 15.1, the chord at the edge of the floor diaphragm must be developed by framing members to sustain the computed tension/compression force of 3 kips. Ordinary framing members may be capable of this action, if attention is paid to splicing for full continuity of the 100-ft-long edge member.

The construction details for the roof, floor, and exterior walls must be carefully studied to ensure that the necessary transfers of force are achieved. These transfers include the following:

1. Transfer of the force from the roof plywood deck (the horizontal diaphragm) to the wall plywood sheathing (the shear wall).
2. Transfer from the second-story shear wall to the first-story wall that supports it.
3. Transfer from the second-floor deck (the horizontal diaphragm) to the first-story wall plywood sheathing (the shear walls).
4. Transfer from the first-story shear wall to the foundation.

In the first-story end shear walls the total lateral load is 5000 lb, as shown in Figure 15.2*d*. For the 21-ft-wide wall the unit shear is

$$v = \frac{5000}{21} = 238 \text{ lb/ft}$$

From Table 13.2 it may be noted that this resistance can be achieved with APA Rated sheathing of 3/8-in. thickness, although nail spacing closer than the minimum of 6 in. is required at the panel edges.

At the first-floor level, the investigation for overturn of the end shear wall is as follows (see Figure 15.13*c*):

Overturning moment = (2)(24)(1.5) = 72 kip-ft
+ (3)(13)(1.5) = 58.5 kip-ft
Total overturning moment with safety factor: 130.5 kip-ft
Restoring moment = (3 + 2 + 11)(21/2) = 168 kip-ft
Net overturning effect = 130.5 − 168 = −37.5 kip-ft

As the restoring moment provides a safety factor greater than 1.5, there is no requirement for the anchorage force *T*.

In fact there are other resisting forces on this wall. At the building corner the end walls are reasonably well attached through the corner framing to the north and south walls, which would need to be lifted to permit overturning. At the sides of the building entrance, with the second-floor framing as described, there is a post in the end of the wall that supports the end of the floor beams. All in all, there is probably no computational basis for requiring an anchor at the ends of the shear walls. Nevertheless, many designers routinely supply such anchors.

15.3 BUILDING THREE: MASONRY AND TIMBER STRUCTURE

This section presents a solution for a structure that utilizes a form of construction known as *mill construction*. This construction evolved during the eighteenth and nineteenth centuries in response to the need for practical buildings for industrial and commerical uses. Some of the typical details for this type of construction are presented here as they were published in popular books on building technology in the early twentieth century.

A common form for mill construction was one that employed heavy masonry exterior walls and timber roof and floor framing. In later times, roofs were developed as clear-spans with timber or steel trusses. Examples of this construction are presented here in a set of illustrations (Figures 15.14 through 15.18) taken from *Construction Revisited* (Ref. 14). These composite figures were assembled from the illustrations in several books published in the early twentieth century, as noted in the figure captions.

Figure 15.14 shows a building form that was used extensively hundreds of years ago and up to the early twentieth century in Europe and

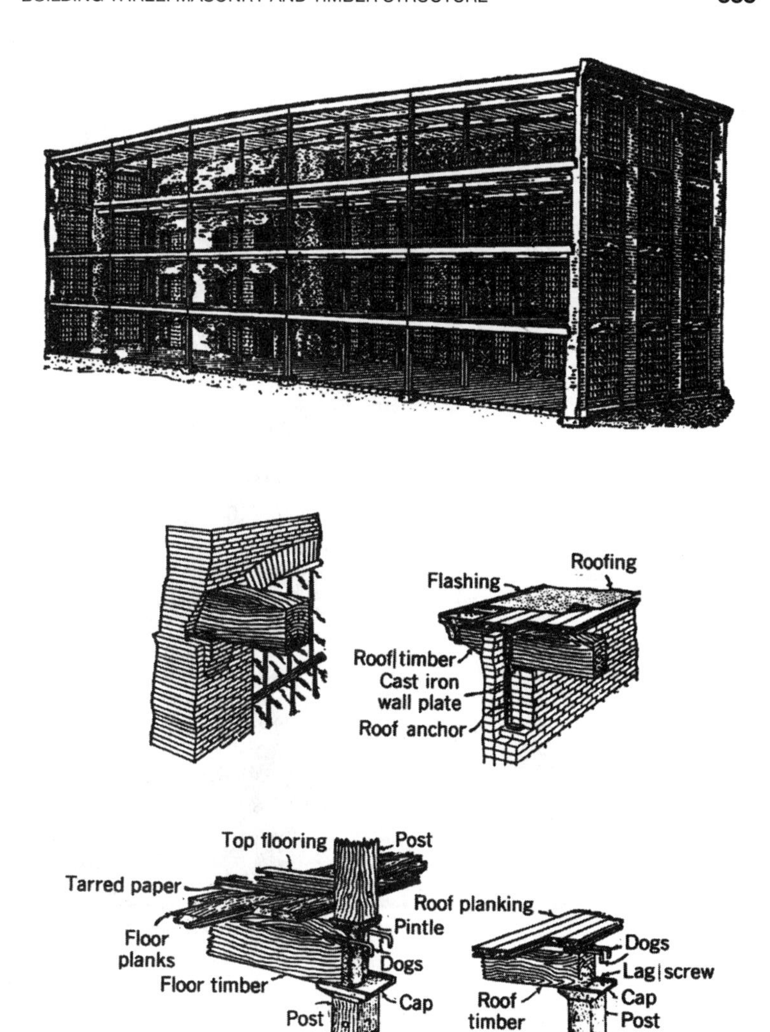

Figure 15.14 Example of mill construction for a multistory building. Reproduced from *Construction Revisited* (Ref. 14) with permission of the publishers, John Wiley & Sons, Inc., New Jersey. This is a composite of illustrations from *Architects' and Builders' Handbook*, 1931, published by John Wiley & Sons, Inc., New Jersey.

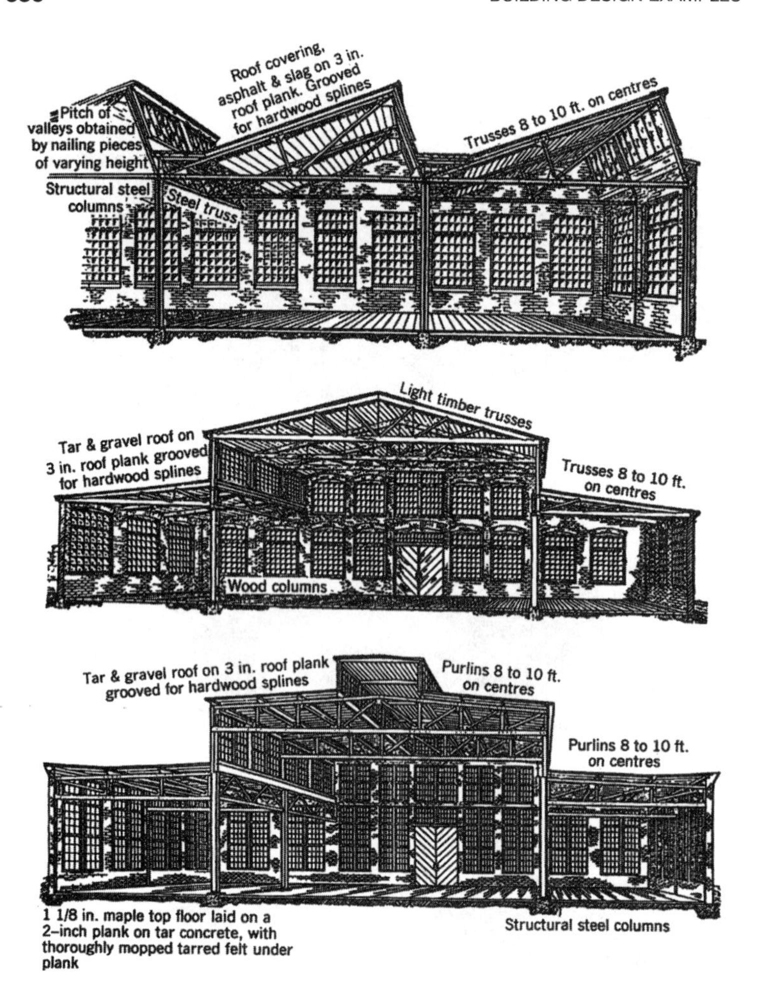

Figure 15.15 Example of mill construction for one-story industrial buildings. Reproduced from *Construction Revisited* (Ref. 14) with permission of the publishers, John Wiley & Sons, Inc., New Jersey. This is a composite of illustrations from *Architects' and Builders' Handbook*, 1931, published by John Wiley & Sons, Inc., New Jersey.

North America. Brick for structural masonry walls and sawn timber for the roof and floor construction were produced by many suppliers. While brick and timber are still available, buildings having this appearance are not likely to have this same form of construction. Structural masonry is now mostly produced with precast concrete blocks, heavy beams will

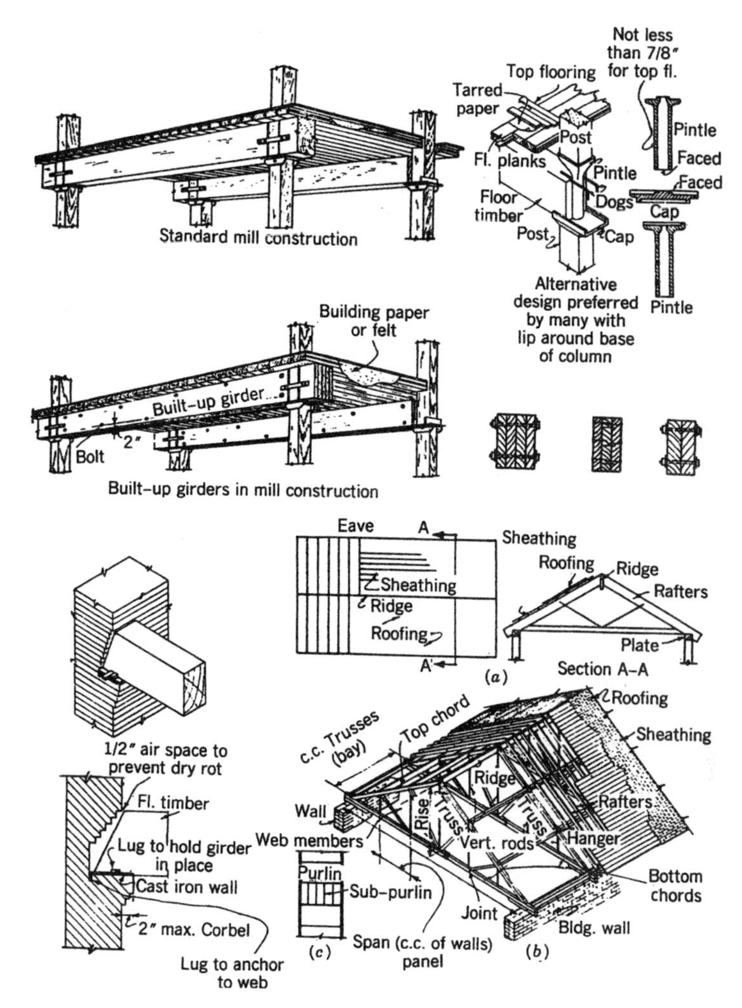

Figure 15.16 Examples of heavy timber construction for roofs and floors. Reproduced from *Construction Revisited* (Ref. 14) with permission of the publishers, John Wiley & Sons, Inc., New Jersey. This is a composite of illustrations from *Architects' and Builders' Handbook*, 1931, published by John Wiley & Sons, Inc., New Jersey.

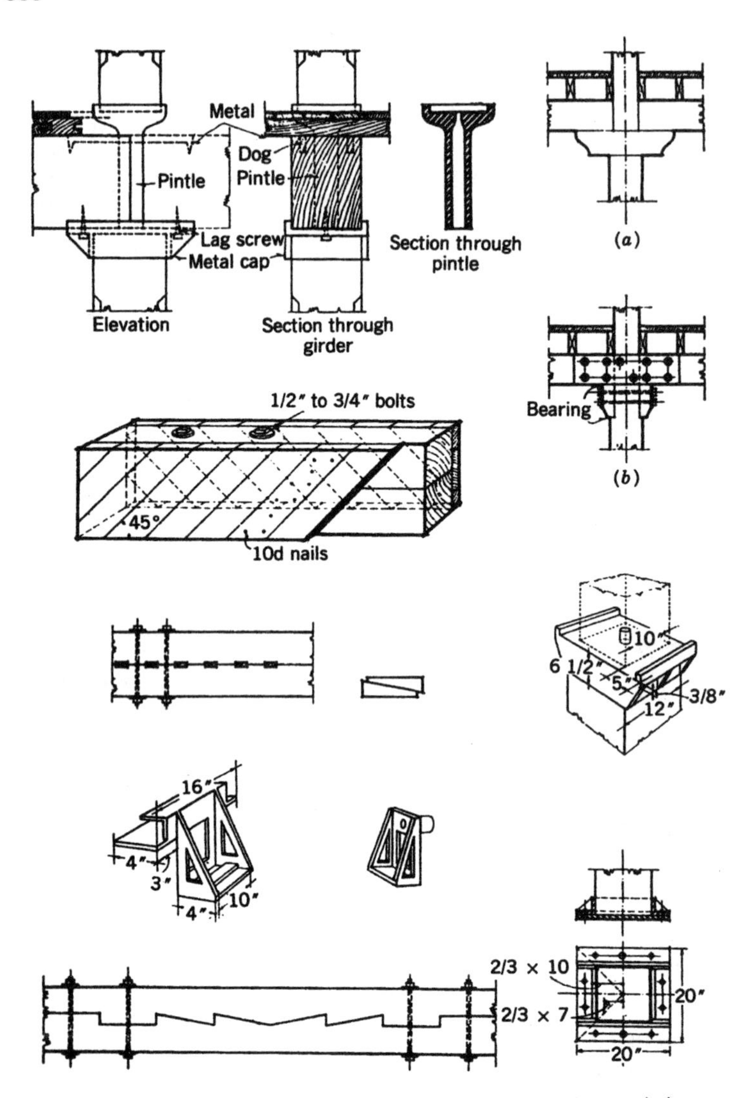

Figure 15.17 Typical elements of early twentieth century heavy timber construction. Reproduced from *Construction Revisited* (Ref. 14) with permission of the publishers, John Wiley & Sons, Inc., New Jersey. This is a composite of illustrations from *Architects' and Builders' Handbook*, 1931, published by John Wiley & Sons, Inc., New Jersey.

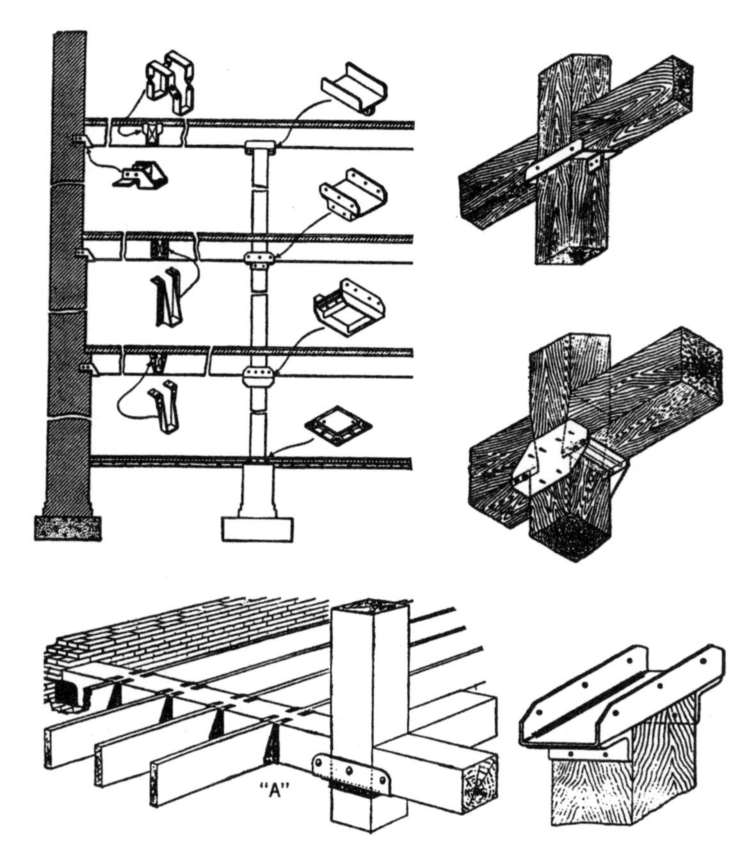

Figure 15.18 Examples of early twentieth century steel framing devices for masonry and timber construction. Reproduced from *Construction Revisited* (Ref. 14) with permission of the publishers, John Wiley & Sons, Inc., New Jersey. This is a composite of illustrations from *Architects' and Builders' Handbook*, 1931, published by John Wiley & Sons, Inc., New Jersey.

most likely be glued-laminated wood or steel, and deck will quite likely be of laminated elements. Indeed, the whole building structure may be of reinforced concrete, with the exterior brick finish produced with thin tiles adhered to framed structural walls. Nevertheless, the timber construction shown is actually still possible with sawn wood members and steel connection devices.

Figure 15.15 shows another ancient development of the masonry-walled, timber-framed building. Here a trussed roof structure provides for relatively large, column-free, interior spaces. This general form is still widely used for industrial and commercial buildings, although now mostly developed with steel framing for the roof structure.

Figure 15.16 shows some additional details for the timber structures of the buildings in Figures 15.14 and 15.15. A possible modification shown here is the replacement of the single large timbers with built-up sections consisting of multiple pieces of thinner lumber, with members held together by bolts or lag screws. This construction produced somewhat better results in terms of dimensional stability (twisting, warping, splitting, etc.), and also reduced the need for obtaining the single large pieces—a problem that developed as the old-growth trees became less available with the elimination of the forests.

Figure 15.17 shows some additional details of the timber beams and columns. A common problem with heavy timber construction is that of the effect of cross-grain shrinkage of the timber beams. Add this to the relatively weak resistance to cross-grain compression (compression perpendicular to the grain), and the development of multistory buildings becomes difficult. Vertical settlement of the interior wood structure becomes a problem as the exterior masonry walls do not settle. Methods are shown here for by-passing the beams to transfer vertical loads directly from column-to-column. Also shown are some means for obtaining large timbers from smaller pieces of wood—now achieved, of course, with glue lamination.

Finally, Figure 15.18 shows some early details for steel connecting devices that are quite similar to many used presently. These ordinary tasks are still required, and classic forms of simple solutions persist in the inventory of available products. Replace the sawn timbers with glued-laminated ones, the sawn joists with I-joists, and the timber deck with plywood, and these connectors are still valid items for the construction.

A general form for Building Three is shown in Figure 15.19. A roof truss is used to provide a column-free, sky-lighted space on the top floor. The interior structure for the second and third floors is developed with elements of heavy timber construction. Some of the details for this construction are shown in Figure 15.19. Additional details are presented in the discussions that follow.

The Timber Floors and Columns

A general system for the interior construction is shown in Figure 15.19. This system uses heavy timber sections for the columns and main beams

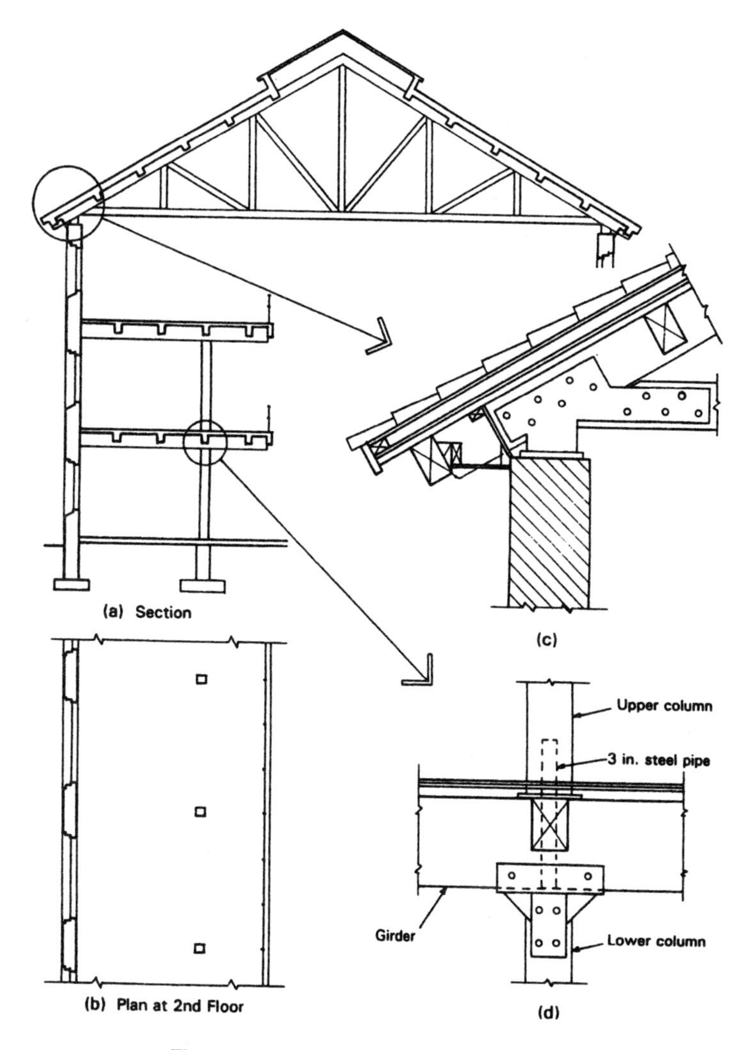

Figure 15.19 Building Three, general form.

and a thick timber deck. Maintaining minimum thickness for these members qualifies the system for a limited fire rating. Various floor finishes may be placed on top of the deck, but a layer of plywood or fiberboard must be used to provide a smooth surface.

A general framing plan for the second and third floors is shown in Figure 15.20. The clear span for the office spaces is provided by

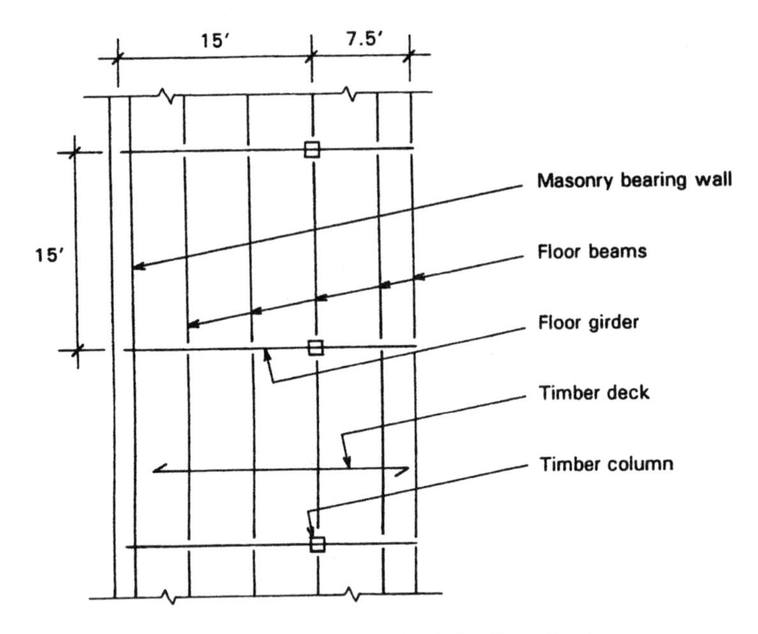

Figure 15.20 Framing for the timber floor structure.

girders at the locations of the interior columns and the exterior masonry wall piers. These girders cantilever to provide support for the balcony corridor. The structural floor deck consists of a timber deck that is exposed to view on the underside. A set of beams at 5-ft centers spans between the girders and supports the deck.

Use of solid-sawn elements for Building Three will be illustrated in the following computations.

Floor Deck. Assume construction with timber deck, fiberboard underlay, carpet and pad, producing approximately 15 psf dead load. Use live load = 100 psf for office + partitions or corridors. Select commercial deck from manufacturer's information. Probably OK for approximately 1.5 in. thick deck.

For beams and girders use Douglas fir-larch, No. 1 grade. From Table 4.1, $F_b = 1350$ psi, $F_v = 170$ psi.

Beams. 15 ft span, 5 ft spacing. Total load $= 5 \times 15 \times 115 = 8625$ lb. Add about 300 lb for the beam and use a total load of 9000 lb.

For the maximum bending moment,

$$M = \frac{WL}{8} = \frac{9000 \times 15}{8} = 16,875 \text{ lb-ft}$$

$$\text{Required } S = \frac{M}{F_b} = \frac{16{,}875 \times 12}{1350} = 150 \text{ in.}^3$$

From Table A.3, use 8 × 12, $S = 165 \text{ in.}^3$

Maximum shear = 9000/2 = 4500 lb, $f_v = 1.5\frac{V}{A} = 1.5\frac{4500}{86.25} =$ 78.3 psi less than 170 psi.

Deflection: see Figure 6.6, 12 in. depth, not critical.

Girder. Area supported = 15 × 22.5 = 337.5 ft^2. Use reduced live load, approximate beam reaction load = 8 kips. See Figure 15.21 for analysis of beam functions.

Maximum $M = 71.4$ kip-ft, Required $S = \frac{71{,}400 \times 12}{1350} = 635 \text{ in.}^3$

From Table A.3, use 10 × 22 or 12 × 20.

For this deep section, a reduction must be made of the effective section modulus based on the factor defined in Section 6.2 and given for various depths in Table 6.1. For the 22 in. member, this factor is 0.927. Thus the effective section modulus for the 10 × 22 is 0.937(732) = 686 in.^{3}, which is still greater than that required.

Maximum shear = 13,135 lb, $f_v = 1.5\frac{V}{A} = 1.5\frac{13,135}{204} = 97$ psi, less than 170 psi. (For 10 × 22.)

Column. 10 ft high to bottom of girder.

In first-story total load = twice the girder reaction, approximately 67 kips.

From Table 9.1, use 10 × 10 or 8 × 12.

Connections for this system can be developed with steel elements much in the same manner as those shown in Figures 15.14 through 15.18. Figure 15.17 shows the use of steel connecting devices that

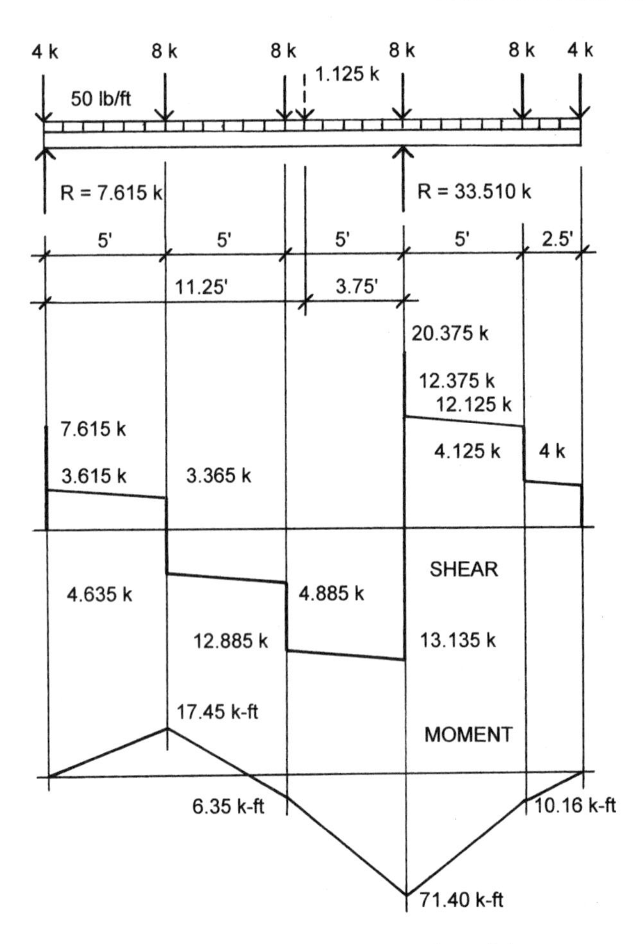

Figure 15.21 Investigation of the floor girder.

have their parallel in standard manufactured products available today. That is to say, the same general tasks (beam-to-beam, beam-to-column, column-to-foundation, etc.) can be achieved with corresponding elements in today's technology—in some cases with little modification.

Two special connection problems are those that occur at the splice joint of the multistory columns and at the end support for the girders at the masonry walls. For the multistory column there are three possibilities for the situation at the splice joint, as follows:

Continuous Columns. This means that the column is not jointed, but rather is continuous through the second-floor joint. Girders would consist of spaced, doubled members, straddling the columns.

End-Bearing Columns. This involves having the upper column bear directly on top of the lower column, using a steel device that is a single-piece combination base and cap in one.

Pintles. This is a device that permits the girder to pass through the joint, while still having the column bear on top of the beam.

Pintles are not much used any more, but either of the other options is possible. A major consideration is what maximum length of the required timber column section can be obtained.

The shorter the required length, the easier it is to find a good solid-sawn timber. Of course, using a glued-laminated column would make the longer piece an easy solution.

Figure 15.19 shows the use of a special connection for the multistory column. In this case the girder is a single piece and bears on a steel cap on the lower column. The upper column is supported by a steel plate welded to a thin steel pipe. The pipe passes through the girder to also bear on the cap plate on the lower column. The advantage gained with this connection is the ability to pass the single-piece girder through the joint to form the cantilever without having the upper column bear on the girder—not a recommended detail. Some lateral shrinkage of the timber girder is likely, with subsequent lowering of the upper column if it bears on the girder.

The Timber Roof Truss

As shown in the building section in Figure 15.19, the roof structure utilizes clear-spanning trusses. There are several options for the layout, materials, and assemblage details of these trusses. In Figure 15.15 the upper figure shows the use of a light steel truss with joints achieved with rivets and steel gusset plates, a common form for moderate spans.

The middle figure in Figure 15.15 shows the use of a truss consisting of a combination of timber and steel elements. This was a common form in earlier times, largely due to the elimination of tension connections between wood members. A somewhat more detailed illustration of this form of truss is shown in Figure 11.17.

Although the composite wood and steel truss is still used, the development of wood tension members is somewhat more feasible today

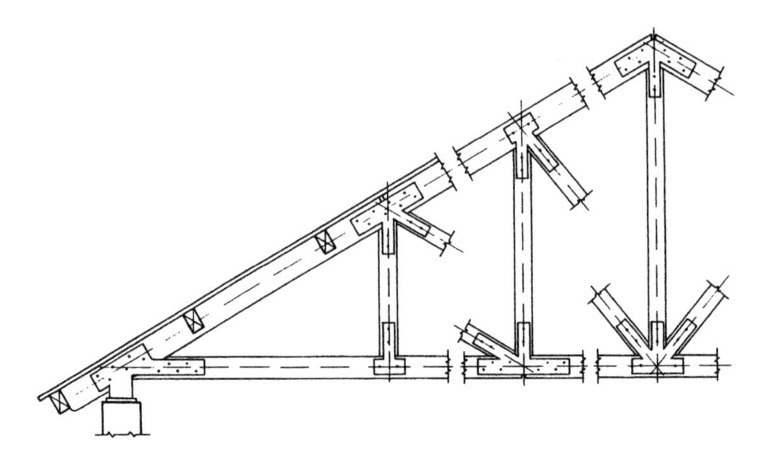

Figure 15.22 Form of the roof truss for Building Three. See also Figure 11.17 for historic form of details.

with modern jointing methods. The example here will utilize all timber elements for the truss, with bolts and steel gusset plates used for all joints. This form is shown in Figure 15.22. Although the span is modest here, available lengths of timber elements will likely require some splicing of the chords for this truss. A possible layout of the separate pieces is indicated in Figure 15.23*b*, which allows for one splice in each of the two top chords and two splices in the bottom chord. The details for the connections in Figure 15.22 provide for these splices.

Figure 15.23 presents an investigation of the truss with results determined for a unit gravity load. These values for internal forces in the members can be used by simple multiplication for various specific loads, as determined by optional forms for the general roof construction. Based on the form of construction shown in Figure 15.19 and a minimum roof live load of 20 psf, the total loads on the individual roof purlins that are supported by the truss top chords will be approximately 1800 lb live load and 3600 lb dead load.

Figure 15.24 presents an investigation for wind load, based on criteria from ASCE 2006 (Ref. 3). The form of loading here is based on the roof slope and a minimum horizontally directed wind pressure of 20 psf at the roof level.

The results of the investigations in Figures 15.23 and 15.24 are summarized in Table 15.1. For design purpose there are three combinations considered, as follows:

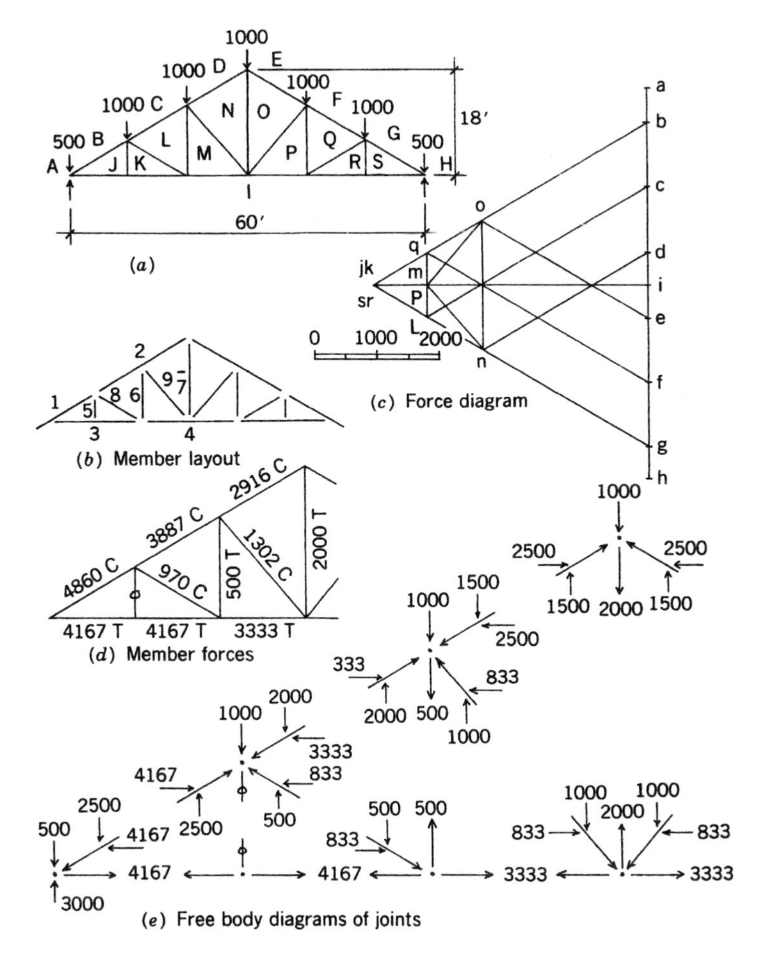

Figure 15.23 Investigation of the truss for gravity loads.

Dead load plus live load.

Dead load plus wind load when the result force is of the same sign as the gravity load force (tension or compression).

Dead load plus wind load when the result is a reversal of the sign of the gravity load member force.

Using the roof live load it is possible to use a stress increase, as described in Section 4.2. However, the live load in this case is already

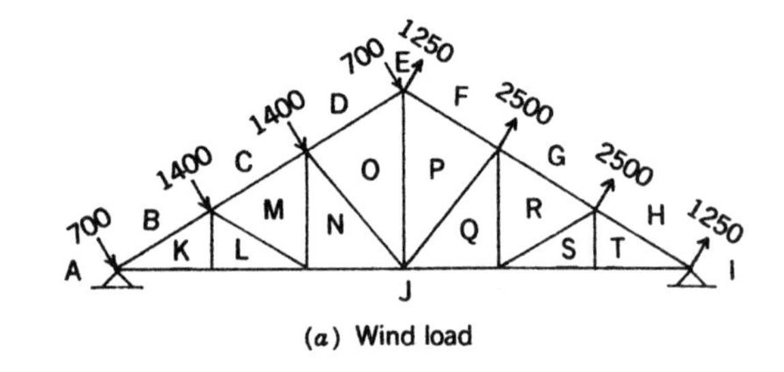

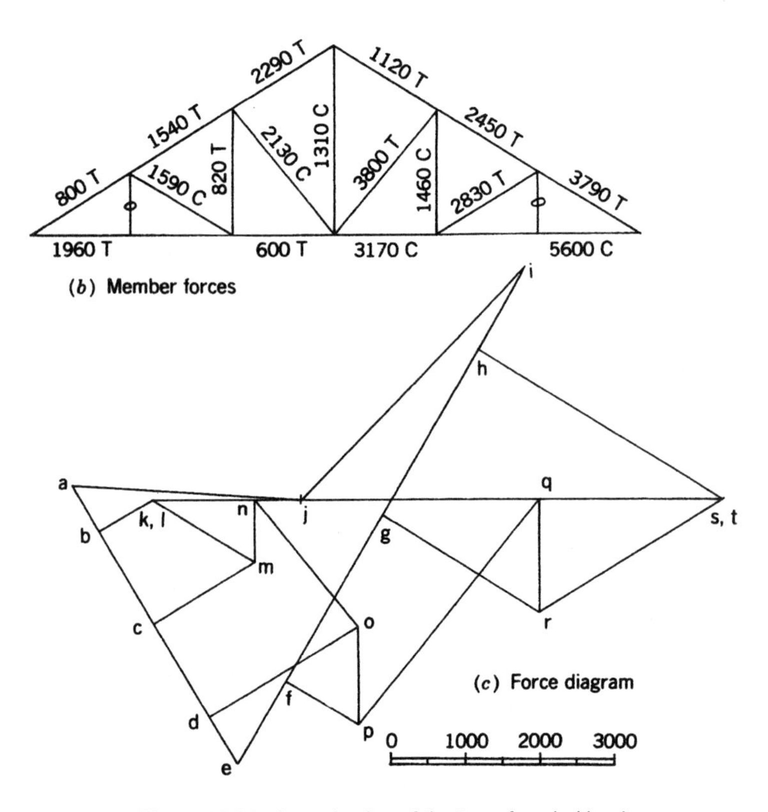

Figure 15.24 Investigation of the truss for wind load.

TABLE 15.1 Design Forces for the Truss (lb)[a]

Member (See Fig. 15.23)	Unit Gravity Load	Dead Load (3.6 × Unit)	Live Load (1.8 × Unit)	Wind Load	*DL* + *LL*
1	4860 C	17496 C	8748 C	3790 T	26244 C
2	3887 C	13994 C	6997 C	2450 T	20990 C
3	4167 T	15000 T	7500 T	1960 T/5600 C	22500 T
4	3333 T	12000 T	6000 T	3170 C	18000 T
5	0	0	0	0	0
6	500 T	1800 T	900 T	820 T/1460 C	2700 T
7	2000 T	7200 T	3600 T	1310 C	10800 T
8	970 C	3492 C	1746 C	1590 C/2830 T	5238 C
9	1302 C	4688 C	2344 C	2130 C/3800 T	7031 C

[a]C indicates compression, T indicates tension, 0 indicates that the member is not stressed for these loading conditions.

reduced due to the area supported by an individual truss, as described in Section 14.6. Thus, use is made of the full values of the dead and live load forces with no consideration for stress modification.

When wind load is included, design values for members can be increased by a factor of 1.6 (Table 4.3 in Section 4.2). Therefore the combinations of dead load plus wind load in Table 15.1 have been reduced by a factor of $1/1.6 = 0.625$ for comparison with the gravity load only combination. In any event, it may be observed that the wind load for this example is not critical. Whereas the wind load itself causes some force reversals in members, the net result of wind plus dead load (always present) indicates no reversals. Thus the character of force in individual members as tension or compression is that shown for gravity loads.

A summary of the design of the truss members for the loads in Table 15.1 is given in Table 15.2. Compression members are obtained from Table 9.1. These selections are probably conservative, as a higher grade of wood may well be used for the truss construction. A critical design decision is the selection of the dimension of members perpendicular to the plane of the truss. This dimension must be the same for all members to achieve the construction as shown in Figure 15.22. The alternative selections in Table 15.2 are based on a common dimension of 6 in. nominal (actually 5.5 in.) or 8 in. nominal (actually 7.5 in.). Considerations for the joint design or for any combined bending and compression in the top chords may affect this decision.

Comparison of the trusses in Figures 11.17 and 15.22 will show some similarities and some differences. The basic truss layout pattern is the

TABLE 15.2 Selection of the Truss Members

Member (See Fig. 15.23)	Member Length (ft)	Design Force (kips)	Member Selection[a]	
			All with 6-in. Nominal Thickness	All with 8-in. Nominal Thickness
1	11.7	26.3 C	6 × 10	8 × 8
2	11.7	21 C	6 × 8	8 × 8
3	10	22.5 T	6 × 6	8 × 8
4	10	18 T	6 × 6	6 × 8
5	6	0	6 × 6	6 × 8
6	12	2.7 T	6 × 6	6 × 8
7	18	10.8 T	6 × 6	6 × 8
8	11.7	5.24 C	6 × 6	6 × 8
9	15.6	7.03 C	6 × 6	8 × 8

[a] Minimum thickness for code qualification as heavy timber is 6 in.
Selections for top chords are without consideration for bending due to purlin loads not at joints.

same, but there are some differences in the manner of development of the roof overhang at the eave. The construction in Figure 11.17 uses projected rafters to achieve this cantilever, while the construction in Figure 15.22 uses the truss chord with only a minor projection of the roof deck.

A major difference in the two trusses has to do with the development of joints. In Figure 11.17 the wood-to-wood compression joints are achieved by direct bearing, using member-to-member custom fitting or intermediate bearing blocks. In Figure 15.22 all the joints use steel bolts and steel plate gussets. Mainly because of the jointing problems, all the tension members in Figure 11.17 consist of steel rods—not a factor for the truss in Figure 15.22, as the bolt and gusset joint is fully capable of transferring either tension or compression.

Alternative Truss Construction

The truss in the preceding design is not likely to be used unless some relationship to historical forms of construction is quite important to the overall building design goals. A slightly more contemporary form for the timber truss is shown in Figure 15.25. This truss uses chords of multiple members with joints achieved by overlapping of members. This joint form permits the use of split-ring connectors and a resulting construction that has much better control of deflections due to much smaller joint deformations.

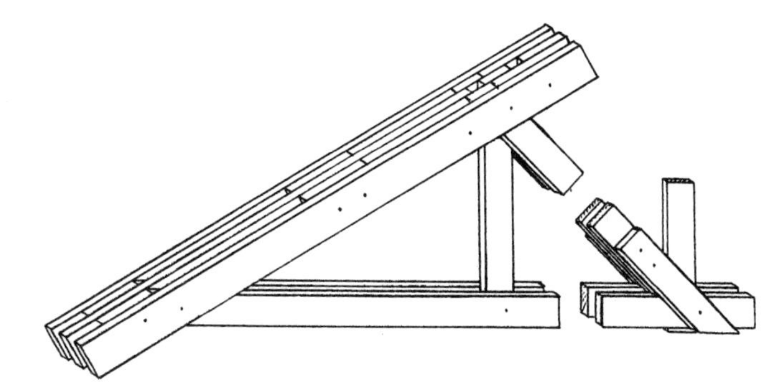

Figure 15.25 Alternative truss construction for Building Three, using multiple-element members.

Feasibility of the scheme for this truss depends on the working out of joint layouts that permit the placing of the required split rings within the space defined by edge limits, spacing limits, and the sizes of the truss members. Also a bit tricky is the choice of members as single-double- or triple-piece combinations. In Figure 15.25 the achieving of the lower chord joint with both a vertical and diagonal member requires the use of two outside splice pieces, since both the bottom chord and the diagonal are two-piece members.

15.4 BUILDING FOUR: STEEL AND WOOD STRUCTURE

Figure 15.26 shows a partial plan for a large warehouse-type structure with a flat roof. Basic elements of the structure are an exterior bearing wall, vertical interior columns (and possibly a column for the girder at the exterior wall), girders, joists, and a roof deck. Under consideration here is a system that uses steel columns and girders with joists and a roof deck of wood.

A major design factor is that of the magnitudes of the two spans, L_1 and L_2. Also of concern is the spacing of the joists as it affects both the span of the deck and the load on individual joists. Over a range of these dimensions, choices for the type as well as the size of elements will vary. Starting with the deck, the wood products most economical here are panels of plywood, composite materials, or particleboard. In reasonable thickness of $^1/_2$ to 7/8 in. these are capable of only 24 to 32 in. span. This then relates to the joist spacing. As the joist span increases, it is not feasible to use joists (actually I-joists or light trusses) at this close spacing. This calls for a coordinated decision.

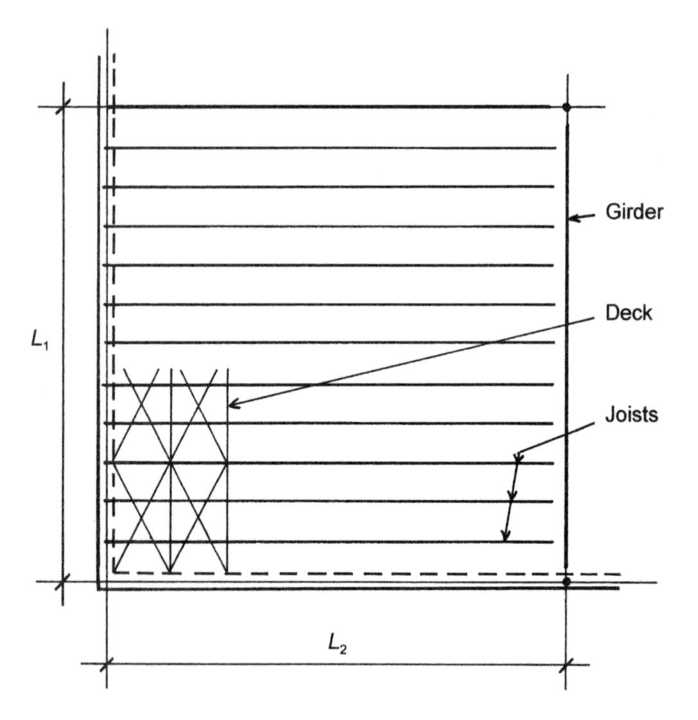

Figure 15.26 Building Four, roof framing, Alternative One.

An alternative is to use the system shown in Figure 15.27. A common dimension for this system is a spacing of the purlins at 8 ft with joists at 24 in., allowing for a panelized unit with one 4 ft by 8 ft panel and two joists. Actually, this system is commonly used for a range of spans, with solid-sawn or glued-laminated members used for the purlins at shorter spans (L_2 in the figures). As the span increases, trusses may be used for the purlins. With the 8 ft spacing these will not be light trusses, but may still be within the choice range for the composite wood and steel manufactured trusses.

The heaviest-loaded members in the system are the girders. These may be steel W shapes or glued-laminated elements, although steel trusses are also a possibility. Relatively heavy custom-made wood trusses may also be possible, probably with chords of glued-laminated members. If planning permits, it would be advisable to reduce the span L_1 to substantially reduce the size of the heavily-loaded girders.

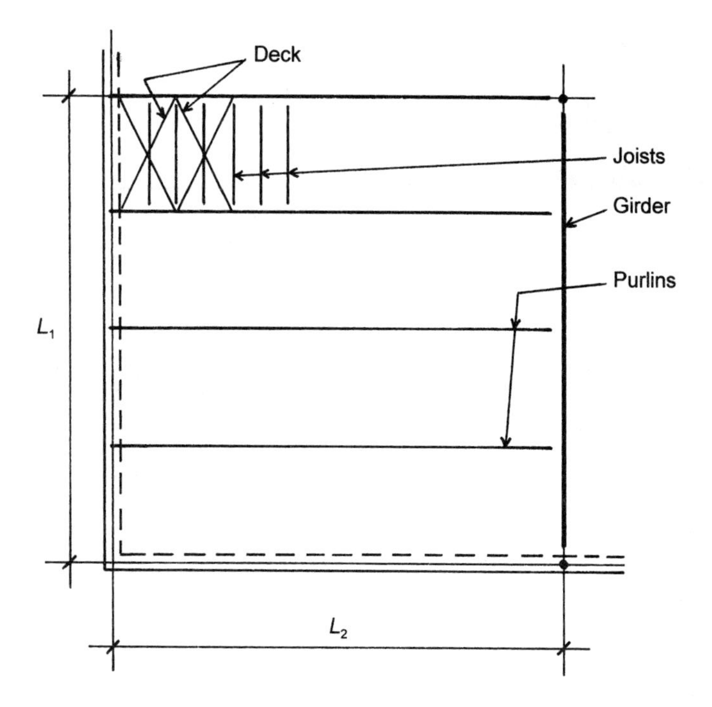

Figure 15.27 Building Four, roof framing, Alternative Two.

The columns for this building may be solid-sawn or glued-laminated timbers. However, they would probably more likely be steel, with round pipe or rectangular tubes as choices.

This is a very common form of building for commercial and industrial uses. A major consideration for design is typically dollar cost, so many solutions may be compared, with choices depending on regional availability of products and regional influences on construction costs.

Appendix A

PROPERTIES OF SECTIONS

This chapter deals with various geometric properties of planar (two-dimensional) areas. The areas referred to are the cross-sectional areas of structural members. These geometric properties are used for the analysis of stresses and deformations in the design of the structural members.

A.1 CENTROIDS

The *center of gravity* of a solid is the point at which all of its weight can be considered to be concentrated. Since a planar area has no weight, it has no center of gravity. The point in a planar area that corresponds to the center of gravity of a very thin plate of the same area and shape is called the *centroid* of the area. The centroid is a useful reference for various geometric properties of planar areas.

For example, when a beam is subjected to a bending moment, the materials in the beam above a certain plane in the beam are in compression, and the materials below the plane are in tension. This plane is the *neutral stress plane*, also called the neutral surface or the zero stress plane (see Section 6.2). For a cross section of the beam, its intersection with the neutral stress plane is a line that passes through the centroid of the section and is called the *neutral axis* of the section. The neutral axis is very important for investigation of bending stresses in beams.

The location of the centroid for symmetrical shapes is located on the axis of symmetry for the shape. If the shapc is bisymmetrical, that is, it has two axes of symmetry, the centroid is at the intersection of these axes. Consider the rectangular area shown in Figure A.1*a*; obviously its centroid is at its geometric center and is quite easily determined. (*Note:* Table A.3, referred to in the discussion that follows, is located at the end of this Appendix.)

For more complex forms, such as those of built-up sections, the centroid will also be on any axis of symmetry. And, as for the simple rectangle, if there are two axes of symmetry, the centroid is readily located.

For simple geometric shapes, such as those shown in Figure A.1, the location of the centroid is easily established. However, for more complex shapes, the centroid and other properties may have to be determined by computations. One method for achieving this is by use of the *statical moment*, defined as the product of an area times its distance from some reference axis. Use of this method is demonstrated in the following examples.

Example 1. Figure A.2 is a beam cross section that is unsymmetrical with respect to a horizontal axis (such as X-X in the figure). The area is symmetrical about its vertical centroidal axis, but the true location of the centroid requires the locating of the horizontal centroidal axis. Find the location of the centroid.

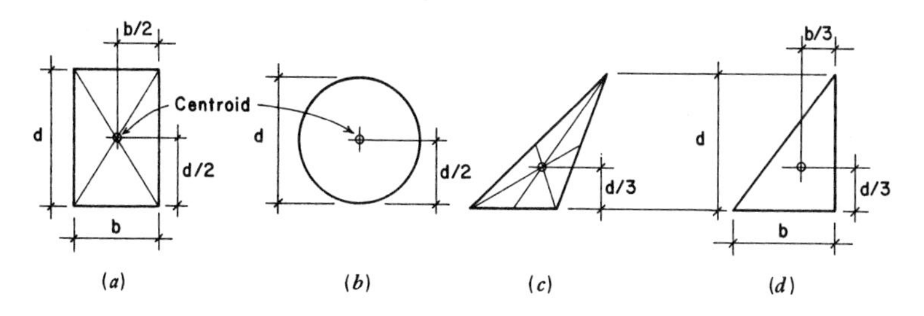

Figure A.1 Centroids of various planar shapes.

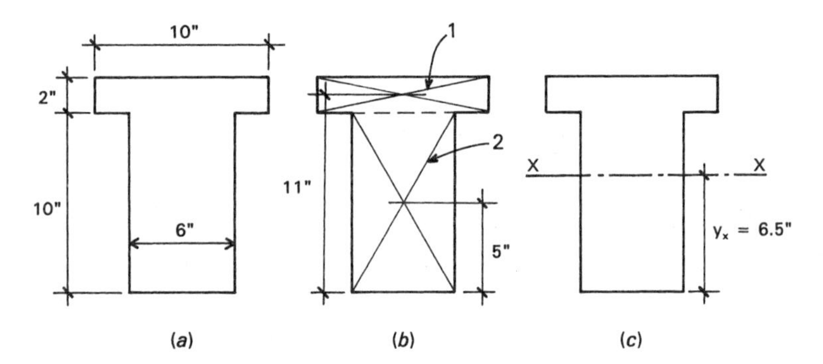

Figure A.2 Reference for Example 1.

Solution: Using the statical moment method, first divide the area into units for which the area and location of the centroid are readily determined. The division chosen here is shown in Figure A.2*b* with the two parts labeled 1 and 2.

The second step is to choose a reference axis about which to sum statical moments and from which the location of the centroid is readily measured. A convenient reference axis for this shape is one at either the top or bottom of the shape. With the bottom chosen, the distances from the centroids of the parts to this reference axis are shown in Figure A.2*b*.

The computation next proceeds to the determination of the unit areas and their statical moments. This work is summarized in Table A.1, which shows the total area to be 80 in.^{2} and the total statical moment to be 520 in.^{3}. Dividing this moment by the total area produces the value of 6.5 in., which is the distance from the reference axis to the centroid of the whole shape, as shown in Figure A.2*c*.

Problems A.1.A-F. Find the location of the centroid for the cross-sectional areas shown in Figure A.3. Use the reference axes indicated

TABLE A.1 Summary of Computations for Centroid: Example 1

Part	Area (in.^{2})	y (in.)	A × y (in.^{3})
1	2 × 10 = 20	11	220
2	6 × 10 = 60	5	300
Σ	80		520

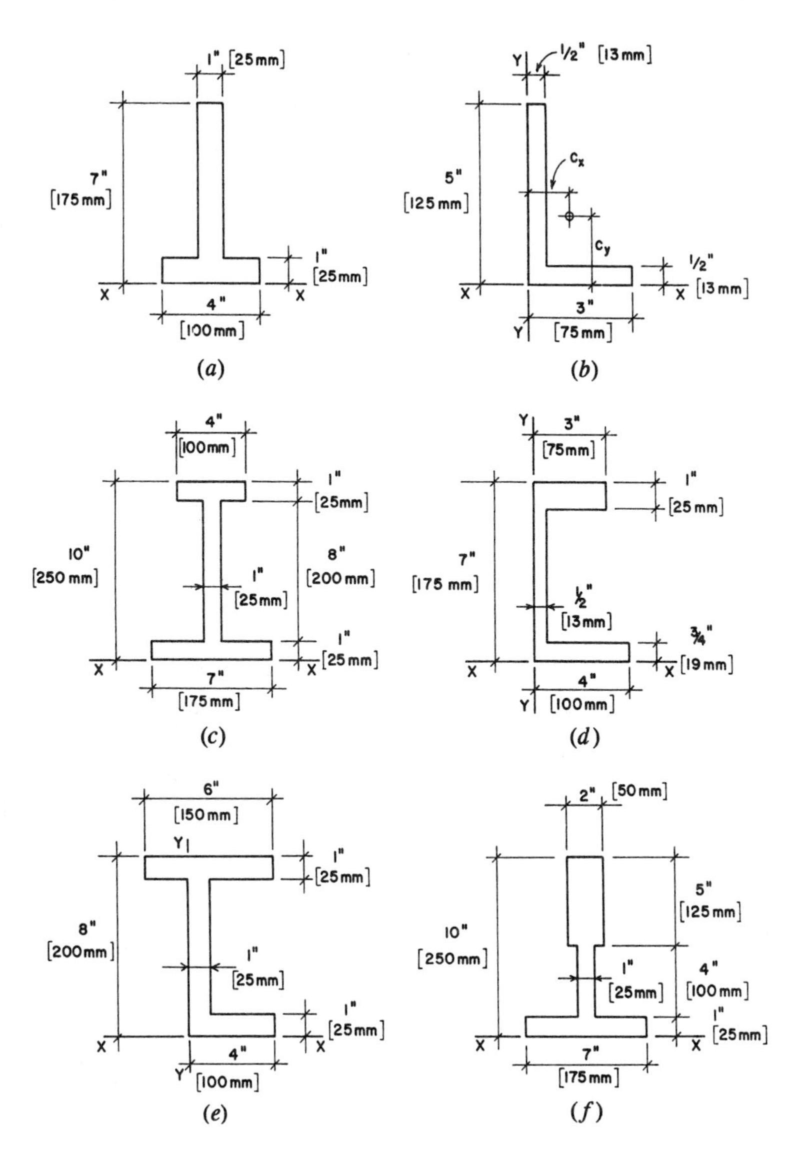

Figure A.3 Reference for Problem A.1.

and compute the distances from the axes to the centroid, designated as c_x and c_y, as shown in Figure A.3*b*.

A.2 MOMENT OF INERTIA

Consider the area enclosed by the irregular line in Figure A.4*a*. In this area, designated *A*, a small unit area *a* is indicated at *z* distance from the axis marked X-X. If this unit area is multiplied by the square of its distance from the reference axis, the result is the quantity az^2. If all of the units of the area are thus identified and the sum of these products is made, the result is defined as the *second moment* or the *moment of inertia* of the area, designated as *I*. Thus

$$\Sigma az^2 = I, \text{ or specifically } I_{X\text{-}X}$$

which is the moment of inertia of the area about the X-X axis.

The moment of inertia is a somewhat abstract item, less able to be visualized than area, weight, or center of gravity. It is, nevertheless, a real geometric property that becomes an essential factor for investigation of stresses and deformations due to bending. Of particular interest is the moment of inertia about a centroidal axis, and—most significantly—about a principal axis for the shape. Figures A.4*b, c, e,* and *f* indicate such axes for various shapes. An inspection of Table A.3 will reveal the properties of moment of inertia about the principal axes of the shapes in the table.

Moment of Inertia of Geometric Figures

Values for moment of inertia can often be obtained from tabulations of structural properties. Occasionally it is necessary to compute values

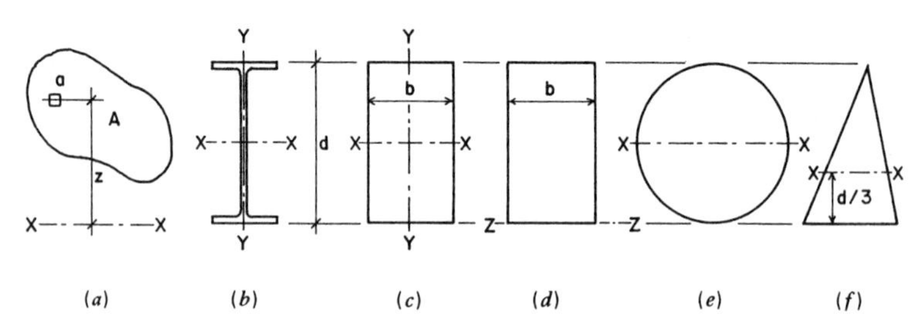

Figure A.4 Consideration of reference axis for the moment of inertia of various shapes of cross sections.

for a given shape. This may be a simple shape, such as a square, a rectangle, a circular area, or a triangular area. For such shapes simple formulas are derived to express the value for the moment of inertia.

Rectangle. Consider the rectangle shown in Figure A.4*c*. Its width is b and its depth is d. The two principal axes are X-X and Y-Y, both passing through the centroid of the area. For this case the moment of inertia with respect to the centroidal axis X-X is

$$I_{X\text{-}X} = \frac{bd^3}{12}$$

and the moment of inertia with respect to the Y-Y axis is

$$I_{Y\text{-}Y} = \frac{db^3}{12}$$

Example 2. Find the value of the moment of inertia for a 6×12 in. wood beam about an axis through its centroid and parallel to the narrow dimension.

Solution: As listed in Table A.3, the actual dimensions of the section are 5.5×11.5 in. Then

$$I = \frac{bd^3}{12} = \frac{5.5(11.5)^3}{12} = 697.1 \text{ in.}^4$$

which is in agreement with the value for $I_{X\text{-}X}$ in Table A.3.

Circle. Figure A.4*e* shows a circular area with diameter d and axis X-X passing through its center. For the circular area the moment of inertia is

$$I = \frac{pd^4}{64}$$

Example 3. Compute the moment of inertia of a circular cross section, 10 in. in diameter, about its centroidal axis.

Solution: The moment of inertia is

$$I = \frac{\pi d^4}{64} = \frac{3.1416(10)^4}{64} = 490.9 \text{ in.}^4$$

Triangle. The triangle in Figure A.4*f* has a height h and a base width b. The moment of inertia about a centroidal axis parallel to the base is

$$I = \frac{bh^3}{36}$$

Example 4. If the base of the triangle in Figure A.4*f* is 12 in. wide and the height from the base is 10 in., find the value for the centroidal moment of inertia parallel to the base.

Solution: Using the given values in the formula

$$I = \frac{bh^3}{36} = \frac{12(10)^3}{36} = 333.3\,\text{in.}^4$$

Open and Hollow Shapes. Values of moment of inertia for shapes that are open or hollow may sometimes be computed by a method of subtraction. The following examples demonstrate this process. Note that this is possible only for shapes that are symmetrical.

Example 5. Compute the moment of inertia for the hollow box section shown in Figure A.5*a* about a centroidal axis parallel to the narrow side.

Solution: Find first the moment of inertia of the shape defined by the outer limits of the box.

$$I = \frac{bd^3}{12} = \frac{6(10)^3}{12} = 500\,\text{in.}^4$$

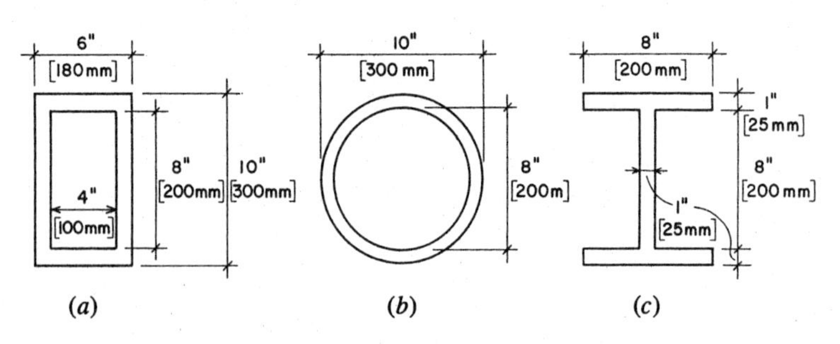

Figure A.5 Reference for Examples 5, 6, and 7.

Then find the moment of inertia for the shape defined by the void area.

$$I = \frac{4(8)^3}{12} = 170.7\,\text{in.}^4$$

The value for the hollow section is the difference, thus

$$I = 500 - 170.7 = 329.3\,\text{in.}^4$$

Example 6. Compute the moment of inertia about the centroidal axis for the pipe section shown in Figure A.5*b*. The thickness of the shell is 1 in.

Solution: As in the preceding example, the two values may be found and subtracted. Or a single computation may be made as follows

$$I = \frac{\pi}{64}(d_o^4 - d_i^4) = \frac{3.1416}{64}(10^4 - 8^4) = 491 - 201 = 290\,\text{in.}^4$$

Example 7. Referring to Figure A.5*c*, compute the moment of inertia of the I-shape section about the centroidal axis parallel to the flanges.

Solution: This is essentially similar to the computation for Example 5. The two voids may be combined into a single one that is 7 in. wide. Thus

$$I = \frac{8(10)^3}{12} - \frac{7(8)^3}{12} = 667 - 299 = 368\,\text{in.}^4$$

Note that this method can only be used when the centroids of the outer shape and the void coincide. For example, it cannot be used to find the moment of inertia for the I-shape about its vertical centroidal axis. For this computation the method discussed in the next section must be used.

A.3 TRANSFERRING MOMENTS OF INERTIA

Determination of the moment of inertia of unsymmetrical and complex shapes cannot be done by the simple processes illustrated in the preceding examples. An additional step that must be used is that involving the transfer of moment of inertia about a remote axis. The formula for achieving this transfer is as follows:

$$I = I_o + Az^2$$

In this formula,

I = moment of inertia of the cross section about the required reference axis

I_o = moment of inertia of the cross section about its own centroidal axis, parallel to the reference axis

A = area of the cross section

z = distance between the two parallel axes

These relationships are illustrated in Figure A.6, where X-X is the centroidal axis of the area, and Y-Y is the reference axis for the transferred moment of inertia. Application of this principle is illustrated in the following example.

Example 8. Find the moment of inertia of the T-shaped area in Figure A.7 about its horizontal (X-X) centroidal axis. (*Note*: the

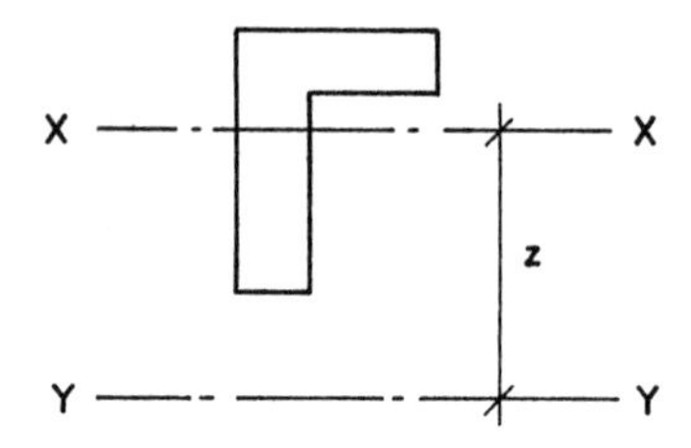

Figure A.6 Transfer of moment of inertia to a parallel axis.

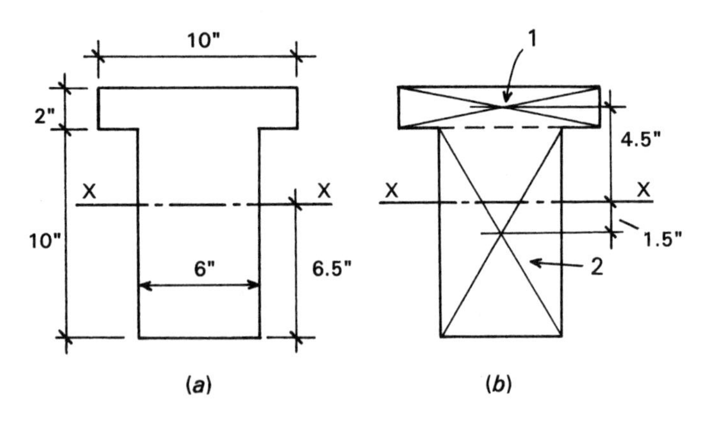

Figure A.7 Reference for Example 8.

TABLE A.2 Summary of Computations for Moment of Inertia: Example 8

Part	Area (in.^{2})	y (in.)	I_o (in.^{4})	$A \times y^2$ (in.^{4})	I_x (in.^{4})
1	20	4.5	$10(2)^3/12 = 6.7$	$20(4.5)^2 = 405$	411.7
2	60	1.5	$6(10)^3/12 = 500$	$60(1.5)^2 = 135$	635
Σ					1046.7

location of the centroid for this section was solved as Example 1 in Section A.1.)

Solution: A necessary first step in these problems is to locate the position of the centroidal axis if the shape is not symmetrical. In this case, the T-shape is symmetrical about its vertical axis, but not about the horizontal axis. Locating the position of the horizontal axis was the problem solved in Example 1 in Sec. A.1.

The next step is to break the complex shape down into parts for which centroids, areas, and centroidal moments of inertia are readily found. As was done in Example 1, the shape here is divided between the rectangular flange part and the rectangular web part.

The reference axis to be used here is the horizontal centroidal axis. Table A.2 summarizes the process of determining the factors for the parallel axis transfer process. The required value for I about the horizontal centroidal axis is determined to be 1046.7 in.^{4}.

Problems A.3.A–F. Compute the moments of inertia about the indicated centroidal axes for the cross-sectional shapes in Figure A.8.

A.4 MISCELLANEOUS PROPERTIES

Elastic Section Modulus

As noted in Section 6.2, the term I/c in the formula for flexural stress is called the *elastic section modulus*, designated S. Use of the section modulus permits a minor shortcut in the computations for flexural stress or the determination of the bending moment capacity of members. However, the real value of this property is in its measure of relative bending strength of members. As a geometric property it is a direct index of bending strength for a given member cross section. Members of

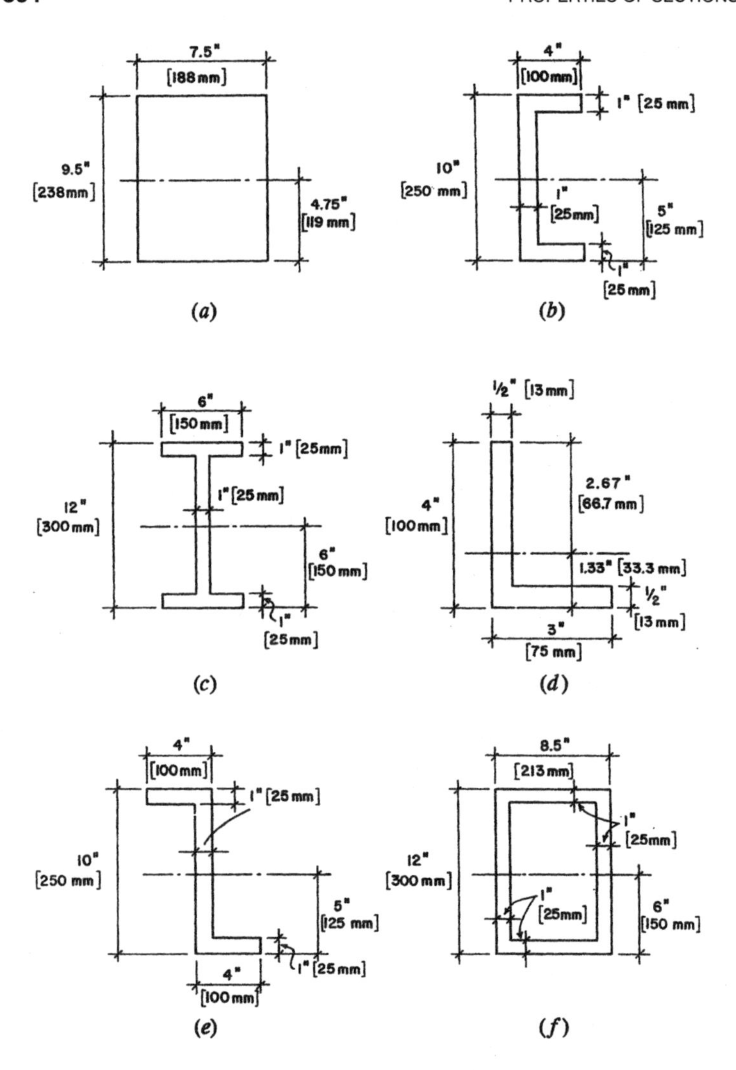

Figure A.8 Reference for Problem A.3.

various cross sections may thus be rank-ordered in terms of their bending strength strictly on the basis of their S values. Because of its usefulness, the value of S is listed together with other significant properties in the tabulations for steel and wood members.

For members of standard form (structural lumber and rolled steel shapes) the value of S may be obtained from tables. For complex forms not of standard form, the value of S must be computed, which is readily done once the centroidal axes are located and moments of inertia about the centroidal axes are determined.

Example 9. Verify the tabulated value for the section modulus of a 6×12 wood beam about the centroidal axis parallel to its narrow side.

Solution: From Table A.3 the actual dimensions of this member are 5.5×11.5 in. And the value for the moment of inertia is 697.1 in.^{4}. Then

$$S = \frac{I}{c} = \frac{697.1}{5.75} = 121.2 \text{in.}^3$$

which agrees with the value in Table A.3.

Radius of Gyration

For design of slender compression members, an important geometric property is the *radius of gyration*, defined as

$$r = \sqrt{\frac{I}{A}}$$

Just as with moment of inertia and section modulus values, the radius of gyration has an orientation to a specific axis in the planar cross section of a member. Thus if the I used in the formula for r is that with respect to the X-X centroidal axis, then that is the reference for the specific value of r.

A value of r with particular significance is that designated as the *least radius of gyration.* Since this value will be related to the least value of I for the cross section, and since I is an index of the bending stiffness of the member, then the least value for r will indicate the weakest response of the member to bending. This relates specifically to the resistance of slender compression members to buckling. Buckling is essentially a sideways bending response, and its most likely occurrence will be on the axis identified by the least value of I or r. For sawn-wood columns with rectangular cross sections, the slenderness of the column is expressed as L/d. For steel columns, use is made directly of r values and slenderness is expressed as L/r.

A.5 TABLES OF PROPERTIES OF SECTIONS

Figure A.9 presents formulas for obtaining geometric properties of various simple plane sections. Some of these may be used for single-piece structural members or for the building up of complex members.

Table A.3 presents the properties of standard lumber shapes. These are sections identified as those of standard industry-produced sections of wood. Standardization means that the shapes and dimensions of the sections are fixed, and each specific section is identified in some way.

Structural members may be employed for various purposes, and thus they may be oriented differently for some structural uses. Of note for any plane section are the *principal axes* of the section. These are the two, mutually-perpendicular, centroidal axes for which the values will be greatest and least, respectively, for the section; thus the axes are identified as the major and minor axes. If sections have an axis of symmetry, it will always be a principal axis—either major or minor.

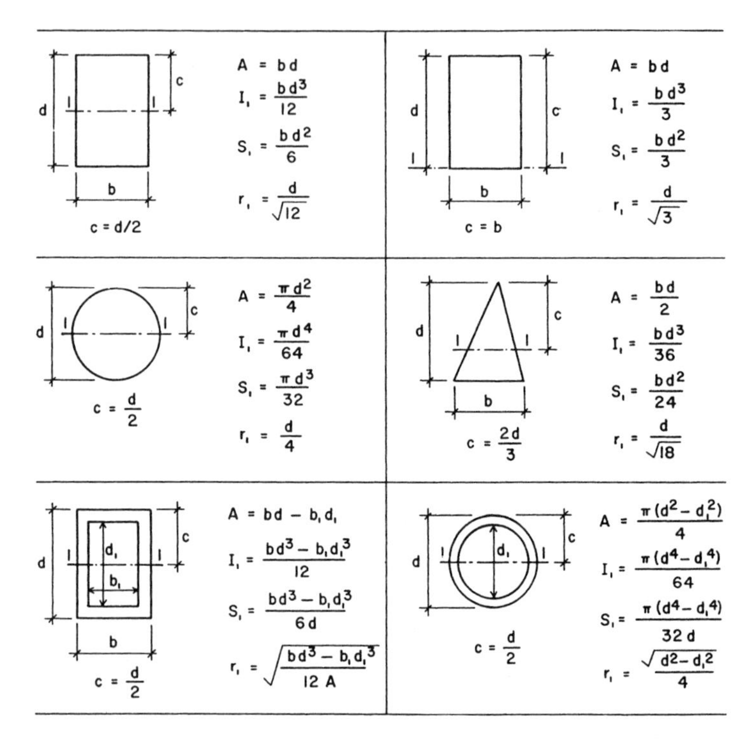

Figure A.9 Properties of various geometric shapes of cross sections.

TABLE A.3 Properties of Structural Lumber

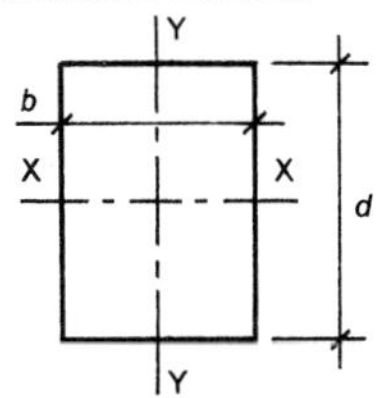

Dimensions (in.)		Area	X-X Axis Section Modulus	X-X Axis Moment of Inertia	Y-Y Axis Section Modulus	Y-Y Axis Moment of Inertia	Weight at 35 lb/ft^3 Density
Nominal $b \times h$	Actual $b \times h$	A (in.^{2})	S (in.^{3})	I (in.^{4})	S (in.^{3})	I (in.^{4})	(lb/ft)
2 × 3	1.5 × 2.5	3.75	1.563	1.953	0.938	0.703	0.911
2 × 4	1.5 × 3.5	5.25	3.063	5.359	1.313	0.984	1.276
2 × 6	1.5 × 5.5	8.25	7.563	20.80	2.063	1.547	2.005
2 × 8	1.5 × 7.25	10.88	13.14	47.63	2.719	2.039	2.643
2 × 10	1.5 × 9.25	13.88	21.39	98.93	3.469	2.602	3.372
2 × 12	1.5 × 11.25	16.88	31.64	178.0	4.219	3.164	4.102
2 × 14	1.5 × 13.25	19.88	43.89	290.8	4.969	3.727	4.831
3 × 4	2.5 × 3.5	8.75	5.104	8.932	3.646	4.557	2.127
3 × 6	2.5 × 5.5	13.75	12.60	34.66	5.729	7.161	3.342
3 × 8	2.5 × 7.25	18.13	21.90	79.39	7.552	9.440	4.405
3 × 10	2.5 × 9.25	23.13	35.65	164.9	9.635	12.04	5.621
3 × 12	2.5 × 11.25	28.13	52.73	296.6	11.72	14.65	6.836
3 × 14	2.5 × 13.25	33.13	73.15	484.6	13.80	17.25	8.051
3 × 16	2.5 × 15.25	38.13	96.90	738.9	15.89	19.86	9.266
4 × 4	3.5 × 3.5	12.25	7.146	12.51	7.146	12.51	2.977
4 × 6	3.5 × 5.5	19.25	17.65	48.53	11.23	19.65	4.679
4 × 8	3.5 × 7.25	25.38	30.66	111.1	14.80	25.9	6.168
4 × 10	3.5 × 9.25	32.38	49.91	230.8	18.89	33.05	7.869
4 × 12	3.5 × 11.25	39.38	73.83	415.3	22.97	40.20	9.570
4 × 14	3.5 × 13.25	46.38	102.4	678.5	27.05	47.34	11.27
4 × 16	3.5 × 15.25	53.38	135.7	1034	31.14	54.49	12.97
5 × 5	4.5 × 4.5	20.25	15.19	34.17	15.19	34.17	4.922
6 × 6	5.5 × 5.5	30.25	27.73	76.26	27.73	76.26	7.352
6 × 8	5.5 × 7.5	41.25	51.56	193.4	37.81	104.0	10.03
6 × 10	5.5 × 9.5	52.25	82.73	393.0	47.90	131.7	12.70
6 × 12	5.5 × 11.5	63.25	121.2	697.1	57.98	159.4	15.37
6 × 14	5.5 × 13.5	74.25	167.1	1128	68.06	187.2	18.05
6 × 16	5.5 × 15.5	85.25	220.2	1707	78.15	214.9	20.72
6 × 18	5.5 × 17.5	96.25	280.7	2456	88.23	242.6	23.39
6 × 20	5.5 × 19.5	107.3	348.6	3398	98.31	270.4	26.07
6 × 22	5.5 × 21.5	118.3	423.7	4555	108.4	298.1	28.74
6 × 24	5.5 × 23.5	129.3	506.2	5948	118.5	325.8	31.41
8 × 8	7.5 × 7.5	56.25	70.31	263.7	70.31	263.7	13.67
8 × 10	7.5 × 9.5	71.25	112.8	535.9	89.06	334.0	17.32
8 × 12	7.5 × 11.5	86.25	165.3	950.5	107.8	404.3	20.96

(*continues*)

TABLE A.3 (*continued*)

Dimensions (in.)		Area	X-X Axis		Y-Y Axis		Weight at 35 lb/ft^3 Density
			Section Modulus	Moment of Inertia	Section Modulus	Moment of Inertia	
Nominal $b \times h$	Actual $b \times h$	A ($in.^2$)	S ($in.^3$)	I ($in.^4$)	S ($in.^3$)	I ($in.^4$)	(lb/ft)
8 × 14	7.5 × 13.5	101.3	227.8	1538	126.6	474.6	24.61
8 × 16	7.5 × 15.5	116.3	300.3	2327	145.3	544.9	28.26
8 × 18	7.5 × 17.5	131.3	382.8	3350	164.1	615.2	31.90
8 × 20	7.5 × 19.5	146.3	475.3	4634	182.8	685.5	35.55
8 × 22	7.5 × 21.5	161.3	577.8	6211	201.6	755.9	39.19
8 × 24	7.5 × 23.5	176.3	690.3	8111	220.3	826.2	42.84
10 × 10	9.5 × 9.5	90.25	142.9	678.8	142.9	678.8	21.94
10 × 12	9.5 × 11.5	109.3	209.4	1204	173.0	821.7	26.55
10 × 14	9.5 × 13.5	128.3	288.6	1948	203.1	964.5	31.17
10 × 16	9.5 × 15.5	147.3	380.4	2948	233.1	1107	35.79
10 × 18	9.5 × 17.5	166.3	484.9	4243	263.2	1250	40.41
10 × 20	9.5 × 19.5	185.3	602.1	5870	293.3	1393	45.03
10 × 22	9.5 × 21.5	204.3	731.9	7868	323.4	1536	49.64
10 × 24	9.5 × 23.5	223.3	874.4	10270	353.5	1679	54.26
12 × 12	11.5 × 11.5	132.3	253.5	1458	253.5	1458	32.14
12 × 14	11.5 × 13.5	155.3	349.3	2358	297.6	1711	37.73
12 × 16	11.5 × 15.5	178.3	460.5	3569	341.6	1964	43.32
12 × 18	11.5 × 17.5	201.3	587.0	5136	385.7	2218	48.91
12 × 20	11.5 × 19.5	224.3	728.8	7106	429.8	2471	54.51
12 × 22	11.5 × 21.5	247.3	886.0	9524	473.9	2725	60.10
12 × 24	11.5 × 23.5	270.3	1058	12440	518.0	2978	65.69
14 × 14	13.5 × 13.5	182.3	410.1	2768	410.1	2768	44.30
14 × 16	13.5 × 15.5	209.3	540.6	4189	470.8	3178	50.86
14 × 18	13.5 × 17.5	236.3	689.1	6029	531.6	3588	57.42
14 × 20	13.5 × 19.5	263.3	855.6	8342	592.3	3998	63.98
14 × 22	13.5 × 21.5	290.3	1040	11180	653.1	4408	70.55
14 × 24	13.5 × 23.5	317.3	1243	14600	713.8	4818	77.11
16 × 16	15.5 × 15.5	240.3	620.6	4810	620.6	4810	58.39
16 × 18	15.5 × 17.5	271.3	791.1	6923	700.7	5431	65.93
16 × 20	15.5 × 19.5	302.3	982.3	9578	780.8	6051	73.46
16 × 22	15.5 × 21.5	333.3	1194	12840	860.9	6672	81.00
16 × 24	15.5 × 23.5	364.3	1427	16760	941.0	7293	88.53
18 × 18	17.5 × 17.5	306.3	893.2	7816	893.2	7816	74.44
18 × 20	17.5 × 19.5	341.3	1109	10810	995.3	8709	82.94
18 × 22	17.5 × 21.5	376.3	1348	14490	1097	9602	91.45
18 × 24	17.5 × 23.5	411.3	1611	18930	1199	10500	99.96
20 × 20	19.5 × 19.5	380.3	1236	12050	1236	12050	92.42
20 × 22	19.5 × 21.5	419.3	1502	16150	1363	13280	101.9
20 × 24	19.5 × 23.5	458.3	1795	21090	1489	14520	111.4
22 × 22	21.5 × 21.5	462.3	1656	17810	1656	17810	112.4
22 × 24	21.5 × 23.5	505.3	1979	23250	1810	19460	122.8
24 × 24	23.5 × 23.5	552.3	2163	25420	2163	25420	134.2

Source: Compiled from data in the *National Design Specification for Wood Construction* (Ref. 1), with permission of the publishers, American Forest and Paper Association.

Appendix B

STUDY AIDS

This appendix provides the reader with some means to measure accomplishment in terms of comprehension of the book materials.

WORDS AND TERMS

Introduction

ASD
LRFD
NDS
IBC
ASCE 2006

Chapter 1

Species
Softwood
Hardwood
Structural lumber
Timber
Annual rings
Sapwood
Heartwood
Knot
Shake
Check
Split

Pitch pocket
Seasoning
Kiln dried
Moisture content
Sawn lumber
Dimension lumber
Light framing
Joists and planks
Beams and stringers
Posts and timbers
Nominal dimension

Chapter 2

Stress design
Service load
Factored load
Resistance factor

Chapter 3

Unit stress
Force
Kip
Concentrated load
Uniformly distributed load
Dead load
Live load
Axial force
Direct stress
Compressive stress
Tensile stress
Shear stress
Deformation
Strain
Deflection
Elasticity
Elastic limit
Ultimate strength
Modulus of elasticity
Design values

Chapter 4

Reference design values
Modification of design values
Load duration
Hankinson formula
Bearing stress

Chapter 5

Beam forms: simple, cantilever, overhanging, continuous, restrained
Moment
Center of moments
Reaction
Beam shear
Shear diagram
Moment diagram
Inflection point
Bending stress sign convention
ETL

Chapter 6

Horizontal shear stress
Statical moment
Resisting moment
Flexure formula
Extreme fiber stress

Size factor for beams
Biaxial bending
Lateral buckling
Torsional buckling

Chapter 7

Joist
Rafter
Bridging
Blocking

Chapter 8

Board deck
Plank deck
Plywood
Plies
Veneer
Fiberboard
OSB (oriented strand board)
Composite panel

Chapter 9

Slenderness of column
Stud construction
Pole construction
Spaced column
Built-up column
Column interaction

Chapter 10

Single shear joint
Double shear joint
Common wire nail
Lag bolt
Split-ring connector
Gusset

Chapter 11

Truss
Panel point
Slope
Maxwell diagram
Method of joints
Heavy timber
Manufactured truss

Chapter 12

Glued-laminated timber
Plywood
Plies
Veneer
OSB (oriented strand board)
Composite panel
I-beam
LVL (laminated veneer lumber)
PSL (parallel strand lumber)
Flitched beam
Built-up beam
Pole construction

Chapter 13

Horizontal diaphragm
Diaphragm chord
Shear wall
Overturn effect
Shear wall tiedown

Chapter 14

Dead load
Live load
Live load reduction
Load periphery
Roof ponding
Load combinations
Factored load

Appendix A

Centroid
Neutral axis
Moment of inertia
Transfer-of-axis equation (also called parallel-axis theorem)
Section modulus
Radius of gyration

GENERAL QUESTIONS

These questions relate to issues that are not so evident in the computational problems or in the simple definitions of words and terms. If the answers to these questions are not readily evident to the reader, it is recommended that the appropriate portions of the book be read again. Answers to these questions are provided at the end of this appendix.

Chapter 1

1. With regard to the classification of species of trees, what does the term *softwood* mean?
2. With regard to the identification of structural lumber, what does the term *timber* indicate?
3. What principal factors determine the weight (unit density) of a particular piece of wood?
4. For structural use, what is the basis for concern for defects in the wood?
5. For what purposes are the following references used: (a) nominal dimensions, (b) dressed or true dimensions?

Chapter 2

1. What are some common goals for the design of building structures?
2. What are some major concerns for a building structure with regard to its performance during a fire?
3. What is the general implication of the statement "this structure has a safety factor of 2 for load resistance"?

4. What load condition is the basis for the strength method of structural design?
5. How is safety generally provided for in the allowable stress design method?

Chapter 3

1. What is the importance of stress as a factor in structural design?
2. What is particularly significant about dead load?
3. How is direct stress different from shear stress?
4. For comparison of different materials, what is significant about the modulus of elasticity?

Chapter 4

1. What load condition is assumed in establishing the reference design values for allowable stress design?
2. What specific property of a load is the basis for the load duration modification factor for design stress?
3. Why must reference design values by adjusted for the LRFD method?

Chapter 5

1. With regard to the nature of deformation of a beam under loading, what is significant about the following support qualifications: (a) Simple support, (b) Restrained (fixed) support, (c) multiple supports (continuous beam)?
2. What is indicated when a moment in a beam is described as being either positive or negative?
3. What constitutes the complete external force system that operates on a beam?
4. What are the significant relationships between the shear diagram and the moment diagram for a beam?
5. What are the significant relationships between the moment diagram and the deflected shape of a beam?

Chapter 6

1. What condition for a rectangular beam section permits the use of a simplified formula for the investigation of beam shear?

2. For what types of wood beams is it necessary to use the general equation for investigation of beam shear stress?
3. What basic relationship must exist between the resisting moment in a beam and the bending moment induced by the loads on the beam?
4. What two properties of a beam influence its resistance to deflection?

Chapter 7

1. Other than the need to perform their structural tasks, what usually determines the choice for spacing for joists and rafters?
2. What are the general conditions that must be considered for the complete structural functioning of a wood beam?

Chapter 8

1. Why is a structural panel layer (usually fiberboard or plywood) often placed on top of a deck of boards or planks?
2. What advantage is gained by use of a placement pattern with controlled random lengths when installing a board or plank deck with short pieces?

Chapter 9

1. Why is strength in resisting compression stress not always the major concern for a very slender wood column?
2. Why is it not possible in most cases to simply invert the direct stress formula ($f = P/A$) for a solution in the design of a column?
3. Although the limitation for maximum slenderness of $L/50$ for a sawn-wood column would limit the height of a 2 × 4 stud to 75 in., what makes it possible to use the studs for taller walls?

Chapter 10

1. Why is the two-member bolted joint not a favored means for the development of major force transfers between wood members?
2. Why is resistance to withdrawal not a reliable form of structural load development in a nailed joint with wood members?

3. What is the reason for the difference of resistance of a bolted joint in wood based on the direction of the load with respect to the grain in the wood?
4. How is structural performance improved when a shear developer is added to a bolted joint between two wood members?

Chapter 11

1. What geometric feature distinguishes a truss as a basic framing system?
2. What basic internal force condition generally exists for the members of a truss?

Chapter 12

1. For structural applications, what are some advantages gained by use of glued-laminated wood elements?
2. Why are odd numbers of plies typically used for plywood panels?
3. What basic property of the materials is significant to the relationship between the separate materials in a composite structural element?
4. What interactive relationship is essential for the separate parts in a built-up structural element?

Chapter 13

1. Horizontal diaphragms perform what basic function in the development of resistance to lateral forces on a building?
2. What are the four primary structural functions of a shear wall?
3. What are the principal variables that determine the shear strength of a plywood diaphragm for resistance to lateral forces?

Chapter 14

1. Of what does the design dead load primarily consist?
2. What primary factor affects the percentage of live load reduction?
3. In what unit is the load periphery for a column expressed?
4. Why is the achievement of optimal structural efficiency not always a dominant concern for the cost of building construction?
5. What are some reasons for the difficulty of making cost estimates during the design process?

6. What are some common rules that designers may apply that typically help to reduce cost of structures?

Appendix A

1. Although there is a single centroid for an area, why is there not a single value for the centroidal moment of inertia for most areas?
2. For what geometric forms are there single values for the centroidal moment of inertia?
3. For a column subject to buckling, what is the significance of the least radius of gyration?

ANSWERS TO THE GENERAL QUESTIONS

Chapter 1

1. It indicates a source tree that is cone-bearing and generally described as "evergreen." The wood from these trees is mostly softer and lighter than that from trees that drop their leaves in the fall.
2. It identifies lumber sizes beyond the classification of *dimension lumber*, generally including pieces with thickness greater than 4 in. nominal.
3. The amount of void (air space) in the wood and the percentage of contained moisture (water).
4. Concern is for the potential effects on strength and durability of the wood piece for structural use. Strength may relate to the location of the defect, as well as to its specific nature. For example, a large knot near the bottom in the center portion of the beam length.
5. (a) For identification or specification of the piece. (b) For determination of properties for structural computations and for dimensions for the development of construction details.

Chapter 2

1. Safety against structural failure; general accommodation of the usage of the building; controlled cost of the construction; satisfaction of established design standards. And maybe: low

use of energy; utilization of renewable sources for materials; no generation of toxic conditions; no creation of hazardous waste; and possibility for recycling of materials or structural elements.

2. Combustibility (burning) of the structure itself; loss of structural capability during the fire; lack of ability to contain the fire and slow its spread.
3. The structure is 2 times as strong as is estimated to be necessary for its tasks.
4. Load sufficient to cause structural failure.
5. By using design values for stresses that are significantly reduced below the failure limits for the materials.

Chapter 3

1. It is the most direct measure of the effort of resistance required by the material of a structure.
2. It is always present; thus, it affects long-time behaviors and is always part of any load combinations for strength evaluation.
3. Direct stress acts normal (perpendicular) to the plane of a cross section (as a push or pull); shear acts in the plane of the cross section (as a slip or slide effect). Direct stress causes deformation of elongation change (shortening or lengthening); shear causes deformations of angular change (wracking, sideways movements).
4. It compares the relative stiffness of the materials.

Chapter 4

1. A condition not modified by moisture content, load duration, or any other special consideration.
2. Modification relates to the time of endurance of the load; the shorter the time, the greater the modification.
3. Values as given relate to service load conditions for the ASD method; for LRFD, values must be adjusted to relate to failure load conditions.

Chapter 5

1. (a) As qualification of a single support, indicates a lack of resistance to rotation of the beam. "Simple beam" is also used to describe a single-span beam with end supports that do not resist rotation.

(b) Support resists rotation and develops negative moment in the end of a single-span beam with downward loads.
(c) If the beam is continuous, it may rotate at the supports, but there will be moment in the beam at supports where the beam is continuous through the support.

2. Refers to the convention of relating sense (sign) of moment to stress conditions in a horizontal beam. A positive moment indicates tension in the bottom of the beam; a negative moment indicates tension in the top.
3. The loads and the reactions.
4. Area of the shear diagram indicates moment changes, both in magnitude and sense (sign). Points of zero shear locate peaks of maximum moment magnitude.
5. Sign of the moment indicates direction of curvature of the beam; zero moment indicates location of inflection (change of direction of curvature).

Chapter 6

1. The specific geometric properties of the rectangular section.
2. For sections with geometries other than a single rectangle.
3. They must be in equilibrium; if the resisting moment is insufficient, the beam will fail.
4. The geometric form and dimensions of the cross section that determine the value of its moment of inertia and the relative stiffness of the material as measured by the modulus of elasticity.

Chapter 7

1. The type, form, and dimensions of units of the decking and ceiling materials used for construction.
2. Bending stress, horizontal shear stress, deflection, bearing, and buckling.

Chapter 8

1. To achieve a smoother surface for application of flooring or roofing, or to improve performance of diaphragm action for lateral loads.

2. A reasonable simulation of a continuous span deck, versus a simple span deck. It reduces both bending stress and deflection for the deck.

Chapter 9

1. Buckling, not stress, is critical. Resistance to buckling is essentially resistance to deflection caused by bending; stiffness is determined by the modulus of elasticity of the material and the radius of gyration of the cross section.
2. Most columns are not designed for compressive stress alone. Design value stress that includes consideration for buckling is also a function of the column stiffness, which is unknown until a cross section is chosen.
3. Wall-covering materials attached to the studs serve to brace the studs on their weak axis (the 1.5 in. direction). The studs are thus critical on the other axis, and the limit for unbraced height is $50 \times 3.5 = 175$ in., or 14 ft − 7 in.

Chapter 10

1. Basic action of the joint produces a twisting of the joint. This is not acceptable for major loads.
2. Shrinkage of the wood with age results in loosening of the wood's grip on the nails, making withdrawal resistance very unreliable.
3. The different design value stresses for resistance to loads parallel to the grain and perpendicular to the grain.
4. The connection is more secure, having minimal slippage during loading. Strength of the joint is also increased.

Chapter 11

1. A triangulated pattern of the framing members.
2. Members resist only direct, axial forces of tension or compression.

Chapter 12

1. Improved dimensional stability (resistance to shape change); reduction of influence of defects in the material; potential for increased strength and stiffness by placing higher-grade materials

in strategic locations; ability to achieve large timbers with use of small-dimension lumber.

2. To obtain face plies with their grain directions the same.
3. Modulus of elasticity.
4. Structural action as a single unit.

Chapter 13

1. They collect lateral loads and distribute them to the vertical bracing elements (shear walls, etc.).
2. Resistance to horizontal shear, to bending moment when working as a vertical cantilever beam, to overturn, and to horizontal sliding at the wall base.
3. Type and thickness of plywood; type, size, and spacing of nails; pattern of placement of the plywood panels on the framing; thickness of framing; presence or absence of blocking.

Chapter 14

1. Weight of the building construction.
2. Total area of the loaded surface being carried by the member being considered.
3. Square feet of supported surface area.
4. Structural efficiency, while an important basic concept for structural engineering design, may be less important economically than the effect of the structure on the cost of the rest of the construction.
5. Unpredictable future market conditions; lack of necessity of the estimator to be the actual seller to the—client—that is, the person who actually gets paid for the finished product; complexity of the building and of the process for its construction.
6. Low-volume use of materials; use of commonly available materials; simplicity and order of the construction; use of materials, products, and processes familiar to builders; reduction of labor time.

Appendix A

1. Except for circular forms—circles and hollow circles (or pipes)—an infinite number of different axes exist.
2. Circular forms—circles and hollow circles (pipes).
3. It indicates the weakest axis of bending resistance and the direction in which buckling is most critical.

Appendix C

ANSWERS TO PROBLEMS

Chapter 4

4.3.A.	746 psi [5.14 MPa]
4.3.B.	943 psi [6.50 MPa]

Chapter 5

5.2.A.	Sample, for point 2: $\Sigma_2 = -(850 \times 4) - (400 \times 2) - (600 \times 6) + (650 \times 12) = 0$
5.3.A.	$R_1 = 3594$ lb [16 kN], $R_2 = 4406$ lb [19.6 kN]
5.3.B.	$R_1 = 8375$ lb [37.26 kN], $R_2 = 10{,}625$ lb [47.26 kN]
5.3.C.	$R_1 = 7667$ lb [34.1 kN], $R_2 = 9333$ lb [41.5 kN]
5.3.D.	$R_1 = 4429$ lb [19.7 kN], $R_2 = 7571$ lb [33.7 kN]
5.3.E.	$R_1 = 7143$ lb [31.8 kN], $R_2 = 11{,}857$ lb [52.7 kN]
5.3.F.	$R_1 = 6750$ lb [30 kN], $R_2 = 5250$ lb [23.4 kN]
5.4.A.	Maximum shear = 10 kips [44.5 kN]
5.4.B.	Maximum shear = 5250 lb [22.97 kN]
5.4.C.	Maximum shear = 1114 lb [4.956 kN]
5.4.D.	Maximum shear = 8.47 kips [37.55 kN]

5.4.E. Maximum shear = 9.375 kips [41.6 kN]

5.4.F. Maximum shear = 4333 lb [19.11 kN]

5.5.A. Maximum M = 60 kip-ft [80.1 kN-m]

5.5.B. Maximum M = 18,375 ft-lb [24.12 kN-m]

5.5.C. Maximum M = 4286 ft-lb [5.716 kN-m]

5.5.D. Maximum M = 61.3 kip-ft [81.7 kN-m]

5.5.E. Maximum M = 18.35 kip-ft [24.45 kN-m]

5.5.F. Maximum M = 20,850 ft-lb [27.62 kN-m]

5.5.G. R_1 = 1860 lb [8.27 kN], maximum V = 1360 lb [6.05 kN], maximum $-M$ = 2000 ft-lb [2.66 kN-m], maximum $+M$ = 3200 ft-lb [4.27 kN-m]

5.5.H. R_1 = 10.32 kips [45.9 kN], maximum V = 7.32 kips [32.1 kN], maximum $-M$ = 4.5 kip-ft [5.9 kN-m], maximum $+M$ = 22.3 kip-ft [29.3 kN-m]

5.5.I. R_1 = 2760 lb [12.28 kN], maximum V = 1760 lb [7.83 kN], maximum $-M$ = 2000 ft-lb [2.67 kN-m], maximum $+M$ = 5520 ft-lb [7.37 kN-m]

5.5.J. R = 10 kips [44.5 kN], maximum V = 7 kips [30.66 kN], maximum $-M$ = 4.5 kip-ft [5.91 kN-m], maximum $+M$ = 20 kip-ft [26.3 kN-m]

5.5.K. Maximum V = 1500 lb [6.67 kN], maximum M = 12,800 ft-lb [17.1 kN-m]

5.5.L. Maximum V = 1500 lb [6.67 kN], maximum M = 9500 ft-lb [12.555 kN-m]

5.5.M. Maximum V = 1200 lb [5.27 kN], maximum M = 8600 ft-lb [11.33 kN-m]

5.5.N. Maximum V = 2700 lb [11.84 kN], maximum M = 12,750 ft-lb [16.8 kN-m]

5.6.A. 32 kip-ft [43.4 kN-m]

5.6.B. 101.25 kip-ft [137.3 kN-m]

5.6.C. 90 kip-ft [122 kN-m]

5.6.D. 108 kip-ft [146.4 kN-m]

5.7.A. $R_1 = R_3$ = 1.25 kips, R_2 = 5.5 kips, maximum V = 2.75 kips, $+M$ = 7.5 kip-ft, $-M$ = 9 kip-ft

5.7.B. $R_1 = R_3$ = 4.8 kips, R_2 = 16 kips, maximum V = 8 kips, $+M$ = 14.4 kip-ft, $-M$ = 25.6 kip-ft

Chapter 6

6.1.A. 70.2 psi [484 kPa]

6.1.B. Maximum shear stress is 62.6 psi [432 kPa], beam is OK

6.1.C. Maximum shear stress is 82.4 psi [568 kPa], beam is OK
6.1.D. 55.94 psi [386 kPa]
6.1.E. 111 psi [765 kPa]
6.2.A. 2.695 ksi [18.6 MPa]
6.2.B. Required $S = 150$ in.^{3}, lightest is 6 × 14
6.2.C. 9.65 kip-ft [13.1 kN-m]
6.3.A. $\Delta = 0.31$ in., allowable is 0.8 in., beam is OK
6.3.B. $\Delta = 0.19$ in., allowable is 0.6 in., beam is OK
6.3.C. $\Delta = 0.23$ in., allowable is 0.75 in., beam is OK
6.3.D. $\Delta = 0.30$ in., allowable is 0.8 in., beam is OK
6.3.E. Required $I = 911$ in.^{4}, lightest is 4 × 16
6.4.A. Stress = 303 psi, allowable is 625 psi, beam is OK
6.4.B. Stress = 480 psi, allowable is 625 psi, beam is OK
6.6.A. Maximum combined stress is 1,899 psi, unmodified allowable stress is 1550 psi, beam is not OK unless roof loading qualifies for modification.
6.6.B. Maximum combined stress is 1,397 psi, unmodified allowable stress is 1600 psi, beam is OK even without modification of stress.
6.7.A. $V_u = 9.504$ kips, $\lambda\phi_v V' = 22.78$ kips, beam is OK
6.7.B. $V_u = 13.81$ kips, $\lambda\phi_v V' = 36.3$ kips, beam is OK
6.7.C. $V_u = 11.136$ kips, $\lambda\phi_v V' = 32.6$ kips, beam is OK
6.7.D. $V_u = 17.28$ kips, $\lambda\phi_v V' = 43.93$ kips, beam is OK
6.7.E. $M_u = 42.77$ kip-ft, $\lambda\phi_v M' = 58.4$ kip-ft, beam is OK
6.7.F. $M_u = 95.62$ kip-ft, $\lambda\phi_v M' = 117$ kip-ft, beam is OK
6.7.G. $M_u = 57.34$ kip-ft, $\lambda\phi_v M' = 108.2$ kip-ft, beam is OK
6.7.H. $M_u = 110.4$ kip-ft, $\lambda\phi_v M' = 162.7$ kip-ft, beam is OK
6.7.I. $M_u = 871.2$ kip-in., required $S = 373.5$ in.^{3}, 6 × 22 is lightest, but is vulnerable to buckling, maybe use 8 × 18 or 10 × 16 if buckling is a problem.
6.7.J. $M_u = 1014$ kip-in., required $S = 434.7$ in.^{3}, 6 × 24 is lightest, but is vulnerable to buckling, maybe use 8 × 20 or 10 × 18 if buckling is a problem.

Chapter 7

7.2.A. Required $S = 127$ in.^{3}, required $I = 683$ in.^{4}, lightest is 4 × 16 if buckling is not a problem, maximum shear stress is 77 psi—not critical, beam weight not a factor.

7.2.B. Required $S = 104\,\text{in.}^3$, required $I = 683\,\text{in.}^4$, lightest is 4 × 16 if buckling is not a problem, $V_u = 5.55$ kips, $\lambda\phi_v V'$ = 12.4 kips—not critical, beam weight not a factor.

7.2.C. Required $S = 96.8\,\text{in.}^3$, required $I = 424\,\text{in.}^4$, lightest is 4 × 14 if buckling is not a problem, maximum shear stress is 53 psi—not critical, beam weight not a factor.

7.2.D. Required $S = 80.6\,\text{in.}^3$, required $I = 424\,\text{in.}^4$, lightest is 4 × 14 if buckling is not a problem, beam weight and shear not critical.

7.3.A. 2 × 10

7.3.B. 2 × 8

7.3.C. 2 × 12

7.3.D. 2 × 12

7.3.E. 2 × 6

7.3.F. 2 × 8

7.3.G. 2 × 10

7.3.H. 2 × 12

Chapter 9

9.2.A. 6800 lbs + or −, using Fig. 9.3

9.2.B. 15,700 lbs, + or −, using Fig. 9.3

9.2.C. 21,300 lbs, + or −, using Fig. 9.3

9.2.D. 53,000 lbs, + or −, using Fig. 9.3

9.2.E. use 6 × 6, allowable load = 24.8 kips

9.2.F. use 10 × 10, allowable load = 79.0 kips

9.2.G. use 10 × 10, allowable load = 52.9 kips

9.2.H. use 12 × 12, allowable load = 111 kips

9.3.A. 10.5 kips, + or −, using Fig. 9.3

9.3.B. 24.9 kips, + or −, using Fig. 9.3

9.3.C. 33 kips, + or −, using Fig. 9.3

9.3.D. 89.0 kips, + or −, using Fig. 9.3

9.6.A. 4660 lbs, + or −, using Fig. 9.3

9.6.B. 10,600 lbs, + or −, using Fig. 9.3

9.8.A. For the combined compression and bending: $0.094 + 0.785 = 0.879 < 1$, OK

9.8.B. For the combined compression and bending: $0.094 + 0.962 = 1.056$, not OK

9.8.C. For the combined compression and bending: $0.056 + 0.931 = 0.987 < 1$, OK

9.8.D. For the combined compression and bending: $0.039 + 0.797 = 0.836 < 1$, OK

9.8.E. Combined action $= 0.038 + 0.751 = 0.789$, less than 1, column OK

9.8.F. Combined action $= 0.026 + 0.643 = 0.669$, less than 1, column OK

Chapter 10

10.1.A. Based on bolts, 1440 lbs

10.1.B. Based on bolts, 1440 lbs

10.1.C. 3200 lbs

10.2.A. 1050 lbs

10.2.B. 1700 lbs

10.6.A. Limit based on rings, 3880 lbs

10.6.B. Limit based on rings, 7180 lbs

10.9.A. 1560 lbs

10.9.B. 2064 lbs

Chapter 11

11.4.A. Sample values: $CI = 2000C$, $IJ = 812.5T$, $JG = 1250T$

11.4.B. Sample values: $CJ = 2828C$, $JK = 1118T$, $KH = 1500T$

11.4.C. Same as 11.4.A

11.4.D. Same as 11.4.B

Chapter 12

12.8.A. Total $W = 21{,}221$ lbs [88 kN]

12.8.B. Required thickness $= 0.483$ in., $1/2$-in. plate OK

Chapter 13

13.2.A. 325 lb/ft

13.2.B. 290 lb/ft

13.2.C. 8d nails at 4 in. spacing at continuous panel edges, 6 in. spacing at other edges

13.2.D. Can't do this – requires 10d nails or 8d nails at 2-1/2 in. spacing, both of which require 3 in. nominal framing.

13.4.A. Possible choice: 3/8 in. thick structural I plywood, 8d nails at 4 in., if studs 16 in. on center or closer; six $3/4$ in. bolts with 2 in. sill, or four $3/4$ in. bolts with 3 in. sill; no tie-down required for overturn

13.4.B. Possible choice: 15/32 in. thick Structural I plywood, 8d nails at 4 in., if studs 16 in. on center or closer; seven 3/4 in. bolts with 2 in. sill, or five 3/4 in. bolts with 3 in. sill; no tie-down required for overturn

Appendix A

A.1.A. $c_y = 2.6$ in. [65 mm]
A.1.B. $c_y = 1.75$ in. [43.9 mm], $c_x = 0.75$ in. [18.9 mm]
A.1.C. $c_y = 4.2895$ in. [107.24 mm]
A.1.D. $c_y = 3.4185$ in. [85.2 mm], c_x 1.293 in. [32.2 mm]
A.1.E. $c_y = 4.4375$ in. [110.9 mm], $c_x = 1.0625$ in. [26.6 mm]
A.1.F. $c_y = 4.3095$ in. [107.7 mm]
A.3.A. $I = 535.86$ in.^{4} [2.11 H 10^6 mm^4]
A.3.B. $I = 205.33$ in.^{4} [80.21×10^6 mm^4]
A.3.C. $I = 447.33$ in.^{4} [175 H 10^6 mm^4]
A.3.D. $I = 5.0485$ in.^{4} [2.034×10^6 mm^4]
A.3.E. $I = 205.33$ in.^{4} [80.21 H 10^6 mm^4]
A.3.F. $I = 682.33$ in.^{4} [267×10^6 mm^4]

GLOSSARY

Every topic area or field has its own private language. To help readers with some of the materials in this book, as well as with that in other publications about wood, the following brief glossary of terms is provided.

Beam Stability Factor, C_L. Factor for modification of design bending stress based on the lateral and torsional buckling stiffness of the beam.

Column Stability Factor, C_p. Factor for modification of design bending stress based on the lateral and torsional buckling stiffness of the column.

Composite Panel. Structural panel with wood veneer faces and a fiberboard core. In thicker panels there is also a center wood veneer.

Dowel-Type Fastener. General term used to describe fasteners that work primarily for lateral (shear) loading. Includes bolts, lag screws, wood screws, nails & spikes, drift bolts, and drift pins.

Engineered Wood. General term for wood products other than single pieces of sawn wood.

Flitched Beam. Built-up beam with combination of elements of lumber and steel plate.

Fiberboard. Panel consisting of compression bonded wood fibers with no orientation of the fibers.

Format Conversion Factor (LRFD only), K_F. Factor for conversion of ASD values for use in the LRFD method.

Grade. Elevation of the top of the ground at a point; or, the slope of a nonflat surface of concrete or earth.

Factored Load (LRFD only). Service load multiplied by a load factor to adjust for use in strength design.

I-Joist. I-shaped joist or rafter, formed with flanges of sawn wood or laminated veneer and a web of plywood or particleboard.

Laminated Veneer Lumber (*LVL*). Lumber produced by glue lamination of very thin wood veneers.

Load Duration Factor, (ASD only) C_D. Factor used to adjust reference design values based on the time of duration of the service load.

Oriented Strand Board, OSB. Structural panel produced by pressure bonding of thin wafers of wood with the wafers placed in layers with the grain direction alternating in adjacent layers

Parallel Strand Lumber (*PSL*). Lumber elements produced by pressure bonding of thin strands of wood with the strands parallel.

Ply. One layer in a multilayered sandwich for a glued-laminated structural panel.

Plywood. Panel consisting of glued-laminated thin wood veneers (called plies), with alternating plies at right angles in terms of grain direction.

Pole. Round wood shaft of tapered form, consisting of a single tree trunk stripped of the outer layers.

Reference Design Value. Basic values for stress and modulus of elasticity for wood, with no modifications for conditions.

Repetitive Member Factor, C_r. Factor for increase of reference design values based on load-sharing in a multiple-element framing system with closely-spaced members.

Resistance Factor (LRFD only), ϕ. Factor for adjustment of the ultimate resistance of a structural element for use in strength design.

Sawn Wood. Linear wood pieces produced by sawing directly from a log.

Service Conditions. Structural concerns (loads mostly) resulting from the usage of the structure.

Split-Ring Connector. Element consisting of a cut steel ring that is inserted between wood members in a bolted joint to enhance the shear resistance of the joint.

Structural Composite Lumber. General class of fabricated wood structural elements that includes *laminated veneer lumber* and *parallel strand lumber.*

Structural Glued-Laminated Timber. Built-up timber sections consisting of multiple layered stacks of 2 in. nominal dimension lumber with the lumber glued together on their flat sides.

Temperature Factor, C_t. Modification factor for reference design values for service conditions involving sustained exposure to temperatures in excess of 100° F.

Time Effect Factor (LRFD only), λ. Factor for modification of reference design values based on duration time of loads.

Veneer. A very thin slice of wood prepared for use in a glued-laminated product.

Volume Factor, C_V. A factor used to modify reference design values for structural glued-laminated timber or structural composite lumber based on the volumetric size of the member.

Wet Service Factor, C_M. A factor for adjustment of reference design values based on sustained moisture conditions.

Wood Structural Panel. A general code classification including plywood, OSB, and composite panels.

REFERENCES

1. *National Design Specification (NDS) for Wood Construction with Commentary and Supplement: Design Values for Wood Construction*, American Forest and Paper Association, Washington, DC, 2005.
2. *Timber Construction Manual*, 5th ed., American Institute of Timber Construction, Wiley, Hoboken, NJ, 2006.
3. *Minimum Design Loads for Buildings and Other Structures*, ASCE/SEI 7-05, American Society of Civil Engineers, Reston, VA, 2005.
4. *International Building Code, Volume 2: Structural Engineering Provisions*, International Code Council, Inc., Country Club Hills, IL, 2006.
5. *Structural Wood Design Solved Example Problems – ASD/LRFD*, American Forest and Paper Association, Washington, DC, 2005.
6. *Manual for Engineered Wood Construction – ASD/LRFD*, American Forest and Paper Association, Washington. DC, 2005.
7. *Engineered Wood Construction Guide*, APA - The Engineered Wood Association, Tacoma, WA, 2005.
8. *Introduction to Lateral Design*, APA – The Engineered Wood Association, Tacoma, WA, 2003.

9. Edward Allen and Rob Thallon, *Fundamentals of Residential Construction*, 2nd ed., Wiley, Hoboken, NJ, 2006.
10. James Ambrose and Patrick Tripeny, *Simplified Engineering for Architects and Builders*, 10th ed., Wiley, Hoboken, NJ, 2006.
11. James Ambrose and Dimitry Vergun, *Simplified Building Design for Wind and Earthquake Forces*, 3rd ed., Wiley, Hoboken, NJ, 1995.
12. James Ambrose, *Simplified Design of Building Foundations*, 2nd ed., Wiley, Hoboken, NJ, 1988.
13. James Ambrose, *Design of Building Trusses*, Wiley, Hoboken, NJ, 1994.
14. James Ambrose, *Construction Revisited*, Wiley, Hoboken, NJ, 1993.
15. *Commercial Design Guide No. 1048: Parallam® PSL for Commercial Applications*, Trus Joist, Boise, ID.

Index

Accuracy of computations, 8
Actual sizes, 17
Adjustment factors, LRFD, 48, 109
Allowable stresses, 35
  modification of, 36
Anchors,
  framing, 192
  horizontal, 192
  shear wall, 274
Angle to grain load, 45, 169
ASD (allowable stress design) method, 2, 24
Axial load, 30

Bay, structural, 29
Beam diagrams and formulas, 81
Beams,
  bearing for, 101
  bending in, 65
    sense of, 71
  bending stress in, 93
  buckling, 103
  built-up, 237, 238
  cantilever, 51, 75
  continuous, 51, 82
  deflection of, 97
  design, 117
  end bearing, 101
  flexure formula, 95
  flitched, 239
  inflection of, 73
  laminated, 237
  laterally unsupported, 103
  moment diagram, 66
  multiple span, 82
  overhanging, 51
  reactions, 54, 57
  restrained, 52
  shear in, 60, 86
  shear diagram, 63
  shear stress in, 86
  simple, 51, 79

Beams (*Continued*)
size factor for, 42, 96
tabulated values for, 78
types, 51, 78
uniformly distributed load, 52
unsymmetrical bending of, 105
Beam stability factor, 44
Bearing,
beam, 101,
bearing area factor, 44
wall, 308
Bending moment, 65
biaxial, 105
in continuous beams, 82
diagrams, 66
formulas for, 95
negative, 71
positive, 71
unsymmetrical, 105
Bending stress, 93
Biaxial bending, 105
Board deck, 129
Bolted joints,166
design values for, 171
edge distance in, 168
Hankinson graph for, 169
modification factors for, 169
spacing of bolts in, 108
two-member, 167
Box-beam, 127
Bridging, 122
Buckling of beams, 103
Buckling stiffness factor, 44
Building codes, requirements, 292
Building systems, general design, 302
Built-up,
beams, 237, 238
columns, 154
plywood beams, 238

Cantilever beams, 51, 75
Center of gravity, 354
Centroid, 354
Chord,
in diaphragm, 254
in truss, 197
Column formulas for,
columns with bending, 158
solid sawn columns, 138
spaced columns, 150
Columns,
built-up, 154
compression capacity, ASD, 136
LRFD, 145
design, 142
design aids for, 142
design values for, 143
eccentrically loaded, 158
interaction, 155
P-delta effect, 156
pole, 147, 244
round, 147
safe loads, 143, 144
slenderness ratio, 135
spaced, 150
stability factor, 44, 138, 140
stud, 148
with bending, 155
Combined flexure and axial load, 158
Compression at angle to grain, 45
Connectors,
bolts, 166
framing elements, 189
lag, 180
nailed, 176
screw, 179
split ring, 182
timber, 189, 222
Continuous span beams, 82
Control joints, 252
Cost of construction, 303

Dead load, 28, 290
Decking, 231
board, 129
plank, 130
plywood, 132
spanning capability, 133
wood fiber, 132
Defects in lumber, 15
Deflection,
allowable, 98
computation of, 97
formulas for, 81, 98
Deformation, 32
Density of wood, 14

Design values, 35
  modification of, 36
Diaphragms,
  chord, 254
  horizontal, 250
  plywood, 233
  vertical, 262
Dimension lumber, 37
Dimensional stability of wood, 17
Direct stress, 30
Dressed sizes, 17
Drift, 299
Duration of load, 41

Eccentrically loaded column, 158
Economics, 303
Edge distance,
  for bolts, 167
  for split-rings, 184
Elasticity, 32
  modulus, 33
End bearing of beams, 101
End distance,
  for bolts, 167
  for split-rings, 184
Engineered wood products, 226
Equilibrium, 28, 53

Fabricated wood products, 18
Fasteners, *see* Connectors
Fiber products, 18, 132
Flat use factor, 43
Flexure formula, for beam, 95
Flitched beam, 239
Floor joists, 121
  span tables for, 124
Force, 28
Formed steel framing elements, 189
Foundations, 326
Framing,
  anchors, 192
  devices, 189

Girder, defined, 50
Glued laminated products, 227
Grade beam, 326
Grading of lumber, 38
Gussets, 192

Hankinson formula, 46
  graph, 169
Horizontal bracing, 279, 281
Horizontal diaphragm, 250, 321
Horizontal shear, 60, 87

I-joist, 127, 237
Incising factor, 43
Inelastic behavior, 23
Inflection point, 73
Interaction, 155
Internal forces in trusses, 203

Joints, method of, 211
Joists, 121
  bridging for, 121
  floor, 121
  span tables for, 124

Kip, defined, 28

Lag bolt, 180
Lateral bracing, special, 284
Lateral buckling,
  beam, 103
  column, 135
  truss, 199
Lateral force, 247, 297
Laterally unsupported beam, 103
Ledger, 192
Let-in bracing, 277
Light wood frame, 150, 307, 327
Live load, 28, 293, 294, 295
  reduction of, 294, 296
Load,
  at angle to grain, 45, 169
  axial, 30
  combinations, 293, 301
  concentrated, 28
  dead, 28, 290
  design, 302
  duration, 41, 293
  factor, LRFD, 301
  floor, 295
  lateral, 247, 297
  live, 28, 293, 294, 295
  of movable partitions, 329

Load (*Continued*)
periphery, 302
roof, 294
seismic, 293, 300
on trusses, 201
uniformly distributed, 28
wind, 290, 293, 298
LRFD (load and resistance factor design), 2, 24
Lumber,
classification of, 17
defects in, 15
density of, 14
grading, 18
laminated veneer, 230
parallel strand, 230
properties of sections, 367
structural composite, 229

Maxwell diagram, 206
Manufactured truss, 224, 315
Measurements, units of, 5
Mechanically-driven fasteners, 181
Method of joints, 203
Methods of design, 23
ASD, 2, 24
choice of, 25
LRFD, 2, 24
Mill construction, 334
Modification of design values, 36
Modulus of elasticity, 33
Moisture content, 42
Moment, 52
bending, 65
diagram, 66
of a force, 52
of inertia, 358
resisting, 93
Movable partitions, 329
Multiple span beams, 82

Nailed joints, 176
Negative bending moment, 71
Net section, 167
Neutral axis, 93
Nominal size, 17
Notation, standard, 9

Occupancy, types of, 292
Oriented strand board (OSB), 231
Overhanging beam, 51
Overturn,
moment, 274, 299, 324, 334
of shear wall, 274

Panel point, of truss, 197
Partitions, movable, 329
Permanent set, 33
P-delta effect, 156
Plank deck, 130
Plywood, 232
deck, 132
design data, 232, 234, 236
diaphragm, 233, 256
gusset, 192
shear wall, 272
stressed-skin panel, 127
types and grades, 232
usage considerations, 233
Pole structure, 244
Poles, 147, 244
Ponding, 294
Principal axes, 366
Properties of,
sections, 366, 367
structural lumber, 367

Radius of gyration, 365
Rafters, 121
span tables for, 129
Reactions, 54, 57
Reference Design values, 35, 36
Adjustment, 41
Reference sources, 4, 390
Relative stiffness of vertical bracing, 251
Repetitive-member uses, 43
Resistance factor (LRFD), 2, 24, 48
Resisting moment, 93
Rigid frame, for lateral bracing, 285
Roof load, 294
Roof pitches and slopes, 197
Round column, 147

Safety, factor, 22
fire, 21
Sandwich panel, 127, 238

Screws, 179
Section modulus, 95, 363
Section properties, 366, 367
Seismic forces, 300
Shear,
  beam, 60, 86
  developers, 181
  diagram, for beam, 63
  formula for rectangular beam, 87, 90
  general formula, 90
  horizontal, 60, 87
  stress, 31, 90
  wall, 262, 272, 323
Simple beam, 51, 79
Size factor for beam, 42, 96
Slenderness ratio, column, 135
Solid column, allowable load for, 136, 145
Spaced column, 150
Span tables for,
  joists, 124
  rafters, 125
Species of wood, 13
Split-ring connector, 182
Standard notation, 9
Statical moment, 90, 355
Stiffness, 33
Strain, 32
Strength method, 2, 24
Stress, 30
  allowable, 35
  bearing, 38, 101
  direct, 30
  kinds of, 31
  shear in beam, 90
  unit, 27
Stress/strain relationships, 33
Stressed-skin panel, 127, 238
Structural design,
  accuracy of, 8
  goals, 20
  methods, 2, 23
  references, 289
  standards, 289
Structural lumber,
  composite, 229
  grading of, 38
  properties for, 367
  use classification, 17
Structural planning, 302
Stud walls, 148
Symbols, 9

Tabulated beam values, 78, 81
Temperature factor, 42
Tension joints, 167
Timber, 13
  connectors, 189, 222
  construction, 335, 340
  trusses, 223, 345
Time effect factor, LRFD, 49, 110
Torsional buckling of beams, 103
Torsional effects on buildings, 250
Transfer axis formula, 361
Transferring moments of inertia, 361
Trusses,
  algebraic analysis, 211
  bracing for, 199, 277
  design considerations, 196
  design forces for, 219, 349
  graphical analysis, 203
  heavy timber, 223, 345
  internal forces in members, 203
  investigation of, 203, 211
  joints in, 222
  loads for, 201
  manufactured, 224, 315
  Maxwell diagram for, 206
  members for, 197, 222, 350
  separated joint diagram, 204
  timber, 223, 345
  types, 199
  weight of, 203
  wind load, 348
Two-member joint, 167

Uniformly distributed load, 52
U. S. units, 5
Units of measurement, 5
Unit stress, 27
Unsymmetrical bending of beam, 105
Uplift, 299
Use classification of lumber, 17

Vertical,
  diaphragm, 262
  shear in beams, 60

Walls,
  shear, 262, 272, 323
  stud, 148
Weight of building materials, 291
Wind,
  design for, 293, 298, 317, 332
  force diagram for truss, 348
  forces, 332
  pressure, 332
Withdrawal of nails, 177
Wood,
  built-up beam, 238
  composite panel, 231
  defects, 15
  density of, 14
  dimensional stability, 17
  flaws in, 15
  flitched beam, 239
  I-joist, 237
  moisture content, 17
  pole structure, 244
  poles, 147, 244
  seasoning of, 16
  sources, 13
  species, 13
  structural panel, 2312
  use, 1

Zoned nailing for plywood diaphragms, 260